2021 34th International Vacuum Nanoelectronics Conference (IVNC 2021)

Lyon, France
5 – 9 July 2021

IEEE Catalog Number: CFP21VAC-POD
ISBN: 978-1-6654-2590-2

**Copyright © 2021 by the Institute of Electrical and Electronics Engineers, Inc.
All Rights Reserved**

Copyright and Reprint Permissions: Abstracting is permitted with credit to the source. Libraries are permitted to photocopy beyond the limit of U.S. copyright law for private use of patrons those articles in this volume that carry a code at the bottom of the first page, provided the per-copy fee indicated in the code is paid through Copyright Clearance Center, 222 Rosewood Drive, Danvers, MA 01923.

For other copying, reprint or republication permission, write to IEEE Copyrights Manager, IEEE Service Center, 445 Hoes Lane, Piscataway, NJ 08854. All rights reserved.

*** This is a print representation of what appears in the IEEE Digital Library. Some format issues inherent in the e-media version may also appear in this print version.

IEEE Catalog Number:	CFP21VAC-POD
ISBN (Print-On-Demand):	978-1-6654-2590-2
ISBN (Online):	978-1-6654-2589-6
ISSN:	2164-2370

Additional Copies of This Publication Are Available From:

Curran Associates, Inc
57 Morehouse Lane
Red Hook, NY 12571 USA
Phone: (845) 758-0400
Fax: (845) 758-2633
E-mail: curran@proceedings.com
Web: www.proceedings.com

34th International Vacuum Nanoelectronics Conference

Schedule

Monday, 5th July, 2021

Time (Lyon)	
12:40-13:00	Opening comments from the two Chairmen

M1 - Tutorials

Chair - tbd

Time(Lyon)	Title and Authors (speaker underlined)	Page
13:00-13:45 Tutorial M1.1	**Micro-Nano Fabrication of Integrated Tip-Based Field Electron Emission Devices** <u>Juncong She,</u> Sun Yat-Sen University	30
13:45-14:30 Tutorial M1.2	**Ion sources and optical charged particles dedicated to FIB technology today. Current trends and challenges in semiconductors, failure analysis and HR SIMS** <u>Arnaud Houël</u>, Anne Delobbe, Justine Renaud, Matthieu Vitteau Orsay Physics	32

14:30-15:00	Pause

M2 - Tutorials

Chair – Richard Forbes

Time(Lyon)	Title and Authors (speaker underlined)	Page
15:00-15:45	**A Tutorial on the Physics And Modeling Of Electron Sources** <u>Kevin L. Jensen</u> Naval Research Laboratory	NA
15:45-16:30 Tutorial M2.2	**Electron emission calculations beyond the classical equations: finite size, space charge and thermal effects in sharp emitters** <u>Andreas Kyritsakis</u> University of Tartu, Estonia	NA

16:30-16:45	Pause

M3 - Tutorials

Chair – Christopher Edgcombe

Time(Lyon)	Title and Authors (speaker underlined)	Page
16:45-17:30	**On the brightness, transverse emittance, and transverse coherence of a field emission beam**	NA

Tutorial M3.1	**Soichiro Tsujino** Paul Scherrer Institut, Switzerland	

17:30-17:45	Pause

M4 - Poster Flashs Time Zone A

Chair S. Purcell and J.-P. Mazellier

Time(Lyon)	Title and Authors (speaker bold underlined, institute of speaker only)	Page
	Industriel Sponsors	
	We wish to thank our industriel sponsors Orsay Physics, Kashiyama Europe GMBH, Hamamatsu France and Slide Pack who all will be available in the poster sessions. Three will present flashes.	
17:50	**Orsay Physics**	
17:52	**Hamamatsu France**	
17:54	**Slide Pack**	
	Microscopy + Spectroscopy	
17:58	**Electron energy analysis in Scanning Field Emission Microscopy using a Bessel box energy analyzer** **M. Bodik**, M. Demydenko, C.G.H. Walker, T. Bähler, T. Michlmayr, A.-K. Thamm, U. Ramsperger, A. Pratt, S.P. Tear, M.M. El Gomati, D. Pescia ETH Zürich, Switzerland	128
18:00	**Fowler-Nordheim Slope Dependence on Pressure in Controlled Poor Vacuum** **Girish Rughoobur**, Olusoji O. Ilori and Akintunde I. Akinwande Massachusetts Institute of Technology, USA	130
18:02	**Collector dependence of field emission in the Scanning Field Emission Microscopy** **H.J. Gotsis**, N.C. Bacalis, and J.P. Xanthakis	132
	Modeling	
18:04	**Study of Self-Heating Effects in Looped Carbon Nanotube Fibers** **Geet Tripathi**, Kartik Sharma, Marc Cahay, Jonathan Ludwick, F. F. Dall' Agnol, T. A. de Assis University of Cincinnati, USA	134
18:06	**Influence of Contact Resistance on the Field Emission Characteristics of a Carbon Nanotube** **Geet Tripathi**, Marc Cahay, Jonathan Ludwick, and Kevin L. Jensen University of Cincinnati, USA	136
18:08	**User-friendly method for testing field electron emission data: Technical report** **Mohammad M. Allaham**, Alexandr Knápek, Marwan S. Mousa, and Richard G. Forbes Institute of Scientific Instruments of CAS, Czech Republic	138

18:10	**Testing the performance of Murphy-Good plots when applied to current-voltage characteristics of Si field electron emission tips** **Mohammad M. Allaham**, Philipp Buchner, Rupert Schreiner and Alexandr Knápek Institute of Scientific Instruments of CAS, Czech Republic	140
18:12	**Estimating the uniformity of nanoscale vacuum channel transistor arrays using space-charge effects** **Jesse M. Snelling**, Gregory R. Werner, John R. Cary University of Colorado, USA	NA
	RF and Xrays from electron beams	
18:14	**Confined Electron Laser** **Arya Fallahi**, Niels Kuster, Lukas Novotny ETH Zurich, Switzerland	143
18:16	**Single - Cycle THz Accelerating Structure with Wave Beam Focusing Lens** **Sergey Antipov**, Sergey Kuzikov, and Alexander Vikharev Euclid Techlabs, Russia	145
18:18	**Magnetron Sputtering Formation of Molybdenum-Copper Alloys for Fabrication of Millimeter-Band Planar Slow Wave Structures** **A.V. Starodubov**, D.A. Nozhkin, A.A. Serdobintsev, I.O. Kozhevnikov, A.M. Pavlov, V.V. Galushka, N.M. Ryskin, G. Ulisse, V. Krozer Saratov Branch, Kotelnikov Institute of Radio Engineering and Electronics, Russia	147
18:20	**A Facile Approach for Surface Quality Improvement of Mm-Band Planar Electromagnetic Structures Fabricated by Laser Ablation** **A.V. Starodubov**, A.A. Serdobintsev, I.O. Kozhevnikov, A.M. Pavlov, V.V. Galushka, N.M. Ryskin Saratov Branch, Kotelnikov Institute of Radio Engineering and Electronics, Russia	149
	Theory of Emission : Classic Quantum Tunneling	
18:22	**Analyses of field electron emission Molybdenum current-voltage data using Fowler-Nordheim and Murphy-Good plots** **Mohammad M. Allaham**, Marwan S. Mousa, Daniel Burda, Mohammad H. AlSa'eed, Sabreen Y. AlJrawen and Alexandr Knápek Institute of Scientific Instruments of CAS, Czech Republic	151
18:24	**Universality of Characteristic Field Enhancement Factor from Arched Carbon Nanofibers** **Thiago A. de Assis**, Fernando F. Dall'Agnol, Marc Cahay Federal University of Bahia, Brazil	153
18:26	**Using the parameter "formal area efficiency" (α_f^{SN}) to analyze current-voltage measurements on large-area field electron emitters** **Richard G. Forbes**, University of Surrey, UK	155
18:28	**A Tutorial Commentary on the Schottky Constant** **Richard G. Forbes,** University of Surrey, UK	157

18:30	**Correction of Conceptual Error in Feynman's Textbook Treatment of Pointed-Conductor Electrostatics** **Richard G. Forbes,** University of Surrey, UK	159
18.32	**Features of the field enhancement factor on blade-type emitters** **S.V. Filippov**, E.O. Popov, A.G. Kolosko, F.F. Dall'Agnol Ioffe Institute, Russia	161
18:34	**Comparison of the effective parameters of single-tip tungsten emitter using FN and MG-plots** **Eugeni O. Popov**, Sergey V. Filippov, Anatoly G. Kolosko, Alexandr Knápek, Ioffe Institute, Russia	163
Vacuum Nano Electronics		
18:36	**High Current Field Emission Arrays for Crossed Field Device Experiments** **Ranajoy Bhattacharya**, Mason Cannon, Rushmita Bhattacharjee, Winston Chern, Nedeljko Karaulac, Girish Rughoobur, Akintunde I. Akinwande and Jim Browning, Boise State University, USA	165
18:38	**Lifetime and Breakdown Mechanisms in Double-Gated Si FEAs** **Girish Rughoobur** and Akintunde I. Akinwande Massachusetts Institute of Technology, USA	167
18:40	**Influence of Geometrical Arrangements of Si Tip Arrays Fabricated by Laser Micromachining on their Emission Behaviour** **Matthias Hausladen**, Vitali Bomke, Philipp Buchner, Michael Bachmann, Alexandr Knápek, and Rupert Schreiner OTH Regensburg, Germany	169
18:42	**Silicon Field Emitter Arrays Fabricated Using a Layout-Independent Process** **Nedeljko Karaulac**, Winston Chern, Girish Rughoobur, and Akintunde I. Akinwande, Massachusetts Institute of Technology, USA	171
18:44	**Current dependent performance test used on different types of silicon field emitter arrays** **Andreas Schels**, Simon Edler, Walter Hansch, Michael Bachmann, Florian Herdl, Felix Düsberg, Magdalena Eder, Manuel Meyer, Markus Dudek, Rupert Schreiner Universität der Bundeswehr München, Germany	173
18:46	**Optimizing current uniformity in nanoscale vacuum channel transistors with space charge feedback** **Gregory R. Werner,** Luke Adams, Jesse M. Snelling, John R. Cary University of Colorado, USA	NA
Nano Emitters		
18:48	***Carbon Nanotube Fiber Cathodes and Saturation of their Field Emission Current*** **Evgenii P. Sheshin**, Ilya N. Kosarev, Bulat I. Masnaviev and D. I. Ozol Moscow Institute of Physics and Technology, Russian Federation	176

18:50	**Field emission properties of sharp tungsten cathodes coated with a thin resilient oxide barrier** **Daniel Burda**, Mohammad M. Allaham, Alexandr Knápek, Dinara Sobola, Marwan Suleiman Mousa Institute of Scientific Instruments of the CAS, Czech Republic	178
18:54	**Using High Aspect Ratio AFM Probe for Digital Twin Development of SiC FEA** **Konstantin Nikiforov**, Nikolay Egorov, Ivan Sokolov, Valery Strebko, Vladimir Mikhailovskiy, Denis Danilov, Vladimir Golubkov, Vladimir Ilyin, and Alexey Ivanov Saint Petersburg State University, Russia	180
18:56	**High Brightness Carbon Nanotube Fiber Field Emission Cathode** **Taha Y. Posos**, Jack Cook, Oksana Chubenko, Steven B. Fairchild, Nathaniel P. Lockwood and Sergey V. Baryshev Michigan State University, Michigan, USA	NA

Tuesday, 6th July, 2021

Time (Lyon)	
12:40-13:00	Opening comments from the two Chairmen

Tu1 - Ultrafast, Ultra-intense Laser Excitation of Free and Bound Electrons

Chairs - Anthony Ayari

Time(Lyon)	Title and Authors (speaker bold underlined, institute of speaker only)	Page
13:00-13:45 Plenary Tu1.1	**Ultrafast electron control with various means: from multiphoton** **physics at needle tips to nanophotonic particle acceleration** Roy Shiloh, Tomáš Chlouba, Ang Li, Philip Dienstbier, Alexander Tafel, Johannes Illmer, Norbert Schönenberger, Peyman Yousefi, Stefanie Kraus, Leon Brückner, Julian Litzel, Bastian Löhrl, **Peter Hommelhoff** Friedrich-Alexander-Universität, Germany	38
13:45-14:15 Invited Tu1.2	**Ultrafast Electron Scattering: Femtosecond Electron Pulses in** **Materials Research** Laurent P. René de Cotret, Martin R. Otto, Jan-Hendrik Pöhls, Tristan Britt, Mark J. Stern, Mark Sutton, **Bradley J. Siwick** McGill University, Canada	40

14:15-14:30	Pause

14:30-14:45 Oral Tu1.3	**Emission of electrons from a metal tip irradiated by femtosecond** **IR lasers at wavelengths of 800 and 1240 nm** A.V. Ovchinnikov, O.V. Chefonov, M.B. Agranat, N.A. Abramovskii, S.B. Bodrov, A.M. Kiselev, A.A. Murzanev, A.V. Romashkin, **A.N. Stepanov** Russian Academy of Sciences (IAP RAS), Russia	42
14:45-15:00 Oral Tu1.4	**Tunable Wavelength One-photon Photoassisted Cold Field Emission** **from W(310)-nanotips** **Rudolf Haindl**, Kerim Köster, Armin Feist and Claus Ropers University of Göttingen, Germany	44
15:00-15:15 Oral Tu1.5	**Photoemission from an ultrabright and ultrafast LaB6 nanowire** **electron emitter studied at atomic scale** **Ang Li**, Han Zhang, Stefan Meier, Alexander Tafel, Peter Hommelhoff Friedrich-Alexander-Universität, Germany	46

15:15-15:45	Pause

Tu2 – Advances in Electron Microscopy and Spectroscopy

Chair Alizera Nojeh

Time(Lyon)	Title and Authors (speaker bold underlined, institute of speaker only)	Page
15:45-16:15 Invited Tu2.1	**Nano-optics with Fast Electrons** **M. Kociak** Université Paris Sud, France	NA
16:15-16:45 Invited Tu2.2	**Longitudinal and transverse modulation of electron wave function with light, and its application to electron microscopy** **Ivan Madan,** Giovanni Vanacore, Gabriele Berruto, Enrico Pomarico, Javier García de Abajo, Ido Kaminer, Fabrizio Carbone École Polytechnique Fédérale de Lausanne, Switzerland	48
16:45-17:00 Oral Tu2.3	**Voltage-controlled three-electron-beam interference by a three-element Boersch phase shifter with top and bottom shielding electrodes** **P. Thakkar,** V.A. Guzenko, P-H. Lu, R.E. Dunin-Borkowski, J.P. Abrahams and S. Tsujino Paul Scherrer Insitut, Switzerland	50
17:00-17:30 Invited Tu2.4	**A standing molecule as a coherent single-electron field emitter** Taner Esat, Marvin Knol, Philipp Leinen, Matthew F. B. Green, Malte Esders, Niklas Friedrich, Michael Maiworm, Nicola Ferri, Pawel Chmielniak, Sidra Sarwar, Torsten Deilmann, Peter Krüger, Hadi H. Arefi, Daniel Corken, James Gardner, Kristof T. Schütt, Jeff Rawson, Paul Kögerler, Michael Rohlfing, Rolf Findeisen, Alexandre Tkatchenko, Klaus-Robert Müller, Reinhard J. Maurer, Christian Wagner, Ruslan Temirov & **F. Stefan Tautz** Peter Grünberg Institut, Germany	52
17:30-17:45 Oral Tu2.5	**Scanning Field Emission Microscopy with Spin and Energy Analysis** **A-K Thamm,** J. Wei, M. Demydenko, C.G.H. Walker, D. Pescia and U. Ramsperger, A. Pratt, S.P. Tear, M.M. El Gomati Eidgenössische Technische Hochschule (ETH) Zürich, Switzerland	54

17:45-18:00	Pause

Tu3 – Poster Session Time Zone A

Chair tbd

Time(Lyon)	Title and Authors (speaker bold underlined, institute of speaker only)	Page
18:00 20:00	Following the M4 - Poster Flashes Time Zone A.	

Wednesday, 7th July, 2021

	W1- Poster Flashes Time ZoneB	

Chair S. Purcell, J.-P. Mazellier

Time(Lyon)	Title and Authors (speaker bold underlined, institute of speaker only)	Page
	Industriel Sponsors	
	We wish to thank our industriel sponsors Orsay Physics, Kashiyama Europe GMBH, Hamamatsu France and Slide Pack who all will be available in the poster sessions. Three will present flashes.	
10:50	**Orsay Physics**	
10:52	**Hamamatsu France**	
10:54	**Slide Pack**	
	Applications	
11:00	**Fabrication of ZnO nanowires cold cathode X-ray source with micro patterned transmission anode** **<u>Song Kang</u>**, Yangyang Zhao, Guofu Zhang, Shaozhi Deng, Ningsheng Xu, Jun Chen Sun Yat-sen University, China	185
11:02	**Cold Cathode X-Ray Flat Panel Detector Based on Ga2O3 Thin Film Photoconductor** **<u>Haojian Huang</u>**, Manni Chen, Zhipeng Zhang, Juncong She, Shaozhi Deng, Ningsheng Xu, Jun Chen Sun Yat-sen University, China	187
11:04	**Development of gated carbon nanotube cold cathode for miniature X-ray source** **<u>Junfan Wang</u>**, Yajie Guo, Haifeng Zhu, Baohong Li, Yu Zhang, Shaozhi Deng, Ningsheng Xu and Jun Chen Sun Yat-sen University, China	189
11:06	**Optimization of Focusing Structure for a Micro-Focus X-ray Source** **<u>Junfan Wang</u>**, Yajie Guo, Haifeng Zhu, Baohong Li, Yu Zhang, Shaozhi Deng, Ningsheng Xu and Jun Chen Sun Yat-sen University, China	191
11:08	**Focal Spot Size Enhancement by Offset control of Triode e-beam Module for High Resolution X-ray Imaging** **<u>Yi Yin Yu</u>** and Kyu Chang Park Kyung Hee University, Korea	193
11:10	**Outgassing during LAFE operation in the diode system** **<u>S.V. Filippov</u>**, A.G. Kolosko, E.O. Popov Ioffe Institute, Russia	195
11:12	**Cathodoluminescent UV Sources for Photocatalytic Disinfection of Air** **<u>Evgenii P. Sheshin</u>**, Ilya N. Kosarev, Bulat I. Masnaviev, Alexander O. Getman, Ilya A. Savichev and Dmitry I. Ozol Moscow Institute of Physics and Technology, Russian Federation	197

11:14	**Concept of a Secondary Emission Converter of the Energy of Fast Electrons and γ-Quanta On the Basis of Carbon Materials (e.g. Graphene)** **Dmitry I. Ozol** Moscow Institute of Physics and Technology, Russian Federation	199
11:16	**Towards a MEMS transmission point X-ray source** **Tomasz Grzebyk**, Krzysztof Turczyk, Anna Górecka-Drzazga, Jan A. Dziuban Wroclaw University of Science and Technology, Poland	201
11:18	**Optimization of Gated ZnO Nanowire Field-Emitter Arrays by Tuning Pixel Density** **Songyou Zhang**, Xiuqing Cao, Guofu Zhang, Shaozhi Deng, Juncong She, Ningsheng Xu and Jun Chen Sun Yat-sen University, China	203
11:20	**Study of Nanoscale Cathodes for Gas Discharge Devices** **Sergey M. Karabanov** Ryazan State Radio Engineering University, Russia	205
11:22	**UV lighting with carbon nanotube based cold cathode electron beam (C-beam) and its characteristics** **Sung Tae Yoo**, and Kyu Chang Park Kyung Hee University, Korea	207
Microscopy + Spectroscopy		
11:24	**Microscope equipped with graphene-oxide-semiconductor electron source** **Yukino Kameda**, Katsuhisa Murakami, Masayoshi Nagao, Hidenori Mimura and Yoichiro Neo Research Institute of Electronics Shizuoka University, Japan	209
Nano Emitters		
11:26	**Nanosphere Lithography to Enhance the Field Emission Properties of a Self Aligned Nanocarbon Based Field Emitters** **Nirupama M.P**, Satyanarayana B.S., O.S. Panwar BML Munjal University, India	211
11:28	**Field Emission Characteristics of ZnO Nanowire Driven by Pulsed Voltage** **Deyi Huang**, Yangyang Zhao, Shuai Wang, Guofu Zhang, Juncong She, Shaozhi Deng, Ningsheng Xu and Jun Chen Sun Yat-sen University, China	213
11:30	**Efficient fabrication of vertical carbon nanotube array cold cathode using laser cutting** **Chuyang Liao**, Jiupeng Li, Xiaoyu Qin, Qi Bo, Baohong Li, Shaozhi Deng, Yu Zhang Sun Yat-sen University, China	215

11:32	**Functionalize of vertically aligned CNTs emitter (C-beam) for surface modification and patterning of self-assembled monolayers (SAM)** **Alfi Rodiansyah**, Kyu Chang Park Kyung Hee University, Korea	217

Novel emitters

11:34	**Field emission properties of line-shape CNT field emitters** **Jun Soo Han**, Sang Heon Lee, Han Bin Go, Si Eun Han and Cheol Jin Lee School of Electrical Engineering, Korea University, Korea	NA
11:36	**Field emission behaviour of fresh and aged Sb2Te3 nanosheets** **Somnath R. Bhopale**, and Mahendra A. More Pune University, India.	221
11:38	**PtSe₂ Nanosheets as Efficient Field Emitter** **Mahendra S. Pawar**, Mahendra A. More, and Dattatray J. Late National Chemical Laboratory, Pune, India	223
11:40	**Electron emission from a solvothermally synthesized ZnS-RGO nanocomposite field emitter** **Sanjeewani R. Bansode**, Mahendra A. More, Rishi B. Sharma Savitribai Phule Pune University, India.	225
11:42	**Low-Macroscopic-Field Electron Emission from Metal Thin Films** **I.S. Bizyaev**, P.G. Gabdullin, M.A. Chumak, V.Ye. Babyuk, S.N. Davydov, A.V. Arkhipov, O.E.Kvashenkina Peter the Great St. Petersburg Polytechnic University, Russia	227

RF and Xrays from electron beams

11:44	**Cold cathode electron gun based on single wall carbon nanotubes field emitters for THz traveling wave tube** **Ruirui Jiang**, Baoqing Zeng, Jianlong Liu, Kaiqiang Yang, and Jing Zhao University of Electronic Science and Technology of China, China	229

Theory of Emision : Ab Initio

11:46	**First-Principle Model of the Electron Field Emission From Silicon Nano-Scale Tip** Gleb D. Demin, Nikolay A. Djuzhev, Nikolay N. Patyukov, and **Ilya D. Evsikov** National Research University of Electronic Technology (MIET), Russia	231

Theory of Emission : Classic Quantum Tunneling

11:48	The notional emission area for cylindrical posts and its variation with local electric field **Rajasree** Ramachandran, Debabrata Biswas Homi Bhabha National Institute, India	233

Vacuum Nano Electronics

11:50	**Cascade Electron Source Based on Horizontal Tunneling Junction** **Zhiwei Li,** Xianlong Wei Peking University, China	235

11:52	**Degradation of an emitter based on VACNT made by DC-PECVD during field emission** **M.A. Chumak,** A.A. Rokacheva, L.A. Filatov, I.S. Bizyaev, E.O. Popov, S.V. Filippov, A.G. Kolosko Peter the Great St.-Petersburg Polytechnical University, Russia	237
11:54	**Analysis of The Field Emission Current From an Array of Silicon Field Nanoemitters For Portable X-Ray Systems** **Petr Yu. Glagolev,** Gleb D. Demin, Nikolay A. Djuzhev, Ilya D. Evsikov, and Nikolay A. Filippov, National Research University of Electronic Technology (MIET), Russia	239
11:56	**Experimental study of the multi-tip field emitter based on the array of silicon pyramidal microstructures** Ilya D. Evsikov, Gleb D. Demin, Tatiana A. Gryazneva, Maksim A. Makhiboroda, Nikolay A. Djuzhev, Oleg V. Pankratov, Eugeni O. Popov, Sergey V. Filippov, Anatoly G. Kolosko and **Maksim A. Chumak** Peter the Great St.-Petersburg Polytechnical University, Russia	241
11:58	**Technology of the fabrication of Mo-based diode and triode structures with nanoscale vacuum gap** **Tatiana A. Gryazneva,** Nikolay A. Djuzhev, Gleb D. Demin, Nikolay A. Filippov, Ilya D. Evsikov and Maksim A. Makhiboroda National Research University of Electronic Technology (MIET), Russia	243

12:00-13:00	Pause

W2 - Novel Emission Mechanisms 1

Chair- Arya Fallahi

Time(Lyon)	Title and Authors (speaker bold underlined, institute of speaker only)	Page
13:00-13:45 Plenary W2.1	**A Plasmon-Mediated Cold-Cathode** **Shaozhi Deng,** Yan Shen, Huanjun Chen, Ningsheng Xu Sun Yat-sen University, China	56
13:45-14:00 Oral W2.2	**Enhancement of thermionic emission and conversion characteristics using polarization and band-engineered n-type AlGaN cathodes** **Shigeya Kimura,** Hisashi Yoshida, Hisao Miyazaki, Takuya Fujimoto, and Akihisa Ogino Toshiba Corporation, Japan	58
14:00-14:30 Invited W2.3	**Planar type electron emission device using atomic layered materials and it applications** **Katsuhisa Murakami,** Naoyuki Matsumoto, Yukino Kameda, Yoshinori Takao, Yoichiro Neo, Yoichi Yamada, Kazutaka Mitsuishi, Masahiro Sasaki, Hidenori Mimura, and Masayoshi Nagao National Institute of Advanced Industrial Science and Technology, Japan	60
14:30-14:45 Oral	**Mechanism of electron emission from graphene/hexagonal boron nitride heterostructure: Implication on MIM planar cathode**	NA

W2.4	**Yicong Chen**, Zhibing Li, Jun Chen Sun Yat-sen University, China	
14:45-15:00 Oral W2.5	**Oxygen Resistance Investigation of Graphene-Oxide-Semiconductor Planar-Type Electron Sources for Low Earth Orbit Applications** **Naoyuki Matsumoto**, Yoshinori Takao, Masayoshi Nagao, and Katsuhisa Murakami Yokohama National University, Japan	64

14:45-15:15	Pause

W3 - Novel Emission Mechanisms 2

Chair: Jun Chen

Time(Lyon)	Title and Authors (speaker bold underlined, institute of speaker only)	Page
15:15-15:45 Invited W3.1	**Development of highly spin-polarized field emitter using Heusler alloy Co2MnGa** **Shigekazu Nagai** Mie University, Japan	66
15:45-16:00 Oral W3.2	**A HfC nanowire field emission point electron source** **Shuai Tang**, Jie Tang, Ta-Wei Chiu, Wataru Hayami, Lu-Chang Qin National Institute for Materials Science, Tsukuba, Japan	68
16:00-16:15 Oral W3.3	**Field Emission from Genuine Graphene: An Experimental Study** **Philippe Poncharal**, Anthony Ayari, Pascal Vincent, Sorin Perisanu, Stephen T. Purcell University Claude Bernard Lyon 1 / CNRS, France	70
16:15-16:30 Oral W3.4	**Combined effect of single-electron charging and quantum confinement on field electron emission from heterostructured nanotips** **Victor I. Kleshch** Moscow State University, Russia	72
16:30-16:45 Oral	**Negative Differential Resistance in Laser-Assisted Field Emission from Si Nanowires** M. Choueib, A. Derouet, P. Vincent, A. Ayari, P. Poncharal, C. S. Cojocaru, **R. Martel**, S.T. Purcell Université de Montréal, Canada	74

16:45-17:00	Pause

W4- Vacuum Nano/Micro Devices

Chair: Rupert Schreiner

17:00-17:30 Invited W4.1	**Vertical Si Nano Vacuum Channel Transistors: Building Blocks for Empty State Electronics** **Akintunde I. Akinwande,** Girish Rughoobur, Nedeljko Karaulac, Winston Chern and Olusoji O. Ilori	76

	Massachusetts Institute of Technology, USA	
17:30-17:45 Oral W4.2	**Ion-Atomic clocks with Spindt type Field Emitter Array** **John D. Prestage**, Christopher Holland, Thai Hoang, Sang Chung, Thanh Le, Nan Yu Jet Propulsion Laboratory, USA	NA
17:45-18:00 Oral W4.3	**Investigation on the Emission Behaviour of p-doped Silicon Field Emission Arrays with Individually Controllable Single Tips** **Philipp Buchner**, Vitali Bomke, Matthias Hausladen, Simon Edler, Michael Bachmann, Rupert Schreiner Ostbayerische Technische Hochschule (OTH) Regensburg, Germany	78
18:00-18:15 Oral W4.4	**Failure Mode of Si Field Emission Arrays based on Emission pattern analysis** **Reza Farsad Asadi**, Tao Zheng, Jaime da Silva, Girish Rughoobur, Akintunde I Akinwande, Bruce Gnade Massachusetts Institute of Technology, USA	80
18:15-18:30 Oral W4.5	**Field Emission Arrays from Graphite Fabricated by Laser Micromachining** **Robert Ławrowski**, Michael Bachmann and Rupert Schreiner Ostbayerische Technische Hochschule (OTH) Regensburg, Germany	82
Oral 18:30-18:45 W4.6	**Effects of Ultra Violet Light Exposure on Gated Silicon Field Emitter Arrays** **Ranajoy Bhattacharya**, Mason Canon, Nedeljko Karaulac, Girish Rughoobur, Winston Chern, Akintunde I. Akinwande and Jim Browning Boise State University, USA	84
18:45-19:00 Oral W4.7	**Emission Behavior of Planar Nano-Vacuum Field Emitters** **Marco Turchetti**, Yujia Yang, Mina R. Bionta, Alberto Nardi, Luca Daniel, Karl K. Berggren, Philip D. Keathley Massachusetts Institute of Technology, USA	86

Thursday, 8th July, 2021

Th1– Poster Session Time Zone B

Chair tbd

Time(Lyon)		
9:45 11:45	Following the W1 - Poster Flashes Time Zone B.	

Th2 – Shoulder Gray Spiçndt Award

Chairman: tbd

Time(Lyon)		
11:45 12:00	Heinz Busta announces SGS award winner.	

Th3 – Presentation IVNC 2022 South Korea

Chair tbd

Time(Lyon)		
12:00 12:30	Professor Park : South Korea attributes for the IVNC 2022	

12:30-13:00	Pause

Th4 - High frequency EM radiation from Electron Beams

Chair: Peter Hommelhoff

13:00-13:45 Plenary Th4.1	**Evolution of traveling wave tubes towards sub-THz frequency** **Claudio Paoloni** Lancaster University, UK	NA
13:45-14:15 Invited Th4.2	**Terahertz Acceleration Technology Towards Compact Light Sources** <u>**Arya Fallahi**</u>, ETH Zurich, Switzerland	89

14:15-14:45	Pause

Th5 – Applications and their modelisation

Chair: tbd

Time(Lyon)	Title and Authors (speaker bold underlined, institute of speaker only)	Page
14:45-15:15 Invited Th5.1	**High performance cold cathode CNT x-ray tube** Sang Heon Lee, Jun Soo Han, Han Bin Go, Si Eun Han and <u>**Cheol Jin Lee**</u> Korea University, South Korea	NA
15:15-15:30 Oral Th5.2	**Direct-Conversion X-Ray Detectors Based on ZnO Nanowire Field Emitters Grown on Ga2O3 Photoconductors** <u>**Zhipeng Zhang**</u>, Manni Chen, Xinpeng Bai, Huanjun Chen, Shaozhi Deng, Jun Chen Sun Yat-sen University, China	NA

Time	Title and Authors	Page
15:30-15:45 Oral Th5.3	**A novel current dependent field emission performance test** <u>**Florian Herdl**</u>, Michael Bachmann, Dominik Wohlfartsstätter, Felix Düsberg, Markus Dudeck, Magdalena Eder, Manuel Meyer, Andreas Pahlke, Simon Edler, Andreas Schels, Walter Hansch, Rupert Schreiner KETEK GmbH, Germany	95
15:45-16:00 Oral Th5.4	**Designing Micro-gap Thermionic Energy Harvesters** **Ehsanur Rahman** and Alireza Nojeh University of British Columbia, Canada	97
16:00-16:15 Oral Th5.5	<u>**Proposal for a Negative Capacitance Vacuum Field Effect Transistors with sub-60mV/dec Subthreshold Swing**</u> <u>**N. Hernandez**</u>, M. Cahay, J. Ludwick, and T. Back University of Cincinnati, Cincinnati, USA	99

16:15-16:30	Pause

Th6 - Nano-Micro Emitters (Nanotubes, Nanowires, Spindt and micro cathodes, etc.)

Chair: Alizera Nojeh

Time(Lyon)	Title and Authors (speaker bold underlined, institute of speaker only)	Page
16:30-16:45 Oral Th6.1 NME2.1	**Direct in situ Electron Microscope Synthesis of CNTs with Applied Electric Field and Field Emission** <u>**P. Vincent**</u>, F. Panciera, I. Florea, M. Ezzedine, M.-R. Zamfir, S. Perisanu, C. Cojocaru, N. Blanchard, D. Pribat, S.T. Purcell, P. Legagneux University Claude Bernard Lyon 1 / CNRS, France	101
16:45-17:00 Oral Th6.2 NME 2.2	**Effect of Substrate Conductivity on Si Self-Assembled Field Emission Arrays** <u>**Shabnam Ghotbi**</u>, Saeed Mohammadi Purdue University, USA	103
17:00-17:15 Oral Th6.4 NME 2.4	**Strongly anisotropic field emission from highly aligned carbon nanotube films** <u>**S. B. Fairchild**</u>, T. A. de Assis, J. H. Park, M. Cahay, J. Bulmer, D.E. Tsentalovich, Y. S. Ang, L. K. Ang, J. Ludwick, P.T. Murray, Y. Zhou, P. Zhang Wright-Patterson Air Force Base, USA	NA
17:15-17:30 Oral Th6.5	**A Universal Multiscale Method for Rapid Determination of Local Emission Current Density from Nanoscale Emitters** <u>**J. Ludwick**</u> and T. C. Back, M. Cahay, N. Hernandez, H. Hall, J. O'Mara, K. L. Jensen, J. H. B. Deane, R. G. Forbes Air Force Research Laboratory,USA	NA

Friday, 9th July, 2021

F1 Theory of Emission : Ab Initio

Chair: Thiago A. de Assis

Time(Lyon)	Title and Authors (speaker bold underlined, institute of speaker only)	Page
13:00-13:30 Invited F1.1	**Field emitters at atomic scale – insights from order-N density functional theory** **C. J. Edgcombe** University of Cambridge, United Kingdom	109
13:30-13:45 Oral F1.2	**Thermal-Field Electron Emission from Three-Dimensional Cd_3As_2** **Wei Jie Chan**, Yee Sin Ang, and L. K. Ang Singapore University of Design and Technology, Singapore	111
13:45-14:00 Oral F1.3	**Field emission from two dimensional materials:a quantum mechanical model and its application to graphene** **Bruno Lepetit** Université Toulouse III Paul Sabatier / CNRS, France	113
14:00-14:15 Oral F1.4	**Tunneling Delay and the Modeling of Electron Emission** **Kevin L. Jensen**, Joel L. Lebowitz, Jeanne M. Riga, Andrew Shabaev, Donald A. Shier, Rebecca Seviour Naval Research Laboratory, USA	NA
14:15-14:30 Oral F1.5	**Theoretical analysis of efficiency of plasmonic photoemission from single silver nanospheres** **Shisong Luo**, Yicong Chen, Zhibing, Jun Chen Sun Yat-sen University, China	NA

14:30-15:00	Pause

Theory of Emission : Classic Quantum Tunneling

Chair: John Xanthakis

Time(Lyon)	Title and Authors (speaker bold underlined, institute of speaker only)	Page
15:00-15:15 Oral F2.1	**General scaling laws of space charge effects in field Emission** **A. Kyritsakis**, M. Veske, V. Zadin and F. Djurabekova University of Tartu, Estonia	117
15:15-15:30 Oral F2.2	**Absence of space-charge-limited current from field emission due to non-FN law** **Cherq Chua**, Chun Yun Kee, Yee Sin Ang, Lay Kee Ang Singapore University of Technology and Design, Singapore	119
15:30-15:45 Oral F2.3	**Behavior of notional cap-area efficiency (gn) for hemisphere-on-plane and related field emitters** **S.V. Filippov**, A.G. Kolosko, E.O. Popov, Richard G. Forbes Ioffe Institute, Russia	121

15:45-16:00 Oral F2.4	**Does a banal tungsten field emitter obey the field emission theory?** <u>Anthony Ayari</u>, Pascal Vincent, Sorin Perisanu, Philippe Poncharal, Stephen T. Purcell University Lyon1/CNRS, France	123
16:00-16:15 Oral F2.5	**A Generalized Formula for Barrier Strength (Gamow Factor), applicable to various field ion and electron emission contexts** <u>Richard G. Forbes</u> University of Surrey, UK	125

16:15-???	Closing statements

TECHNICAL DIGEST

2021 34th International Vacuum Nanoelectronics Conference (IVNC)

2nd IVNC virtual conference hosted in Lyon, France, July 5-9, 2021.

TECHNICAL DIGEST

2021 34th International Vacuum Nanoelectronics Conference (IVNC)

July 5-9, 2021.

Editors and Chairmen
Stephen Purcell
Jean-Paul Mazellier

Welcome to the 2021 IVNC Virtual Conference

Dear Colleagues,

On behalf of the International Steering Committee (ISC) and the 2021 IVNC Committees, we are happy to welcome all of you to the 34th IVNC, virtual version.

We were constrained to hold this meeting in a virtual format, as with last year, due to the world-wide sanitary crisis. However, rather than the reduced scientific offering in 2020 we have aimed to create a full conference with leading plenary, invited and tutorial speakers. The response of our scientific community has been outstanding with around 100 submissions for oral and poster presentations covering a large swath of free electron physics going well beyond our core base of field emission tip sources.

This Technical Digest contains over 100 Extended Abstracts spanning all the main topics treated by the IVNC for electron sources and their applications in the vacuum: industriel applications, new and refined nano and micro emitters, theory of emission, vacuum devices, new forms of emission, new developments in source metrology and modeling, and in closing a special mention for the growing interest in the interaction of em radiation and electrons that may allow us to also play a role in the second quantum revolution.

The conference will be held over five days based on Lyon time. The first tutorials day begins with two technological subjects in electron and ion sources rooted in modern microfabrication, followed by three complementary and quite complete theoretical courses on tunneling emission and source physics within the framwork of quantum mechanics.

Tuesday is the official opening day with our lead plenary speaker being Peter Hommelhiof of the Friedrich-Alexander-Universität in Erlangen, chair for Laser Physics, well known for his pioneering work in photon assisted electron emission with femto second lasers. Professor Shaozhi Deng of Sun Yat Sen university will open the main wednesday session with a talk on the rising subject of plasmon mediated photoemissin which is currently sparking interest for a new form of solar energy recuperation. Professor Claudio Paoloni of Lancaster University will open the main Thursday session with a talk on pushing tube electron RF amplifiers to the THz domain where there are great demands from the communication industry. The Friday is dedicated to theoretical work with a bigger offer of ab intio models than in the past but also an equal effort on the challenges of classic tunneling theory particulary applied to nano emitters.

The conference is broken into defined sessions that, together with the invited talks, contain 10 minute live talks, all drawn from the submitted short abstracts, each followed by 5 minute live question periods. As well, two poster sessions and accompanying two min flash presentations are planned which contain the largest number of submissions. The attendees are encouraged to ask questions in the public forum and in private. The posters may be available to participants in a separate pdf file as a simple way for participants to access them particularly during the poster sessions. The platform to be used for the conference is Zoom.

Looking forward to a fulfilling conference and interesting scientific exchanges within a wonderful spirit of international cooperation.

Stephen Purcell and Jean-Paul Mazellier

IVNC 2020 and 2021 Chairmen. Lyon and Paris, France, July 2021.

International Steering Committee (ISC)

Prof. Akintunde I. Akinwande, Massachusetts Institute of Technology, USA

Dr. Heinz H. Busta, Prairie Prototypes and University of Illinois at Chicago, USA

Prof. Marc Cahay, University Cincinnati, Ohio, USA

Prof. Shaozhi Deng, Sun Yat-sen University, P. R. China

Prof. Jan Dziuban, Wroclaw University of Technology, Poland

Prof. Yasuhito Gotoh, Kyoto University, Japan

Dr. Christopher Holland, SRI International, USA

Prof. Charles Hunt, University of California, Davis, USA

Dr. Hans W. P. Koops, HaWilKo GmbH, Germany

Prof. Cheol Jin Lee, Korea University, Republic of Korea

Prof. Hidenori Mimura Shizuoka, University, Japan

Prof. Günter Müller, University of Wuppertal, Germany, honorary - deceased.

Prof. Alireza Nojeh, Quantum Matter Institute, University of British Columbia, Canada

Prof. Rupert Schreiner, OTH Regensburg, Germany

Dr. Jonathan Shaw, Naval Research Laboratory, USA

Dr. Charles Spindt, SRI International, USA

Prof. Mikio Takai, Osaka University, Japan

Dr. Soichiro Tsujino, Paul Scherrer Institute, Switzerland

Dr. Stephen Purcell, University Claude Bernard Lyon1, France

Prof. Ningsheng Xu, Fudan University, P. R. China, Chairman IVNC-ISC

IVMC/IVNC Conference History and Chair Persons

IVMC 1988 Williamsburg, USA Dr. Henry Gray and Dr. Capp Spindt
IVMC 1989 Bath, U.K. Dr. Rosemary Lee, Dr. Cyril Hilsum,
and Prof. Johannes Mitterauer
IVMC 1990 Monterey, USA, Dr. Capp Spindt and Dr. Henry Gray
IVMC 1991 Nagahama, Japan, Prof. Susumu Namba and Dr. Takao Utsumi
IVMC 1992 Vienna, Austria, Prof. Johannes Mitterauer
IVMC 1993 Newport, USA, Dr. Mark Hollis
IVMC 1994 Grenoble, France, Dr. Robert Baptist and Dr. Robert Meyer
IVMC 1995 Portland, USA, Prof. Bill Mackie and Dr. Tony Bell
IVMC 1996 St. Petersburg, Russia, Prof. Georgiy Furgey and Dr. Vladimir Makov
IVMC 1997 Kyongju, Korea, Prof. Jong Duk Lee
IVMC 1998 Asheville, USA, Prof. John Hren
IVMC 1999 Darmstadt, Germany, Dr. Hans Koops
IVMC 2000 Guangzhou, P. R. China, Prof. Ningsheng Xu
IVMC 2001 Davis, USA, Prof. Charles Hunt
IVMC 2002 Lyon, France, Prof. Vu Thien Binh (Joint with IFES)
IVMC 2003 Osaka, Japan, Prof. Mikio Takai
IVNC 2004 Cambridge, USA, Prof. Akintunde Akinwande
IVNC 2005 Oxford, U.K., Prof. Ejaz Huq
IVNC 2006 Guilin, P. R. China, Prof. Ningsheng Xu (Joint with IFES)
IVNC 2007 Chicago, USA, Dr. Heinz Busta
IVNC 2008 Wroclaw, Poland, Prof. Jan Dziuban
IVNC 2009 Hamamatsu, Japan, Prof. Hidenori Mimura
IVNC 2010 Palo Alto, USA, Dr. Chris Holland
IVNC 2011 Wuppertal, Germany, Prof. Günter Müller
IVNC 2012 Jeju, Korea, Prof. Cheol Jin Lee, Dr. Yong Churl Kim,
and Kyu Chang Park
IVNC 2013 Roanoke, USA, Dr. Jonathan Shaw and Dr. Kevin Jensen
IVNC 2014 Engelberg, Switzerland, Dr. Soichiro Tsujino and Prof. Jens Gobrecht
IVNC 2015 Guanzhou, P. R. China, Prof. Shaozhi Deng, Prof. Juncong She,
and Prof. Jun Chen
IVNC 2016 Vancouver, Canada, Prof. Alireza Nojeh
IVNC 2017 Regensburg, Germany, Prof. Rupert Schreiner
IVNC 2018 Kyoto, Japan, Prof. Yasuhito Gotoh
IVNC 2019 Cincinnati, USA, Professor. Marc Cahay
IVNC 2020 Virtuel, Lyon France, Stephen Purcell, Jean-Paul Mazellier

Sponsors of IVNC 2021

Gap in pagination due to formatting issues.

Pages 1-29

Micro-Nano Fabrication of Integrated Tip-Based Field Electron Emission Devices

Juncong She
State Key Laboratory of Optoelectronic Materials and Technologies,
Guangdong Province Key Laboratory of Display Material and Technology,
School of Electronics and Information Technology,
Sun Yat-sen University, Guangzhou 510275, People's Republic of China
*Contact: shejc@mail.sysu.edu.cn

Abstract— **This talk present the state of the art processes in micro-nano fabrication of well-defined integrated metal and Si tip-based vacuum micro-nano electronic devices, the field electron emission performance, and the related mechanisms.**

I. INTRODUCTION

Field electron emission sources possess merits of radiation hardness, temperature tolerance, and fast-switching, which is potential for modern vacuum micro/nano electronic applications. The integrated addressable and uniform tip-based field electron emission devices with low driving voltage and high emission current is regarded as one of the promising candidates. Nowadays, tip-based cathode is the most studied on-chip integrated device structure. The fabrication of the integrated tip-based field emitters needs specific procedures for precisely defining the tip-shape and the integrated driving/focus electrodes, by using the IC compatible processes of lithography, etch, and metallization. Also, rational device structures are needed in achieving the improvement on electron emission performance. This tutorial talk will introduce the rational processing details for achieving well defined integrated tip-based (metal and Si) vacuum micro-nano electronic devices, together with the field electron emission performance and the related mechanisms.

II. RESULTS AND DISCUSSION

The first part of the talk will introduce the fundamentals of micro-nano fabrication techniques, i.e., high-resolution lithography, isotropic/anisotropic etch, and thin film deposition (metallization). The second part will review the state of the art fabrication of the integrated Spindt-type metal tip cathode. Typically, the Spindt-type metal-tip cathode is fabricated by depositing the metal-flux into the pre-defined micro-cavities to form the sharp tips by self-shrinking the cavity aperture. [1] Discussions will focus on the rational methods been developed in reducing the "internal stress" of the deposited films which may cause the distortion of the tips. They are, (i) heating up the sample (~200 °C) and performing a glancing angle co-deposition of alumina to reduce the internal stress in the cathode-metal film. The compressive stressed alumina can offset the tensile stress in the cathode-metal. [2] (ii) Use double-layer photoresist as sacrificial layer and employ a high power pulsed magnetron sputtering system with controlled positive plasma potential and efficiently accelerate ion species to realize the stress-releasing. [3] (iii) Employed metals (nickel or titanium) with lower melting point (corresponding to a weak internal-stress). [4] In addition, a newly developed in-situ stress-released fabrication by employing the patterned photoresist micro-cavities with surrounding trenches will present. Uniform gated Cr-tips with high current density field emission were achieved. The third part of the talk will introduce our recent development in micro/nano fabrication of gated Si-tips with individually integrated hyperbolic nano-channels. [5] Discussions will focus on the non-uniaxial stress assisted fabrication of the integrated hyperbolic nano-channel on vertically aligned Si-tips, [6] followed by the rational processing in fabrication of gated devices. The integrated device of p-type Si showed stable non-saturated high current density field electron emission. No tip profile change was observed after the tests, suggested a well reliability. The findings were interpreted by the mechanism of extension of depletion layer underneath the gate-metal to provide generation current for non-saturated emission, while the hyperbolic nano-channel impeding the current overloading and prevent the tip-emitters from a breakdown. Accordingly, the emission uniformity in an array and thus the total emission current was improved.

III. SUMMARY

This tutorial talk aims to provide useful technical guidelines for developing homogeneous and reliable tip-based field electron micro/nano-emitters for modern vacuum electronic applications.

ACKNOWLEDGMENT

The work was supported by the NSFC (Grant No. 61874144, 51702374), the Science and Technology Department of Guangdong Province (Grant No. 2018B030311045), and the Fundamental Research Funds for the Central Universities.

REFERENCES

[1] C. A. Spindt, I. Brodie, L. Humphrey, and E. R. Westerberg, J. Appl. Phys. 47, 5248 (1976).

[2] C. A. Spindt, I. Brodie, C. E. Holland, and P. R. Schwoebel, Chapter 4, Vacuum Microelectronics, edited by W. Zhu (Wiley, New York, 2004).

[3] T. Nakano, T. Narita, K. Oya, M. Nagao, and H. Ohsaki, J. Vac. Sci. & Technol. B, 35 (2), 022204 (2017).

[4] M. Nagao, Y. Gotoh, Y. Neo, and H. Mimura, J. Vac. Sci. Technol. B 34, 02G108 (2016).

[5] Z. J. Huang, Y. F. Huang, Z. X. Pan, J. C. She, S. Z. Deng, J. Chen, and N. S. Xu, Appl. Phys. Letts. 109, p. 233501 (2016).

[6] M. X. Zeng, X. M. Li, Y. F. Huang, Z. J. Huang, R. Z. Zhan, J. Chen, N. S. Xu, J. C. She, S. Z. Deng, Nanotechnology 30, 365601 (2019).

Ion sources and optical charged particles dedicated to FIB technology today. Current trends and challenges in semiconductors, failure analysis and HR SIMS.

Arnaud HOUËL, Anne DELOBBE, Justine RENAUD, Matthieu VITTEAU

Orsay Physics, Fuveau 13710, France

Contact: arnaud.houel@orsayphysics.com, phone +33 442 538 090 Web site: https://www.orsayphysics.com/

I. INTRODUCTION

The focused ion beam (FIB) market is expected to grow from $820 million in 2019 to $1,185 million in 2024 and is already a well-known technology for electronics and semiconductor industry, materials science and it becomes more and more present in life sciences area.

In this tutorial, will be presented main properties of a charged particle optics for FIB applications.

Firstly, basic notions of charges particles will be reviewed. This will allow to identify the performances and limitations of ion sources combined with optics elements.

Secondly an overview of the different types of ions sources will be presented in the particular example of FIB. Because specifications obtainable in terms of spot size or beam current are very much dependant of the sources characteristics.

After then we will discuss the trends in FIB technologies and major challenges in fields of material science, failure analyses (and semiconductor in general) or high resolution secondary mass spectrometry.

II. FIB AND OPTICAL COMPONENTS

Electrostatics deflectors are the simplest elements used in charged particle optics. One of the characteristics of this type of deflectors is that particles of different masses and charges are all deflected with the same amplitude of deflection if their initial velocity is obtained by the same potential of acceleration as shown Fig. 1.

This is why all ion beam transportation use electrostatic deflectors or electrostatic lenses to be independent of q/m ratios during beam transportation, beam correction or focalization. Only brightness, energy spread and emittance of source are affecting the final spot size of ion beam considering constant the transfer function of an optical system [1], [2].

A FIB column consists of an ion source block, a condenser lens, motorized movable apertures, a blanking, some steerers, a double deflector stage to scan the beam, a stigmator octopole and an objective lens. All these components are of electrostatic type.

$$y = \frac{q.D.l.E}{m.v_0^2} \qquad \frac{2q.E_0}{m} = v_0^2 \qquad \boxed{y = \frac{D.l.E}{2.E_0}}$$

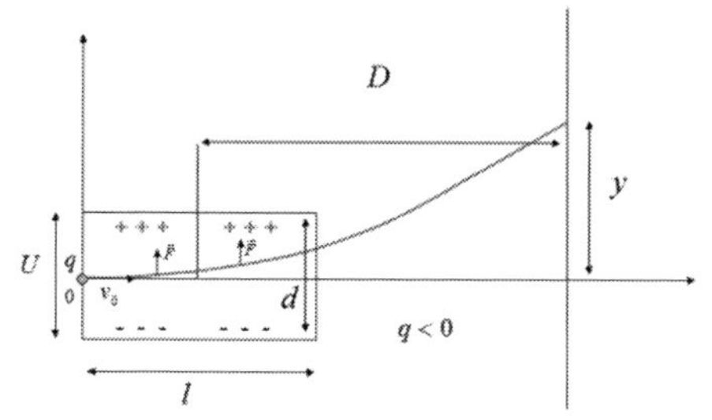

Fig. 1 Properties of electrostatic deflexion

Fig. 2 shows principle of electrostatic Einzel lens. This type of lens consists of three rotationally symmetric electrodes. The focal power of such lenses is adjusted by the value of bias voltage applied to the medium set of electrode.

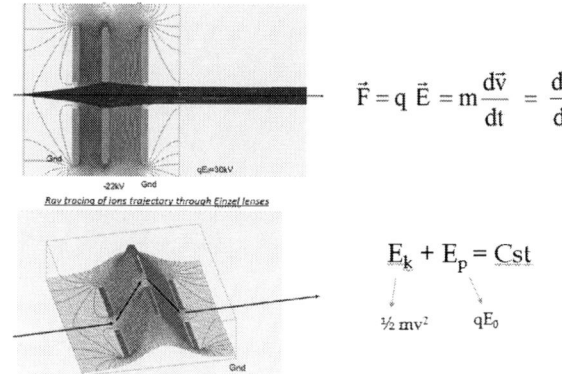

$$\vec{F} = q\,\vec{E} = m\frac{d\vec{v}}{dt} = \frac{d\vec{p}}{dt}$$

$$E_k + E_p = Cst$$

$\tfrac{1}{2}mv^2 \qquad qE_0$

Fig. 2 Properties of electrostatic Einzel lens

All optical elements constituting FIB column allow to correct, to focus charged particles and to play with the optical transfer function.

Final spot size is then the root mean square of the Gaussian diameter, spherical aberration diameter and chromatic aberration diameter as shown in following equation.

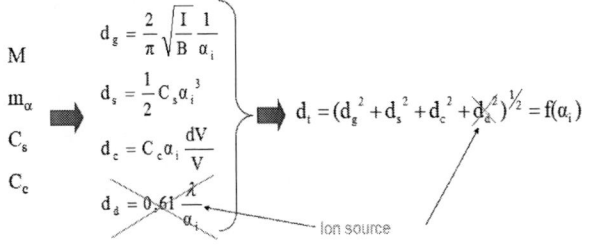

$$d_g = \frac{2}{\pi}\sqrt{\frac{I}{B}}\frac{1}{\alpha_i}$$

$$d_s = \frac{1}{2}C_s\alpha_i^3$$

$$d_c = C_c\alpha_i\frac{dV}{V}$$

$$d_d = 0{,}61\frac{\lambda}{\alpha_i}$$

$$d_t = (d_g^2 + d_s^2 + d_c^2 + d_d^2)^{1/2} = f(\alpha_i)$$

We can find an optimization of α_i (image angle) giving us better density of current in final spot size. [1]

Geometries of sources and ionization process involved are also two important factors affecting final optical features. The existing features of a source are (1) Emission characteristic I(V) (2) Virtual source size (3) Angular density (4) Brightness (5) Energy spread (6) Life time.

III. IONS SOURCES FOR FIB TECHNOLOGY

All charged particles source requires a material supply function, an excitation energy supplied on a surface element or in a volume and finally an extraction system as shown in Fig. 6.

The combination of these parameters influences and affects drastically the optical features of the ion source.

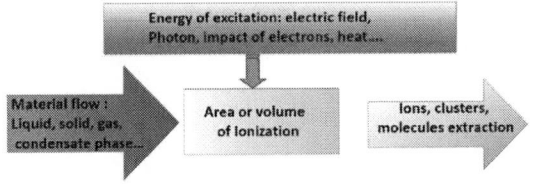

Fig. 6 Principle an of ions source

Operating conditions can affect also specifications of ions sources. E.g. current of emission used affect energy spread.

The existing ionization mechanisms are (1) Surface ionization (2) Ionization by impact (3) Field ionization (4) Field desorption (5) Field evaporation

It is difficult and even impossible to find a single ion source able to satisfy all the applications existing today in the field of FIB, because the needs are numerous and different.

Source type	Ions	Brightness, A.m^{-2}.sr^{-1}.V^{-1}	Energy spread (FWHM), eV	Ion current, nA max	Beam diameter (FWHM), nm
LMIS	Ga, Bi, In	10^6	[5;10]	100	2,5
LMIAS	Au, Ge, Si, Li, Be, B, Co and others	10^6	[5;25]	100	2,5
LICIS	[R]$^+$,[A]$^-$, {[R]$^+$[A]$^-$}$_m$[R]$^+$...	10^5-10^6	[5;15]	<100	5-50
ECRIS ICPIS RFPIS	Xe, Kr, Ar, O, N and others	10^4	[1;5]	<10^4	20
MOTIS	Li, Cs, Rb, Cr	10^7	0,35	0,5-1	5-30
CAIS	Li, Cs, Rb, Cr	10^7	0,3	0,5-1	1
NAIS	Inert gases, H	10^4	[1;2]	<0,23	50-60
GFIS	Inert gases	10^9	1	<0,025	0,35

Fig. 8 Types of Ions sources ([3], [4])

The table shows in Fig. 8 the different ions sources existing with their characteristics:

Another important factor is the angular density of the ions source. For example the RFPIS source (Radio-Frequency Plasma Ions Source) are used today in FIB technology for the realization of TEM lamella preparations because of their high angular density and low solid angle of emission. This type of source allows to reach ionic currents of the order of 1µA in a spot size diameter of 10µm, which is very useful for addressing the failure analysis of 3D technologies where big volume of material must be removed. This final density of current is absolutely not reachable with LMIS having a half angle of emission closed to 300mr [4]. Involving pretty high diameter of spherical aberration. Nevertheless LMIS Ga has however a higher brightness.

IV. CONCLUSIONS

The next challenge for ion sources are multiple and directly related to the scope of FIB use. For circuit editing, high brightness and low energy dispersion are needed to address <5nm technology and 2D devices where the size of the region of interest is closed to several nanometer in width and thickness. For these types of applications, users want to maintain ultra-high milling resolution, but with acceptable sputtering rate to allow time to stop milling in the layer of interest. In this case, it is absolutely necessary to reduce the accelerating bias voltage of ions while maintaining milling resolution. Low energy spread is absolutely necessary for such applications.

Today Gas field ion sources emitting light ions such as Helium are used for imaging and for etching heavier ions such as Neon. Other species can also be emitted, such as Ar, Kr, which are heavier than Ne and therefore more desirable for sputtering applications [5]. For high resolution SIMS (Secondary Ions Mass Spectroscopy) applications, LICIS (Liquid Ionic Compounds Ions Source) sources are very interesting and promising, having the possibility to combine high brightness, low energy dispersions and the possibility to emit reactive ions. The lifetime of this type of source must however be increased in order to meet industrial applications.

ACKNOWLEDGMENT

I would like to thank Pierre SUDRAUD and Bernard RASSER with whom I had the chance to learn a lot by having the opportunity to work closely with them for several years.

REFERENCES

[1] A. Septier, *Focusing of Charged Particles,* Vol.II, 1967
[2] J. Orloff, *Handbook* of *Charged Particle Optics*, 1997
[3] M. Viteau, M. Reveillard, L. Kime, B. Rasser, P. Sudraud, Y. Bruneau, G. Khalili, P. Pillet, D. Comparat, I. Guerri, A. Fioretti, D. Ciampini, M.Allegrini, and F. Fuso, Ultramicroscopy164, 70,2016
[4] P. Mazarov, V G Dudnikov and A B Tolstoguzov, *Electrohydrodynamic emitters of ion beams, Phy.-Usp.* **63** 1219, 2020
[5] Hong-Shi, Kuo,aIng-Shouh, Hwang, Tsu-Yi Fu, Yi-Hsien Lu, Chun-Yueh Lin, and Tien T. Tsong, *Gas field ion source from an Ir/W‹111‹single-atom tip,* Applied Physics Letters **92**, 63106200, 2008

Gap in pagination due to unavailable papers.

Pages 34-37

Ultrafast electron control with various means: from multiphoton physics at needle tips to nanophotonic particle acceleration

Roy Shiloh, Tomáš Chlouba, Ang Li, Philip Dienstbier, Alexander Tafel, Johannes Illmer, Norbert Schönenberger, Peyman Yousefi, Stefanie Kraus, Leon Brückner, Julian Litzel, Bastian Löhrl, <u>Peter Hommelhoff</u>*

Department of Physics, Friedrich-Alexander-Universität Erlangen-Nürnberg (FAU), Erlangen, Germany, EU
*Contact: peter.hommelhoff@fau.de, phone +49-9131-85-27090, www.laser.physik.fau.eu

Abstract— **Ultrafast control of electrons has evolved tremendously in recent years. In the first part, we will give an overview over the interaction of ultrashort laser pulses with various needle tips. Starting with new types of nanometric emitters and a characterization of their photoemission properties, we will continue with two-color photoemission from needle tips, both in the perturbative quantum-path interference regime as well as in the strong-field regime. In the second part, we will give an overview of nanophotonics-based particle acceleration. Having demonstrated all individual components needed for a laser-based accelerator, including optical acceleration, deflection and complex electron phase space control, we are now able to build the accelerator on a nanophotonic chip.**

I. TWO-COLOR PHOTOEMISSION FROM A NEEDLE TIP

When ultrashort laser pulses with moderate intensities are focused at metal needle tips, electrons can be released from the tip by multiphoton electron emission: An electron, separated in energy by at least the work function from the vacuum level, is lifted above the barrier by the energy imparted to it from several photons with energies smaller than the work function (Fig. 1a). Interestingly, two photons with (angular) frequency ω can be replaced by one photon with energy 2ω (Fig. 1b). Hence, another quantum path exists in addition to the multiphoton channel consisting of ω photons only. The two paths can interfere quantum-coherently, meaning that they may lead to an increase or decrease in photoemission. The exact interference condition can be straightforwardly controlled by the delay of the two colors.

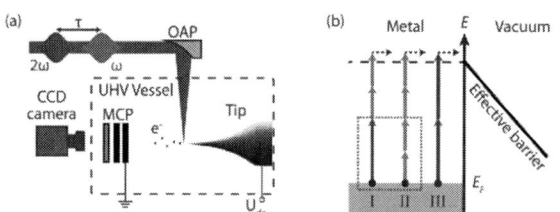

Fig. 1: (a) Femtosecond laser pulses at around 800 nm (ω) and at their second harmonic wavelength or around 400 nm (2ω) are focused at a metal needle tip (W or Au). (b) For intensities smaller than 2×10^{11} W/cm², we observe quantum path interference in multiphoton emission: one 2ω photon quantum-coherently replaces two ω photons. Taken from [1].

We have investigated the quantum-path interference from tungsten tips in a large parameter range. Most importantly, we have scanned the wavelength of the fundamental pulse over almost an octave from 1180 to 2000 nm (while always generating the second harmonic from these fields). We find that the contrast of the quantum path interference as function of delay between the two colors is always larger than 72%, implying a high level of coherence.

Based on these results, we have come up with an insightful theoretical model explaining the large contrast, but also the observed decrease in visibilities for certain intensity ratios. We will discuss the experimental data and the model.

We will also show experimental results with larger intensities and few-cycle two color laser pulses. It is well known that strong-field effects can arise at metal needle tips [2]. We will additionally show how we use our excellent control of the two-color fields ratio to gain deep insights into attosecond dynamics at the surface of a tungsten needle tip.

II. DIAMOND-COVERED NEEDLE EMITTERS

Tungsten needle tips are known to deteriorate over time so that the emission current decreases along with it. We have come 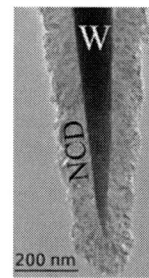 up with a technique to coat a tungsten needle tip with nanodiamonds [3], which is, in addition, a negative electron affinity material. With the help of wavelength-tunable femtosecond laser pulses, we can identify various single-photon and multi-photon emission channels. We find furthermore that the emission is much more stable over time than that from a naked tungsten tip [4].

III. NANOPHOTONICS-BASED PARTICLE ACCELERATION

Particle accelerators are used in science, medicine and industry. They provide deep insights into the foundations of matter, and they are instrumental for radiotherapy and other industrial applications. So far, most particle accelerators operate with microwave cavities, which are rather large and heavy. Nanophotonics-based schemes have been proposed immediately following the invention of the laser that take advantage of the same physical principle, namely to excite an electromagnetic mode that continuously imparts momentum to the charged particle [5; 6]. In 2013, two groups have demonstrated this scheme at optical frequencies and with the help of transparent dielectric structures [7; 8]. Transparent structures allow accelerating gradients much larger than those attainable in conventional RF accelerators, so that much smaller particle accelerators can be conceived, yielding a comparable final electron energy to large RF machines. For example, a 1 m long medical LINAC might be shrunk to 1 cm, which could make it attractive for catheter-based treatment tools [9]. A gradient close to 1 GeV/m has already been reached with relativistic electrons, which is a factor of 40 larger than what conventional accelerators usually operate at [10]. With sub-relativistic electrons, we have reached 210 MeV/m [11].

In recent years, we have demonstrated all elements needed to build a nanophotonic version of a particle accelerator. Next to acceleration, we have shown deflection and collimation structures, all driven with laser fields [12; 13; 14; 15; 16; 17].

Recently, we have also shown that we can generate attosecond electron pulse trains. For this, we imprint a periodic energy modulation on a ~500 fs long electron pulse. After propagation, faster electrons catch up with slower ones, so that electron bunching takes place. In a second laser interaction region, we can measure the electron energy as function of time delay between the laser pulses, which allows us to measure a bunch duration as short as 270 ± 80 as [18].

Last, we will report on a scheme to continuously transport an electron beam through a long nanophotonic structure. With a channel width of just 225 nm and a length of 78 μm, this structure entails phase jumps (gaps in the SEM image above), which make an electron experience repeatedly alternating forces: transversally focusing and, at the same time, longitudinally defocusing, and following a phase jump

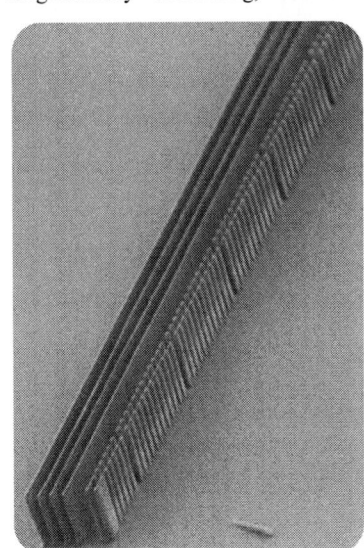

transversally defocusing and longitudinally focusing. This way, the electron beam can be transported through the structure. This alternating phase focusing scheme represents the first demonstration of complex electron phase space control at optical frequencies.

We are now at a point to build a particle accelerator on a chip. We expect to reach electron energies reaching 1 MeV soon.

ACKNOWLEDGMENT

Funding is gratefully acknowledged from the Gordon and Betty Moore Foundation (ACHIP), ERC Grants NearFieldAtto and AccelOnChip, BMBF, DFG and Max Planck Society. We thank the members of the ACHIP collaboration for discussions.

REFERENCES

[1] A. Li, Y. Pan, P. Dienstbier, and P. Hommelhoff, "Quantum Interference Visibility Spectroscopy in Two-Color Photoemission from Tungsten Needle Tips," *Phys. Rev. Lett.*, **126**, 137403 (2021)

[2] M. Krüger, C. Lemell, G. Wachter, J. Burgdörfer, and P. Hommelhoff, "Attosecond physics phenomena at nanometric tips" *J. Phys. B: At. Mol. Opt. Phys.*, **51**, 172001 (2018).

[3] A. Tafel, M. Wu, E. Spiecker, P. Hommelhoff and J. Ristein, "Fabrication and structural characterization of diamond-coated tungsten tips". *Diam. A. Rel. Mat.*, **97**, 107446 (2019)

[4] A. Tafel, S. Meier, J. Ristein and P. Hommelhoff, "Femtosecond laser-induced electron emission from nanodiamond-coated tungsten needle tips", *Phys. Rev. Lett.*, **123**, 146802 (2019)

[5] Shimoda, K. "Proposal for an electron accelerator using an optical maser" *Appl. Opt.* **1**, 33 (1962)

[6] Lohmann, A. W. "Electron acceleration by light waves" *Tech. Rep.*, IBM (1962)

[7] Breuer, J. & Hommelhoff, P. "Laser-based acceleration of nonrelativistic electrons at a dielectric structure." *Phys. Rev. Lett.* **111**, 134803 (2013)

[8] E. A. Peralta, K. Soong, R. J. England, E. R. Colby, Z. Wu, B. Montazeri, C. McGuinness, J. McNeur, K. J. Leedle, D. Walz, E. B. Sozer, B. Cowan, B. Schwartz, G. Travish, and R. L. Byer, "Demonstration of electron acceleration in a laser-driven dielectric microstructure." *Nature* **503**, 91-94 (2013)

[9] England, R. J. et al. "Dielectric laser accelerators." *Reviews of Modern Physics* **86**, 1337-1389 (2014).

[10] D. Cesar, S. Custodio, J. Maxson, P. Musumeci, X. Shen, E. Threlkeld, R. J. England, A. Hanuka, I. V. Makasyuk, E. A. Peralta, K. P. Wootton, and Z. Wu, "High-field nonlinear optical response and phase control in a dielectric laser accelerator." *Commun. Phys.* **1**, 46 (2018)

[11] M. Kozák, M. Förster, J. McNeur, N. Schönenberger, K. Leedle, H. Deng, J.S. Harris, R.L. Byer, P. Hommelhoff, "Dielectric laser acceleration of sub-relativistic electrons by few-cycle laser pulses." *NIMA* **865**, 84-86 (2017)

[12] J. McNeur, M. Kozak, D. Ehberger, N. Schönenberger, A. Tafel, A. Li, and P. Hommelhoff, "A miniaturized electron source based on dielectric laser accelerator operation at higher spatial harmonics and a nanotip photoemitter." *J. Phys. B: At. Mol. Opt. Phys.* **49**, 034006 (2016)

[13] P. Yousefi, N. Schönenberger, J. McNeur, M. Kozák, U. Niedermayer, and P. Hommelhoff, "Dielectric laser electron acceleration in a dual pillar grating with a distributed Bragg reflector." *Optics Letters* **44**, 1520-1523 (2019)

[14] R. Shiloh, T. Chlouba, P. Yousefi, and P. Hommelhoff, "Particle acceleration using top-illuminated nano-photonic dielectric structures." *Optics Express* **29**, 14403-14411 (2021)

[15] J. McNeur, M. Kozák, N. Schönenberger, K. J. Leedle, H. Deng, A. Ceballos, H. Hoogland, A. Ruehl, I. Hartl, R. Holzwarth, O. Solgaard, J. S. Harris, R. L. Byer, and P. Hommelhoff, "Elements of a dielectric laser accelerator." *Optica* **5**, 687-690 (2018)

[16] M. Kozak, J. McNeur, K. J. Leedle, H. Deng, N. Schönenberger, A. Ruehl, I. Hartl, J. S. Harris, R. L. Byer, and P. Hommelhoff, "Optical gating and streaking of free electrons with sub-optical cycle precision," *Nature Communications* **8**, 14342 (2017)

[17] U. Niedermayer, T. Egenolf, O. Boine-Frankenheim, and P. Hommelhoff, „Alternating-Phase Focusing for Dielectric-Laser Acceleration." Phys. Rev. Lett. **121**, 214801 (2018)

[18] N. Schönenberger, A. Mittelbach, P. Yousefi, J. McNeur, U. Niedermayer, and P. Hommelhoff, "Generation and Characterization of Attosecond Microbunched Electron Pulse Trains via Dielectric Laser Acceleration." *Phys. Rev. Lett.* **123**, 264803 (2019)

Ultrafast Electron Scattering:
Femtosecond Electron Pulses in Materials Research

Laurent P. René de Cotret[1], Martin R. Otto[1], Jan-Hendrik Pöhls[1], Tristan Britt[1], Mark J. Stern[1], Mark Sutton[1], Bradley J. Siwick[1,2,*]

[1]Department of Physics, Center for the Physics of Materials, McGill University, 3600 rue Université, Montréal, Québec, Canada, H3A 2T8
[2]Department of Chemistry, McGill University, 801 rue Sherbrooke Ouest, Montréal, Québec, Canada, H3A 0B8
*bradley.siwick@mcgill.ca

Abstract— **Combining ultrafast lasers and electron microscopes in novel ways makes it possible to directly 'watch' the time-evolving structure of condensed matter on the fastest timescales open to atomic motion. By combining such measurements with complementary (and more conventional) spectroscopic probes one can develop structure-property relationships for materials under even very far from equilibrium conditions and explore how light can be used to control material properties. Several examples of the remarkable new kinds of information that can be gleaned from ultrafast electron scattering will be given. For example, it is possible to make 'molecular movies' of atomic-level structural dynamics. It is also possible to directly probe the strength of the coupling between electrons and phonons in materials across the entire Brillouin zone and to probe nonequilibrium phonon dynamics (or relaxation) in exquisite detail. Femtosecond electron pulses are having broad impacts in materials research.**

I. INTRODUCTION

Ultrafast electron scattering is based on a pump-probe technique, whereby a femtosecond (10^{-15} s) laser-pulse 'pumps' (or excites) the sample and an ultrashort electron pulse 'probes' the subsequent dynamics in one of several of ways[1] (Fig. 1). My talk will focus on methods of femtosecond electron pulse generation and important applications of these pulses in materials research.

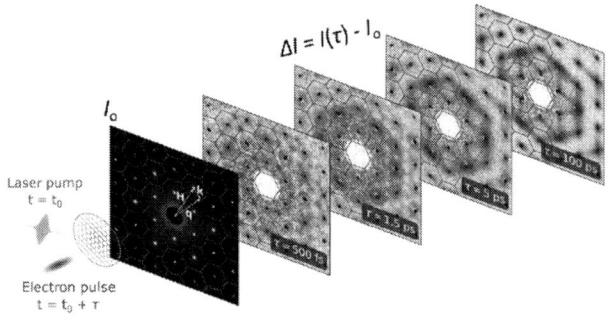

Fig. 1 Schematic of an ultrafast laser-pump, electron-probe experiment in which the time-evolution of photoinduced changes in electron scattering, ΔI, are measured. Data shown is for graphite [refs. 4,5]. The dynamic atomic and electronic structure of the unit cell is encoded in the time-dependent intensity of Bragg peaks (blue dots) [ref 6], and changes in phonon occupancy at all wavevectors/momenta is encoded in the time-dependent diffuse scattering intensity (red) [refs. 4,5].

First, it is well established that detailed information on atomic and charge reorganization can be followed with femtosecond time-resolution using ultrafast electron diffraction (UED)[1-3]. That is, it is now possible to watch the directed motions of atoms and certain details of the associated electronic reorganization in materials on their natural timescales in response to optical excitation. UED enables 'molecular movies' of fundamental dynamics in materials through the information contained in the time-dependent intensity of the Bragg peaks in a diffraction pattern. Second, one can obtain both time and momentum-resolved information on the vibrational excitation of the lattice (phonons) following optical excitation using the information contained in the time-dependent diffuse scattering signals that appear between the diffraction peaks of a UED pattern (Figs. 1). This new technique has been called ultrafast electron diffuse scattering (UEDS).

II. UED: PHOTO-INDUCED IMT IN VO₂

Using UED, we have uncovered a new mechanism by which optical excitation can be used to control both the structure and properties a strongly correlated material. Driving the insulator-to-metal (IMT) transition in VO_2 with light, a new meta-stable, charge-ordered metallic phase of VO_2 was discovered that has no equilibrium analog[2]. This phase is structurally similar to the insulating monoclinic (M1) phase of the material, but with a conductivity similar to the rutile (R) metal. This monoclinic metal ($\mathcal{M}$) appears to only be accessible through photoexcitation[3]. Employing a multi-modal approach we were able to determine the impact that the observed optically induced structural changes have on electronic transport properties[3], determining non-equilibrium structure-property relationships and shed new light on this benchmark problem in materials physics (Fig. 2)

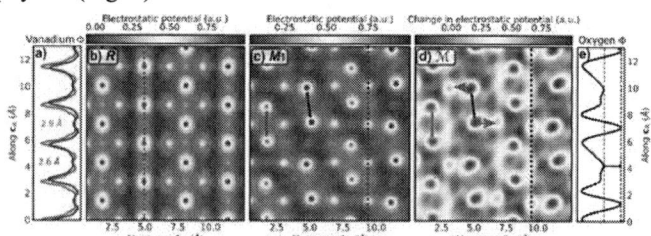

Fig. 2) UED maps changes in the electrostatic potential of the unit cell in VO_2 following photoexcitation, and reveals two qualitatively distinct photoinduced insulator to metal transitions [ref. 3]. First, from equilibrium monoclinic insulator (M1) to the equilibrium rutile metal phase (R). Second, from M1 to a monoclinic metal phase ($\mathcal{M}$) that has no equilibrium analog.

978-1-6654-2590-2/21 $31.00 © 2021 IEEE

III. UEDS: TIME AND MOMENTUM RESOLVED PHONON DYNAMICS

Ultrafast electron diffuse scattering (UEDS) can be used to determine the non-equilibrium populations of phonon modes across the entire Brillouin zone (BZ) in a single crystal material with femtosecond time resolution[4,5,7]. These capabilities are entirely due to the profound sensitivity of UEDS signals to nonequilibrium phonon dynamics and occupancies. This scattering appears between the Bragg peaks (*phonon diffuse scattering)*, and is described by the following equations[5]:

$$I_1(\boldsymbol{q}) = \sum_j \frac{n_j(q)+1/2}{\omega_j(q)} \left| F_{1j}(\boldsymbol{q}) \right|^2 \qquad (1)$$

$$\left| F_{1j}(\boldsymbol{q}) \right|^2 = \left| \sum_s exp\left(-W_s(q)\right) \frac{f_s(q)}{\sqrt{\mu_s}} \left(\boldsymbol{q} \cdot \mathbf{e}_{j,s,k}\right) \right|^2 \qquad (2)$$

Equation (1) states that $I_1(\boldsymbol{q})$, the diffuse intensity at scattering vector $\boldsymbol{q}$, depends on the occupancy of phonon modes $n_j(\boldsymbol{q})$ divided by the mode frequencies $\omega_j(\boldsymbol{q})$ summed over all phonon branches j (see Fig. 1 for a definition of the key vectors). The intensity of scattering at $\boldsymbol{q}$ also depends on the *one-phonon structure factor, F_{1j},* defined in Equation (2). $F_{1j}(\boldsymbol{q})$ depends only on the polarization vectors $\mathbf{e}_{j,s,k}$ of each basis atom s for the phonon mode in branch j with momentum/wavevector $\mathbf{k}$ (where $\mathbf{k}$ depends on the position relative to the closest Bragg peak, $\mathbf{H}$, according to $\mathbf{q} = \mathbf{H} - \mathbf{k}$. See Fig. 1); i.e. only phonon modes with wavevector $\mathbf{k}$ contribute to the scattering intensity at $\mathbf{q}$. $F_{1j}(\boldsymbol{q})$ can be robustly computed using density functional theory methods[5]. UEDS probes the ultrafast changes to $I_1(q)$ and therefore directly measures phonon creation/annihilation in response to laser excitation at all wavevectors via $n_j(q)$, as well as the possibility of measuring the renormalization of mode frequencies $\omega_j(q)$.

We have performed UEDS measurements on graphite[4,5] and TiSe$_2$[7] following photoexcitation (or 'photo-doping' of carriers) to unravel electron-phonon and phonon-phonon interactions in momentum and time. These studies show that UEDS can determine:

1. Which phonon modes are most strongly coupled to charge carriers; a full map of the wave-vector dependent electron-phonon coupling strength[4,5,7].

2. A microscopic view of non-equilibrium phonon dynamics and relaxation channels at all momenta $\mathbf{q}$ including anharmonic couplings, and hot phonon equilibration pathways[4,5] (Fig. 3)

3. Renormalization of phonon frequencies following photoexcitation/photodoping, which reports directly on the many-body phonon self energy in a material[7].

UEDS is like time and angle resolved photoelectron spectroscopy, but for the lattice. This technique will shine new light on phenomena that emerge from the coupling within and between lattice and charge degrees of freedom, which contribute to the fundamental microscopic basis of carrier mobility, photovoltaic efficiency and thermal transport/power dissipation in devices. It has not previously been possible to directly visualize/measure this physics at the level of detail provided by UEDS. Thus, the approach is poised to make significant contributions to our understanding of superconductors, CDW materials, thermoelectrics, and transport phenomena.

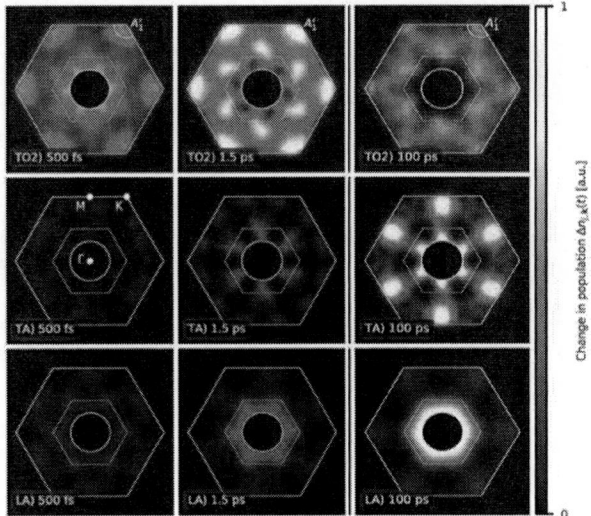

Fig. 3) From the UEDS signals shown in Fig. 1, it is possible to extract the full time, wavevector and band-dependent changes in phonon population following photodoping of carriers [ref 5]. Those changes, everywhere in the hexagonal BZ of graphite, are shown for three phonon bands (TO, TA and LA)4. The strong increase in TO mode population at the K-point at 500fs indicates that this mode is strongly coupled to the photodoped carriers (note that there is little or no increase in TA and LA mode populations by 500fs). By 1.5 ps, vibrational excitation is in the acoustic branches. By 100ps the LA branch appears to be close to thermalized, but the TA branch is still far-from equilibrium. The hot-acoustic phonon thermalization has generated a large population of high wavevector TA phonons near the M-point.

IV. CONCLUSIONS

Ultrafast electron scattering has made enormous strides over the last decade, primarily driven by enhancements in ultrafast electron source capabilities. The field will continue to advance along with instrument performance, and one promising route to higher brightness electron beams is through laser and/or THz field emission from nanotips[8].

ACKNOWLEDGMENT

This work was supported by the Natural Sciences and Engineering Research Council of Canada (NSERC), the Fonds de Recherche du Quebec—Nature et Technologies (FRQNT) and the Canada Foundation for Innovation (CFI).

REFERENCES

[1] G. Sciaini, R. J. D. Miller, Reports on Progress in Physics, 74 (2011) 096101

[2] V. Morrison, R. P. Chatelain, K. Tiwari, A. Hendaoui, M. Chakker and B. J. Siwick, Science 346 (2014) 445 – 448.

[3] M. R. Otto, L. P. Rene de Cotret, K. Tiwari, D. Valverde-Chavez, N. Emond, M. Chakker, D. Cooke and B. J. Siwick, Proceedings of the National Academy of Sciences, 116 (2019) 450-455.

[4] M. J. Stern, L. P. René de Cotret, M. R. Otto, R. P. Chatelain, J.-P. Boisvert, M. Sutton, B. J. Siwick, Phys. Rev. B 97 (2018) 165416.

[5] L. P Rene de Cotret, J. H. Pohls, M. Stern, M. R. Otto, M. Sutton and B. J. Siwick, Phys. Rev. B 100 (2019) 214115.

[6] R. P. Chatelain, V. Morrison, Bart L. M. Klarenaar and B. J. Siwick, Phys. Rev. Lett. 113 (2014) 235502.

[7] M. R. Otto, J. H. Pöhls, L. P. Rene de Cotret, M. Sutton and B. J. Siwick, Science Advances 7 (2021) eabf2810.

[8] D. Matte, N. Chamanara, L. Gingras, L. P. René de Cotret, T. L. Britt, B. J. Siwick, and D. G. Cooke, Phys. Rev. Res. 3 (2021) 013137.

Emission of electrons from a metal tip irradiated by femtosecond IR lasers at wavelengths of 800 and 1240 nm

A.V. Ovchinnikov[1], O.V. Chefonov[1], M.B. Agranat[1], N.A. Abramovskii[2,3], S.B. Bodrov[2,3], A.M. Kiselev[2], A.A. Murzanev[2], A.V. Romashkin[2], A.N. Stepanov[2,*]

[1]Joint Institute for High Temperatures of Russian Academy of Sciences, Moscow, Russia
[2]Federal research center Institute of Applied Physics of the Russian Academy of Sciences (IAP RAS), Nizhny Novgorod, Russia
[3]Lobachevsky State University, Nizhny Novgorod, Russia
*Contact: step@ufp.appl.sci-nnov.ru

Abstract— **The use of IR laser radiation (800 and 1240 nm) with femtosecond pulses with a low repetition rate made it possible to provide non-destructive operation with a metal tip as an electron emitter at an increased intensity of laser radiation on the tip. As a result, the emission of electrons with a charge in the range of tens of picocoulombs per laser pulse was obtained from a metal tip with a characteristic radius of curvature of the order of 1 μm. The power-law dependence of the electron pulse charge proportional to the fourth and sixth powers of the laser pulse energy indicates a multiphoton mechanism of electron emission.**

I. INTRODUCTION

The creation of new short-pulse sources of electron emission using metal tips with a small radius of curvature as an emitter, irradiated by laser radiation with a pulse duration lying in the femtosecond range, seems important due to their possible application in studies of fundamental processes requiring high temporal resolution and high spatial coherence. Electronic pulses of subpicosecond and femtosecond duration are used in free electron lasers [1], allow tracking the movement of atoms in the course of chemical reactions [2], etc. In this work, we present the results of studies in which, with a decrease in the repetition rate of laser pulses in comparison with studies carried out earlier [3, 4], it was possible to increase the intensity of laser radiation on the tip without destroying it to such an extent that it made it possible to obtain electron bunches with a charge of tens of picoculons.

II. EXPERIMENTAL RESULTS AND DISCUSSION

The experiments were carried out on two laser installations [5, 6]. In one, the radiation source was a Ti:Sapphire femtosecond laser complex generating femtosecond laser pulses at a wavelength of 800 nm with a duration of 70 fs, energy up to 10 μJ, and a repetition rate of 10 Hz. In the second, a femtosecond Cr:Forsterite laser system was used, which generated pulses at a radiation wavelength of 1240 nm, a duration of 100 fs, an energy of up to 50 μJ, and a repetition rate of 10 Hz. The metal tips used as the emitter were made of tungsten by electrochemical etching in a potassium hydroxide (KOH) solution. The characteristic size of the tip was on the order of 1 μm. Laser radiation was focused on a metal tip placed in a vacuum chamber at a pressure $p=10^{-4}-10^{-5}$ Torr on

a precision three-coordinate stage. The direction of the vector of the electric field of the laser radiation coincided with the axis of the tip. A negative voltage was applied to the tip in the range U=0-100 V from a separate source. Diagnostics of electron emission from the tip using microchannel plates made it possible to study the spatial distribution of electron emission and to carry out time of flight measurements. For quantitative diagnostics of the magnitude of the charge of electron bunches emitted from the tip, a Faraday cup was used, which was a hollow cylinder made of brass with an inner diameter of 15 mm and a length of 60 mm. The dependence of the electron bunch charge, measured with a Faraday cup, on the laser radiation intensity for two laser radiation wavelengths is shown at Fig. 1.

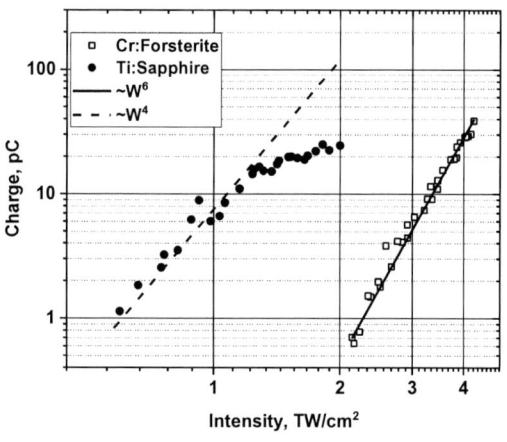

Fig. 1 Dependence of the charge of electron pulses on the intensity of laser radiation, empty squares - Cr:Forsterite, λ=1240 nm, filled circles - Ti:Sapphire, λ=800 nm, solid line - proportional to the sixth degree, dashed line - proportional to the fourth degree

The charge emitted from the cathode rapidly increased with increasing optical pulse energy. For a wavelength of 800 nm at high energies, the effect of saturation of the charge value was observed; at a wavelength of laser radiation of 1240 nm, this effect was not observed. The growing parts of the dependences could be approximated by a power function with

978-1-6654-2590-2/21 $31.00 © 2021 IEEE

an exponent equal to four (λ = 800 nm) and six (λ = 1240 nm), which, in all likelihood, indicates a multiphoton mechanism of electron emission [7]. The saturation of the dependence at high laser pulse energies for a wavelength of 800 nm is apparently associated with the screening effect of the space charge field created by the emitted electrons near the tip. The maximum value of the charge of electrons emitted from the tip reached a value of several tens of picocoulombs in both experiments.

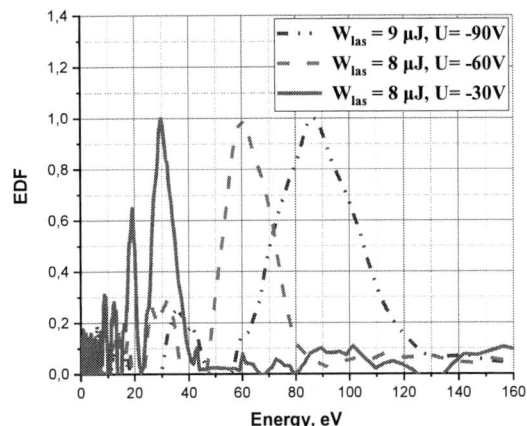

Fig. 2 The electron energy distribution function, λ=1240 nm, pulse energy W = 8-9 μJ

The energy distributions of electrons (EDF) emitted from the tip is shown at Fig. 2. It was measured using a secondary electron multiplier VEU-7M, consisting of a chevron connection of two microchannel plates, and the time-of-flight reconstructed energy distributions of electrons emitted from the tip when it is exposed to laser pulses with an energy of about 9 μJ for different voltages applied to the tip (λ=1240 nm).

III. CONCLUSIONS

Thus, it has been demonstrated that the use of micron-sized metal tips as an emitter of electrons, irradiated by femtosecond laser pulses of radiation with wavelengths of 800 and 1240 nm, makes it possible to obtain electron pulses with a charge of several tens of picocoulombs at microjoule energy in the laser pulse. The experimentally observed power-law dependence of the charge of an electron pulse with different exponents, depending on the wavelength of laser radiation, indicates a multiphoton mechanism of electron emission. The short nanosecond duration of electron pulses synchronized with laser radiation and the micron size of the source allow them to be used in fundamental research requiring high spatial coherence and large electron fluxes.

ACKNOWLEDGMENT

The research was carried out within the framework of the Russian Science Foundation, grant no. 19-42-04133.

REFERENCES

[1] Ayvazyan V., Baboi N., Bohnet I., et al. Generation of GW Radiation Pulses from a VUV Free-Electron Laser Operating in the Femtosecond Regime // Phys. Rev. Lett. 2002. V. 88. P. 104802

[2] Miller R.J. Dwayne. Mapping Atomic Motions with Ultrabright Electrons: The Chemists' Gedanken Experiment Enters the Lab Frame // Annu. Rev. Phys. Chem. 2014. V. 65. P. 583.

[3] Yanagisawa H., Hengsberger M., Leuenberger D., Klöckner M., Hafner C., Greber T., Osterwalder J. Energy Distribution Curves of Ultrafast Laser-Induced Field Emission and Their Implications for Electron Dynamics // Phys. Rev. Lett. 2011. V. 107. P. 087601.

[4] Hilbert S.A., Neukirch A., Uiterwaal C.J.G.J., Batelaan H. Exploring temporal and rate limits of laser-induced electron emission // J. Phys. B: At. Mol. Opt. Phys. 2009. V. 42. P. 141001.

[5] Abramovsky N.A., Bodrov S.B., Kiselev A.M., Murzanev A.A., Romashkin A.V., Stepanov A.N. Generation of Picocoulomb-Level Electron Bunches from a Metal Tip on Femtosecond Ti:Sapphire Laser Irradiation // High Temp. 2020. V. 58. P. 938.

[6] A.V. Ovchinnikov, O.V. Chefonov, M.B. Agranat, A.N. Stepanov. Emission of electrons from a metal tip irradiated with a femtosecond IR laser at a wavelength of 1240 nm // High Temp. 2021. To be published.

Tunable Wavelength One-photon Photoassisted Cold Field Emission from W(310)-nanotips

Rudolf Haindl[†‡], Kerim Köster[†‡], Armin Feist[†‡] and Claus Ropers[†‡*]

[†]*4th Physical Institute - Solids and Nanostructures, University of Göttingen*
[‡]*Max Planck Institute for Biophysical Chemistry*
37077 Göttingen, Germany
[*]Corresponding author: cropers@gwdg.de

Abstract—In this study, we investigate the wavelength-dependent linear photoemission from W(310)-nanotips operated in a Schottky-type emitter assembly. Fowler-Nordheim plots of the electron current are recorded for both static field emission and photoemission with a varying laser wavelength. For all wavelengths, one-photon photoassisted field emission is observed, as demonstrated by the perfect agreement of the experimental data with Fowler-Nordheim tunneling theory.

Index Terms—electron sources, electron guns, cold field emission, transmission electron microscopy, Fowler-Nordheim plot

I. INTRODUCTION

Laser-triggered photoemission from nanoscale field emitters enables high-brightness electron sources [1] and is utilized in photocathodes for ultrafast transmission electron microscopy (UTEM) [2]. Cold field emitters promise the generation of high coherence and low energy spread pulsed electron beams, paving the way towards ultrafast local probing and time resolved electron holography [3].

Here, we report on the observation of one-photon photoassisted field emission from a W(310)-nanotip for variable optical wavelengths.

II. EXPERIMENTAL SETUP

The experimental setup is shown in Fig. 1A. Electron emission is generated from a W(310)-nanotip operated in a Schottky-type emitter assembly (see Fig. 1B) mounted in a vacuum chamber at a base pressure of $5 \cdot 10^{-11}$ mbar. The tip and suppressor are kept at a fixed negative voltage of 4 kV, the vacuum chamber is kept fixed at ground potential and the extractor is negatively biased relative to the tip potential. The electron emission pattern is captured by a multi-channel plate (MCP) with subsequent phosphor screen and camera. The electron current is proportional to the intensity of the MCP-signal that we record for field and photoelectron emission. In the latter case, the frequency-doubled output of a Ti:Sa-oscillator-pumped optical parametric oscillator (OPO) generates frequency-tunable femtosecond optical pulses (100-fs pulse duration, 80-MHz repetition rate). The fs-pulses are focused onto the apex of the 120-nm-radius nanotip with a focal spot size of 80x20 μm^2.

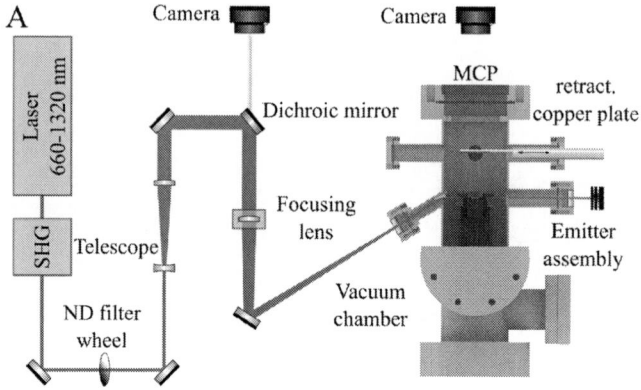

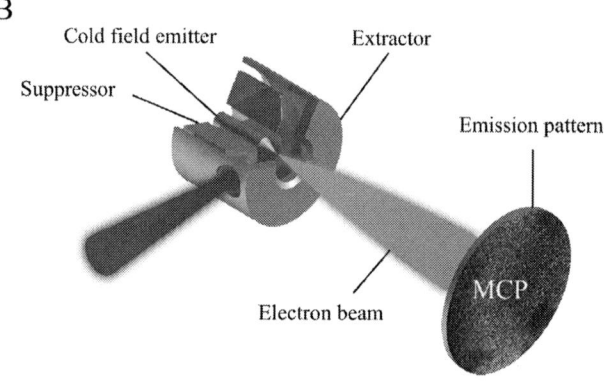

Fig. 1. **A:** A pulsed, frequency doubled laser is focused onto a cold-field tip emitter inside of a vacuum chamber. The emitted electrons are detected at a multi-channel plate for photoemission measurements. **B:** The cold-field emitter is mounted in a Schottky-type emitter assembly consisting of a suppressor and an extractor electrode. Optical access to the emitter is realized by holes in the suppressor and extractor caps. Figure is adapted from [4].

III. WAVELENGTH-DEPENDENT PHOTOASSISTED FIELD EMISSION

The electric field at the nanotip is given by $E_{\mathrm{tip}} = \frac{U_{\mathrm{ext}}}{kr}$, where U_{ext} is the applied extractor voltage, r the tip radius and k a geometric field enhancement factor dependent on the tip and electrode geometry. Field emission from the nanotip is described by the Fowler-Nordheim (FN) equation. The tunneling current density j is related to the local field strength

978-1-6654-2590-2/21 $31.00 © 2021 IEEE

E_{tip} and the work function Φ [5]

$$j = \frac{e^3 E_{\text{tip}}^2}{8\pi h \Phi} \exp\left[-\frac{8\pi\sqrt{2m}\Phi^{3/2}}{3heE_{\text{tip}}}v(w)\right]. \qquad (1)$$

Here, e is the elementary charge, h Planck's constant, m the electron mass, and $v(w) \approx 1 - w + (1/6)w\ln w$ a slowly varying function of $w = (e^3/4\pi\epsilon_0)(E_{\text{tip}}/\Phi^2)$. In the case of photoassisted field emission, electrons are emitted by tunneling through the potential barrier after transient photoexcitation to a higher energy, resulting in an effective work function Φ_λ [6, 7]. We recorded FN curves for field emission and photoassisted field emission at laser wavelengths $\lambda = 400$ nm and $\lambda = 500$ nm (see Fig. 2).

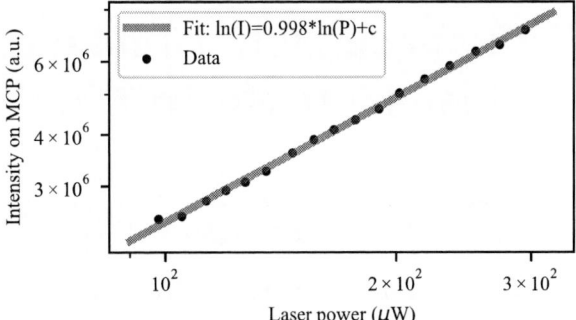

Fig. 3. The photoemission intensity is plotted for varying the laser power between 100 µW and 300 µW on a double logarithmic plot. From the linear fit, we extract a nonlinearity coefficient of 0.998 ± 0.013.

V. Outlook

Future work on one-photon photoassisted field emission will be focused on improving the lifetime of laser-driven cold field emitters, thus making them readily accessible in ultrafast transmission electron microscopy. Hereby, combining the flexibility of linear photoemission with the excellent coherence properties of cold field emitters will be in reach.

Acknowledgment

We thank the Göttingen UTEM team for constant support. This work was funded with resources from the Gottfried Wilhelm Leibniz prize.

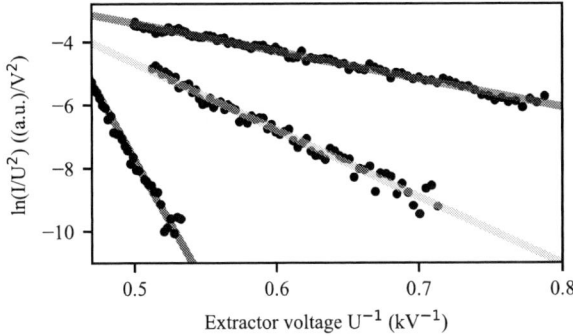

Fig. 2. The photoemission current is shown in a Fowler-Nordheim type plot for no-laser/field emission (grey line), photoemission at $\lambda = 500$ nm (turquoise line) and photoemission at $\lambda = 400$ nm (violet line). Straight lines indicate perfect agreement to (1).

Fitting the FN curves using Eq. (1), we retrieve the respective work function Φ for each dataset. The field emission data is well reproduced by (1), and we attribute a work function of $\Phi = 4.35$ eV [8] to the data. Considering the slope of the FN curves for the field emission data, we extract an effective work function of $\Phi_{400\text{nm}} = 1.00$ eV for $\lambda = 400$ nm (photon energy of 3.1 eV) and $\Phi_{500\text{nm}} = 1.78$ eV for $\lambda = 500$ nm (2.48 eV), respectively.

IV. One-photon Photoassisted Field Emission

For both laser wavelengths, the sum of the effective work function and the photon energy ($\Phi_{400\text{nm}}^{\text{exp}} + \hbar\omega_{400\text{nm}} = 4.15$ eV and $\Phi_{500\text{nm}}^{\text{exp}} + \hbar\omega_{500\text{nm}} = 4.26$ eV) roughly match the literature value of the W(310) work function Φ. This indicates that observed photoassisted field emission is a one-photon process. Further evidence is shown in Fig. 3. The photoemission current is plotted against the applied laser power in a double logarithmic plot for a fixed laser wavelength of 400 nm and an extractor voltage of 1750 V. A linear fit to the data yields a slope of 0.998 ± 0.013, i.e. we observe a linear dependence between photocurrent and laser power.

References

[1] D. Ehberger et al., "Highly Coherent Electron Beam from a Laser-Triggered Tungsten Needle Tip," *Physical Review Letters*, vol. 114, no. 22, p. 227601, Jun. 2015.

[2] A. Feist et al., "Ultrafast transmission electron microscopy using a laser-driven field emitter: Femtosecond resolution with a high coherence electron beam," *Ultramicroscopy*, vol. 176, pp. 63–73, May 2017.

[3] F. Houdellier, G. Caruso, S. Weber, M. Kociak, and A. Arbouet, "Development of a high brightness ultrafast Transmission Electron Microscope based on a laser-driven cold field emission source," *Ultramicroscopy*, vol. 186, pp. 128–138, Mar. 2018.

[4] R. Bormann, "Development and characterization of an electron gun for ultrafast electron microscopy," Ph.D. dissertation, Georg-August-Universität Göttingen, Göttingen, Dec. 2015. [Online]. Available: http://hdl.handle.net/11858/00-1735-0000-0028-867D-4

[5] R. G. Forbes and J. H. Deane, "Reformulation of the standard theory of Fowler–Nordheim tunnelling and cold field electron emission," *Proceedings of the Royal Society A: Mathematical, Physical and Engineering Sciences*, vol. 463, no. 2087, pp. 2907–2927, Nov. 2007.

[6] P. Hommelhoff, C. Kealhofer, and M. A. Kasevich, "Ultrafast Electron Pulses from a Tungsten Tip Triggered by Low-Power Femtosecond Laser Pulses," *Physical Review Letters*, vol. 97, no. 24, p. 247402, Dec. 2006.

[7] C. Ropers, D. R. Solli, C. P. Schulz, C. Lienau, and T. Elsaesser, "Localized Multiphoton Emission of Femtosecond Electron Pulses from Metal Nanotips," *Physical Review Letters*, vol. 98, no. 4, p. 043907, Jan. 2007.

[8] H. Kawano, "Effective work functions for ionic and electronic emissions from mono- and polycrystalline surfaces," *Progress in Surface Science*, vol. 83, no. 1-2, pp. 1–165, Feb. 2008. [Online]. Available: https://linkinghub.elsevier.com/retrieve/pii/S0079681607000779

978-1-6654-2590-2/21 $31.00 © 2021 IEEE

Photoemission from an ultrabright and ultrafast LaB$_6$ nanowire electron emitter studied at atomic scale

Ang Li[1,*], Han Zhang[2], Stefan Meier[1], Alexander Tafel[1], Peter Hommelhoff[1]

1. *Department of Physics, Friedrich-Alexander-Universität Erlangen-Nürnberg (FAU), Erlangen, Germany, EU*
2. *Research center for advanced material characterization, National institute for materials science, Tsukuba, Ibaraki, Japan*
*Contact: ang.li@fau.de, phone +49-9131-85-28318, www.laser.physik.fau.eu

Abstract— **Lanthanum hexaboride (LaB$_6$) nanowire emitters have been demonstrated to work as monochromatic ultrabright field emitters. A small atomic array consisting of only a few La atoms at the foremost (100) facet generates a stark contrast of the surrounding work function. With the help of femtosecond laser pulses, we will show that this special surface morphology is also directly reflected in the photoemission as a function of wavelength. By comparing the measured total absorption energy with the density of states (DOS), emission channels with different initial and final states are revealed to explain the observed results. Moreover, an electron biprism interference measurement is performed, showing excellent transverse beam quality with a normalized emittance of ε_n = 1.41 pm rad and a normalized peak brightness of $B_{n,p}$ = 9 × 10^{13} A/(m^2 sr). These results show that LaB$_6$ nanowire emitters are ultrafast and ultrabright, which could benefit novel applications like dielectric laser acceleration (DLA) and ultrafast electron diffraction (UED).**

I. Introduction

LaB$_6$ single crystals have been one of the most popular choices of thermionic electron sources benefiting from its low work function of 2.6 eV [1]. Recently, a field emitter based on a nanowire of a LaB$_6$ single crystal has been demonstrated as a stable direct current (DC) emitter with an ultrahigh beam brightness and low energy spread [2]. A special treatment of the emitter actuates the La atoms from inside the crystal to the surface at the foremost (100) facet, forming a small atomic array consisting of only a few La atoms. This unique crystallographic profile on the nanowire apex gives rise to a high contrast in the local work function between the foremost La-terminated (100) facet (~2.07 eV) and the adjacent surfaces (4 – 6 eV) [3], confining the physical emission size to an atomic scale. We here investigate the femtosecond laser-triggered electron emission from such a source.

II. Multiphoton Emission from LaB$_6$ Nanowire Emitters

Our experimental setup is based on a side-illuminated tip-based emitter in an ultra-high vacuum (UHV) (Fig. 1a). The LaB$_6$ nanowire consisting of single crystal in (100) orientation has an apex shaped as a half-sphere with a typical radius of curvature of 17 nm. The nanowire emitter is triggered with ultrashort ~67 fs laser pulses at a 1 kHz repetition rate and a variable central wavelength. Inside the UHV vessel, photoemission from the emitter is detected by a microchannel plate (MCP) detector with a phosphor screen, where the emission patterns are inspected via a CCD camera. The emitter is applied with a negative voltage of ~−50 V that is chosen to

be below the field emission threshold at U_{th} = −250 V to ensure that only laser-induced emission is detected.

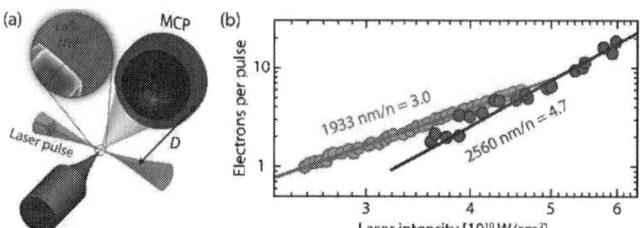

Fig. 1: (a) Femtosecond laser pulses are focused at the LaB$_6$ nanowire (NW) emitter, which is mounted on a trimmed tungsten needle tip. (b) Double-logarithmic plot of the power-scaling results shows photon order of 3 for 1933 nm (orange circles) and 4.7 for 2650 nm (blue circles).

Two typical power-scaling measurements where the photocurrent from the LaB$_6$ nanowire emitter is measured as a function of the laser intensity are plotted for wavelengths of 1933 nm and 2650 nm (Fig. 1b). The slopes of the photocurrent in the double-logarithmic plot reveal the photon orders of the multiphoton processes, namely n = 3 for 1933 nm and n = 4.7 for 2650 nm. The corresponding effective barrier of (2.05 ± 0.25) eV is consistent with the work function obtained by previous studies (2.07 eV [3] and 2.09 eV [2]), implying a small physical emission size, hence the excellent field emission properties can likely be preserved in the photoemission regime.

III. Emission Channel Dependent Beam Divergence

Furthermore, we have investigated the root mean square (rms) divergence angle as a function of the wavelength varied from 354 nm to 2650 nm (Fig. 2). A minimum of the beam divergence angle θ = 82 mrad (4.7°) is observed at central wavelengths of 1024 nm and 2650 nm, showing a factor of 2 higher θ than that measured in field emission of 40 mrad, yet 77% lower than the photoemission from a tungsten nano-scale tip [4]. More interestingly, a ~40% monotonic increase with the decreasing wavelength is found from 1024 nm to 354 nm, while it remains constant at 1024 nm and 2650 nm.

This divergence growth is further studied by measuring the photon order and the total absorption energy $nh\nu$ for the investigated wavelengths (Fig. 3a). Here, we find an increase of n with decreasing wavelength from 1933 nm to 512 nm, deviating from a typical step-like photon order transition for increasing photo energy [5]. Correspondingly, the total absorption energy shows a monotonic increase from 1933 nm to 354 nm with decreasing wavelength and a maximum of 9 eV at 512 nm. The data points spaning over 1.5 octave (354 nm -

1024 nm) lie above the highest local work function on the nanowire apex. This can be attributed to the photoemission process with its initial state lying below the Fermi level.

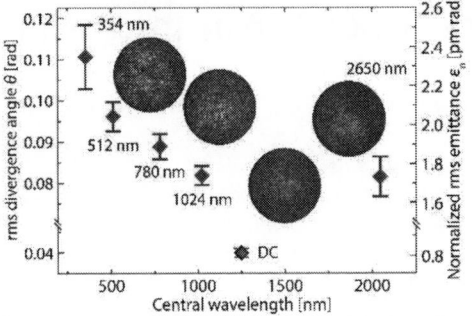

Fig. 2: Root-mean-square (rms) divergence angle plotted as a function of the laser wavelength. Insets display the recorded emission patterns on the MCP detector for various wavelengths from 354 nm to 2650 nm. A clear increase of the beam divergence angle with decreasing laser wavelength is observed with an onset at 1024 nm.

We will also show that by comparing the above experimental results and the density of states (DOS), different emission channels can be identified (Fig. 3b). The gap-like feature in the DOS distribution located around E_F causes different initial states. For long wavelengths (e.g. $\lambda > 1560$ nm), emission channels are predominantly from the Fermi level to the foremost La(100) facet (La atomic array).

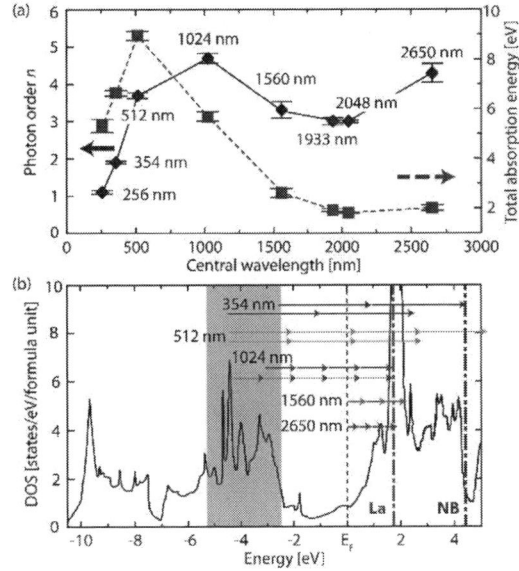

Fig. 3: (a) Scaling of photon order n and the total absorption energy plotted as a function of the central laser wavelength. (b) Different emission channels for different wavelengths (coloured arrows) are recognized by comparing the results in (a) with the DOS of LaB$_6$ crystal. The black and blue dash-dot lines show the effective barrier of the La(100) facet (La) and the neighbouring surfaces (NB).

For shorter triggering wavelengths ($\lambda \sim 1024$ nm), photoemission from the strong peaks below E_F (violet shade in Fig. 3b) to the La(100) facet seem primarily addresses, explaining the increasing total absorption energy yet low beam divergence at 1024 nm. As the wavelength further decreases, emission channels at neighboring surfaces are activated, which further leads to different physical emission areas on the emitter apex with atomic scales, matching the observed large total absorption energy and divergence growth with decreasing wavelength.

IV. VIRTUAL SOURCE SIZE AND BEAM BRIGHTNESS

Moreover, an electron biprism interference measurement is performed to determine the effective source size of the emitter illuminated by few-cycle femtosecond laser pulses (6.1 fs FWHM). A free-standing Ag nanowire with a diameter of $60 - 80$ nm splits the electron wavefront and the interference fringes are detected on the MCP detector located at a distance of D= 73.3 mm downstream of the emitter (inset in Fig. 4).

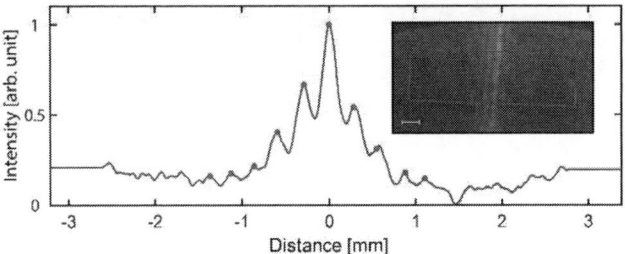

Fig. 3: Line profile of the interference pattern of the photoelectrons spilt by a nanowire biprism. Inset: interference fringes on the MCP detector with the range of interest marked by a red frame. The blue scale bar is 1 mm on the screen

As stated by the van Cittert-Zernike theorem [6], the effective source size is given by $r_{\mathrm{eff}} = \lambda_{\mathrm{db}} D / (\pi \xi_\perp)$, with the far-field interference width $\xi_\perp$, which can be determined by the lateral width of the visible interference fringes. The line profiles of the 9 recorded interference fringes correspond to an interference width $\xi_\perp = (2.5 \pm 0.3)$ mm (Fig. 4). The obtained effective source size $r_{\mathrm{eff}} = (1.5 \pm 0.2)$ nm indicated an excellent transverse beam quality with a normalized rms emittance of $\varepsilon_{\mathrm{n}} = 1.41$ pm rad and a normalized rms peak brightness of $B_{\mathrm{n,p}} = 9 \times 10^{13}$ A/(m^2 sr). Note that only 1 electron per pulse was considered in the calculation to avoid charge density-induced broadening. These results are comparable to those of tungsten tip-based emitters [7]. Hence, LaB$_6$ nanowire represent an ultrafast and ultrabright nanowire emitter that fulfills the requirements of novel applications like dielectric laser acceleration (DLA) and ultrafast electron diffraction (UED) [8].

ACKNOWLEDGMENT

Funding is gratefully acknowledged from the Gordon and Betty Moore Foundation (ACHIP), ERC Grants NearFieldAtto and AccelOnChip, and BMBF.

REFERENCES

[1] M. Gesley and L. W. Swanson, "A determination of the low work function planes of LaB$_6$," *Surface Science,* vol. 146, pp. 583-599, 1984.

[2] H. Zhang, *et al.*, "An ultrabright and monochromatic electron point source made of a LaB$_6$ nanowire," *Nature Nanotech.*, **11,** 273–279, 2016.

[3] M. A. Uijttewaal, G. A. de Wijs, and R. A. de Groot, "Ab initio and work function and surface energy anisotropy of LaB$_6$," *The Journal of Physical Chemistry B*, vol. 110, pp. 18459-18465, 2006.

[4] D. Ehberger, *et al.*, "Highly coherent electron beam from a laser-triggered tungsten needle tip," *Phys. Rev. Lett.*, vol. 114, p. 227601, 2015.

[5] A. Tafel, S. Meier, J. Ristein, and P. Hommelhoff,. "Femtosecond laser-induced electron emission from nanodiamond-coated tungsten needle tips," *Phys. Rev. Lett.*, vol. 123, p. 146802, 2019.

[6] F. Zernike, "The concept of degree of coherence and its application to optical problems," *Physica*, vol. 5.8, pp. 785-795, 1938.

[7] F. Houdellier, G. M. Caruso, S. Weber, M. Kociak, and A. Arbouet, "Development of a high brightness ultrafast transmission electron microscope based on a laser-driven cold field emission source," *Ultramicroscopy*, vol. 186, pp. 128-138, 2018.

[8] A. Li, H. Zhang, S. Meier, A. Tafel, and P. Hommelhoff, MS in preparation.

978-1-6654-2590-2/21 $31.00 © 2021 IEEE

Longitudinal and transverse modulation of electron wave function with light, and its application to electron microscopy

Ivan Madan[1*], Giovanni Vanacore[1], Gabriele Berruto[1], Enrico Pomarico[1], Javier García de Abajo[2], Ido Kaminer[3], Fabrizio Carbone[1]

[1] École Polytechnique Fédérale de Lausanne, Lausanne, Switzerland
[2] ICFO-Institut de Ciencies Fotoniques, Castelldefels (Barcelona), Spain
ICREA-Institucío Catalana de Recerca i Estudis Avançats, Barcelona, Spain
[3] Technion—Israel Institute of Technology, Haifa, Israel

*Contact: ivan.madan@epfl.ch, phone +41 21 693 61 20

Abstract— **Conventional electron microscopy operates with flat electron wave fronts. Employing attenuative or electrostatic phase plates allows transversally modulating the wave front, enhancing cross-sections of particular scattering channels. Utilization of light from coherent ultrafast sources allows modulating electrons phase *both* transversally and longitudinally. This results in a plethora of phenomena, as well as conceptually new techniques such as non-local electron holography of light fields and generation of ultrafast electron vortex beams, which we present in this contribution.**

I. INTRODUCTION

Modulation of longitudinal and phase profiles of particles (photons, electrons, neutrons and even atoms) has a profound impact on their fundamental properties as well as drastically increasing the variability of potential applications. Some of the most celebrated examples include so-called vortex beams (Laguerre-Gaussian beams with non-zero azimuthal number) [1], Bessel beams used in bar-code scanners and optical tweezers [2], beams adjusted by spatial phase modulators (SPM) for ultrafast pulse compression [3] and many others.

Most approaches to particle phase modulation imply transversal modulation. This can be achieved by quasi-static means, most commonly using phase-plates − nano-fabricated structures that use amplitude modulation to obtain phase modulation in the conjugate plane. Longitudinal phase modulation is more challenging and can be achieved either by coupling longitudinal to transverse degrees of freedom as in SPMs for ultrafast laser applications or by employing modulators that operate on the particle coherence time-scale, with, to the best of our knowledge, unique example of electron longitudinal phase modulation by light described below.

In this contribution, we focus on electron phase modulation by light and demonstrate few prominent examples of applications.

II. PHOTON INDUCED NEAR FIELD ELECTRON MICROSCOPY (PINEM)

The basis of electron wave function modulation by light is the so-called PINEM interaction – the coherent electron light

interaction mediated by a third body to ensure momentum-energy conservation [4–6].

The high-energy electron pulse can be well approximated by $\psi(z,t) = g(z - v_e t, t)\psi_0(z, t)$, that is a monochromatic carrier $\psi_0(z, t) = \exp[i(k_e z - \omega_e t)]$ with a Gaussian envelope $g(z - v_e t, t)$ with v_e being the electron group velocity along propagation direction z, and k_e and ω_e are the electron spatial and temporal frequencies.

The PINEM interaction can be described solely as phase modulation of the electron envelope function by the laser field of frequency ω, and complex electric field E_z:

$$g(z', t) = g(z', t_0) \exp\left[\frac{iq_e}{\hbar\omega}\sin\left[\frac{i\omega}{v_e}z' + \arg(\beta)\right]|\beta\left(\frac{\omega}{v_e}\right)|\right],$$
(1)

with the interaction strength

$$\beta\left(x, y, \frac{\omega}{v_e}\right) = \left(\frac{e}{\hbar\omega}\right)\int dz' E_z(x, y, z')e^{-\frac{i\omega z'}{v_e}}$$
(2)

that defines the depth of electron phase modulation. Importantly, $\beta\left(x, y, \frac{\omega}{v_e}\right)$ follows the spatial distribution of the electric field.

These equations provide the basis for phase modulation of electron by light. Engineering transverse and temporal distribution of optical fields allows achieving the desired phase modulation.

III. HOLOGRAPHY OF OPTICAL FIELDS WITH ELECTRONS.

Eq. (1) implicitly includes optical phase as a parameter of the electron phase modulation profile through the phase of the interaction strength. This essentially is the key to all the interference phenomena using optically modulated electron wave function.

The first example includes consecutive interaction of electron with two spatially separated optical fields in a fashion similar to Ramsey control experiment, in the set-up originally explored by Echternkamp et al. [7]. The final interaction strength after the two interactions is given by $\beta = \beta_1 + e^{i\phi}\beta_2$, where ϕ is the phase offset given by the spatial separation of two regions and becomes significant over few microns distance.

In a set-up where the two fields are separated by a nanoscopic distance, the total interaction strength is essentially the complex sum of two interaction strengths. Since the integral in Eq. (2) is a linear operation, the resulting interaction strength is defined by the "interference" of two fields even if these never overlap in space. This allows for a very special holography of nanoscopic optical fields, where the electron serves as a phase-sensitive recording medium traveling through space [8].

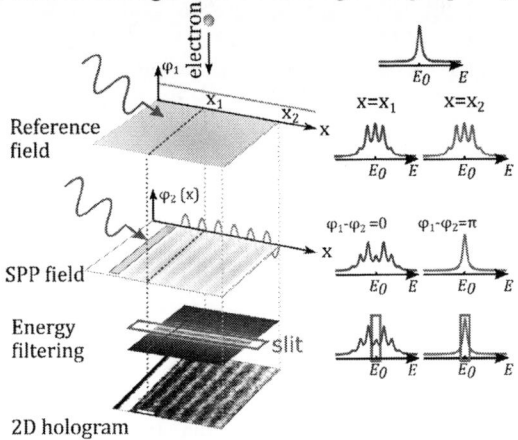

Fig. 1 Principle of non-local holography with electrons. Left-hand side depicts consecutive stages of electron propagation – initial propagation, interaction with a reference field, signal field, energy filtering and the detection of an energy-filtered image. Right-hand-side shows respective variation in electron spectrum at two different positions in space. From [8] © distributed by author under a CC BY-NC 4.0 License

Schematically the technique is represented in Fig. 1, where we apply to surface plasmon polaritons (SPPs) holography. The reference spatially homogeneous field is prepared by simply reflecting the beam from the surface of the sample and employing inverse transition radiation effect [9]. Further down the trajectory of electron, at the interface between metal and insulator, the electron interacts with the spatially inhomogeneous field of the SPP. The resulting electron spectrum depends on the relative phase of two fields. Employing energy filtered imaging we obtain TEM contrast defined by the SPP phase. In case the two pulses are independent and their relative delay is varied, the technique allows extracting group and phase velocities of the modes. Full Fourier analysis of the contrast allows to discriminate between the modes with different momentum [10].

IV. GENERATION OF ELECTRON VORTEX BEAMS

The spatial dependence of the coupling constant in Eq. (2) allows imprinting desired phase distributions on the electron wave function. An example of practical interest involves an electron vortex beam and an electron beam with a π-shift. The former is of interest for electron magnetic dichroism measurement, while the latter allows for discrimination between optically allowed and forbidden nanoscopic excitations by the electron beam. Both cases can be achieved in a rather simple set-up involving optical illumination of microscopic circular aperture for electron beam. The illumination by circularly polarised light results in the spin to

orbital angular momentum (OAM) transfer in the near field around the edges of the aperture. Such spiral distribution is consequently transferred to the electron upon interaction: an absorption of m photons with $l=+1$ photon increases the electron OAM by lm. Although an electron exists in the superposition state of all possible OAMs, the OAM can be eventually selected employing an electron spectrometer, since the OAM variation is directly proportional to the energy exchange $\Delta E_{lm} = lm\hbar\omega$.

We realised such configuration and have shown the control between Gaussian, vortex and a π-beams simply by changing the polarization of the incident optical beam [11] (see Fig.2).

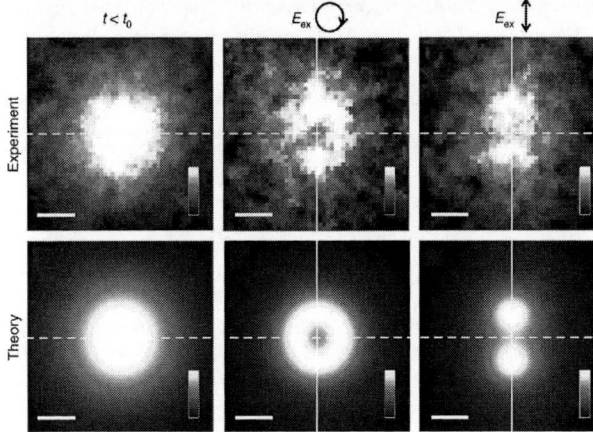

Fig. 2. Experimental and theoretical profiles of electron beam in far field after being modulated by optically controlled phase plate. From [11], distributed by author.

V. CONCLUSIONS

We presented two examples of respectively longitudinal and transverse modulation of electron wave function. These results are placed among many others [12–17] (a more complete list of references can be found in [18]) that demonstrate the rich capabilities of phase-controlled electron measurements and the perspectives for future quantum microscopies [10].

REFERENCES

[1] Y. Shen, et al. Light Sci. Appl. **8**, (2019).
[2] M. R. LaPointe, Curr. Dev. Opt. Des. Opt. Eng. **1527**, 258 (1991).
[3] A. M. Weiner, et al., IEEE J. Quantum Electron. **28**, 908 (1992).
[4] B. Barwick, D. J. Flannigan, and A. H. Zewail, Nature **462**, 902 (2009).
[5] S. T. Park, M. Lin, and A. H. Zewail, New J. Phys. **12**, (2010).
[6] F. J. Garcia De Abajo, A. Asenjo-Garcia, and M. Kociak, Nano Lett. **10**, 1859 (2010).
[7] K. E. Echternkamp, et al., Nat. Phys. **12**, 1000 (2016).
[8] I. Madan, et al.,Sci. Adv. **5**, eaav8358 (2019).
[9] G. M. Vanacore, et al., Nat. Commun. **9**, 2694 (2018).
[10] I. Madan, et al., Appl. Phys. Lett. **116**, 230502 (2020).
[11] G. M. Vanacore, et al Nat. Mater. **18**, 573 (2019).
[12] E. Pomarico, I et al, ACS Photonics **5**, 759 (2018).
[13] X. Fu, et al, Nat. Commun. **11**, 5770 (2020).
[14] O. Kfir, et al., 1 (2019).
[15] R. Dahan et al., Nat. Phys. **16**, 1123 (2020).
[16] M. Kozák, et al., Nat. Phys. **14**, 121 (2018).
[17] Y. Morimoto and P. Baum, Nat. Phys. **14**, 252 (2018).
[18] G. M. Vanacore, I. Madan, and F. Carbone, Riv. Del Nuovo Cim. **43**, 567 (2020).

Voltage-controlled three-electron-beam interference by a three-element Boersch phase shifter with top and bottom shielding electrodes

P. Thakkar[1,2,*], V.A. Guzenko[3], P-H. Lu[4,5], R.E. Dunin-Borkowski[4], J.P. Abrahams[1,2,6] and S. Tsujino[1,2,†]

[1]*Division of Biology and Chemistry, Paul Scherrer Insitut, Forschungsstrasse 111, 5232 Villigen PSI, Switzerland*
[2]*Swiss Nanoscience Institute, University of Basel, Klingelbergstrasse 82, 4056 Basel, Switzerland*
[3]*Photon Science Division, Paul Scherrer Institut, Forschungsstrasse 111, 5232 Villigen PSI, Switzerland*
[4]*Ernst Ruska-Centre for Microscopy and Spectroscopy with Electrons and Peter Grünberg Institute, Forschungszentrum Jülich, Wilhelm-Johnen-Strasse, 52425 Jülich, Germany*
[5]*RWTH Aachen University, Ahornstraße 55, 52074 Aachen, Germany*
[6]*Biozentrum, University of Basel, Klingelbergstrasse 70, 4056 Basel, Switzerland*
Contact: *pooja.thakkar@psi.ch, †soichiro.tsujino@psi.ch

Abstract—**We report a three-element Boersch phase shifter for the wavefront manipulation of coherent electron beam. A device with five-layer structure is fabricated by the state-of-the-art lithography. In a recently reported simplified three-layer device, parasitic beam deflection by contacting wires was observed, which poses a critical obstacle for device upscaling. To minimize the effect, a five-layer structure including a top-contact-shielding electrode is adopted in this work. Despite the mechanical stresses caused by the additional layers, we successfully produced the five-layer device on a suspended thin silicon nitride membrane and tested its performance of the voltage-controlled phase shift of coherent electron beam. The experiment confirmed that the parasitic beam deflection was suppressed. Our work paves a way toward the realization of a multi-element Boersch phase shifter for a programmable holographic synthesis of electron waves in two- and three-dimensions.**

I. INTRODUCTION

The electron beam produced by a cold field emission gun provides a high brightness, a narrow energy spread, and a large transverse coherence length comparing to the beam emitted from Schottky emitters or thermionic emitters. Recent advances in the Cs-corrected transmission electron microscopes fully exploits such advantage of the field emission electron beam to routinely achieve single atom resolution. To further improve the imaging capability of such electron microscopes, manipulating the wavefront of coherent electron beam analogous to beam shaping in light optics, has been actively studied recently, e.g., for the phase modulation imaging, the aberration correction, and other applications [1]. The use of static phase plates or voltage-controlled phase shifters have been reported. However, to realize devices with a large number of phase shifter elements for voltage-controlled phase manipulation, nanofabrication methods compatible with the upscaling are desired.

Using a Boersch phase shifter device [2]–[4], the phase of the electron wave transmitting through its aperture is shifted proportionally to the applied voltage. When a coherent electron beam is incident on a three-element device, a hexagonal interference pattern is formed in far-field (see the TEM image on the right of Fig. 1a). A bright central spot is formed by the constructive interference of three in-phase electron beams. When a voltage corresponding to the π phase shift is applied to one of the phase shifter elements, the central spot of the interference pattern becomes dark due to destructive interference [Fig. 1b].

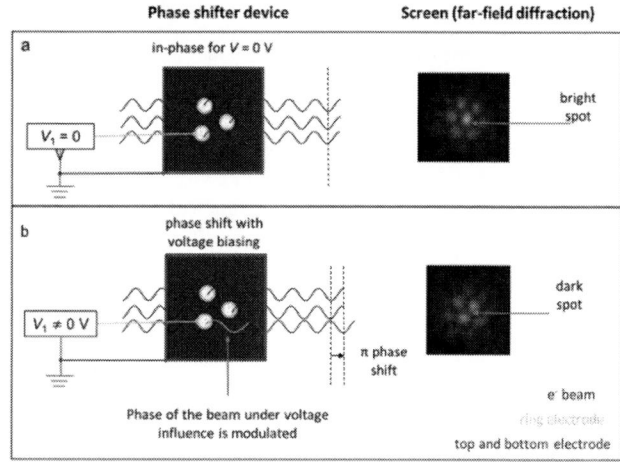

Fig. 1 A schematic of voltage-controlled three-beam interference. The modulation of electron interference pattern is observed in far-field proportional to the voltage bias to the phase shifting element, (a) in-phase condition, (b) π phase shift.

Recently, we reported the voltage-controlled phase shift of a three-element Boersch phase shifter device with a metal-insulator-metal (MIM), three-layer structure [3]. However, contact wires exposed to the electron beam caused parasitic beam deflection. This is undesired since the crosstalk between neighbouring phase shifter elements caused by this effect obstructs the programming capability of a multi-element device. In this work, we study the fabrication of a three-element Boersch phase shifter device with the metal-insulator-metal-insulator-metal (MIMIM), five-layer structure, including the top electrode to shield the contacting

978-1-6654-2590-2/21 $31.00 © 2021 IEEE

wires.

II. DEVICE FABRICATION

A five-layer Boersch device with three individually controllable phase shifter elements, is fabricated on a suspended low-stress silicon nitride membrane with the thickness of 200 nm. First, a three-layer Boersch device is fabricated by the method described in [3], and subsequently the top insulating layer and the top metal layer are added on the top. To minimize the additional stress by these layers, we developed a method using an electron-beam negative resist (hydrogen silsesquioxane) that after cross-linking (electron-beam exposure) attains insulating properties. The top metal layer is then patterned by metal (gold) evaporation and lift-off with a mask patterned by the electron beam lithography.

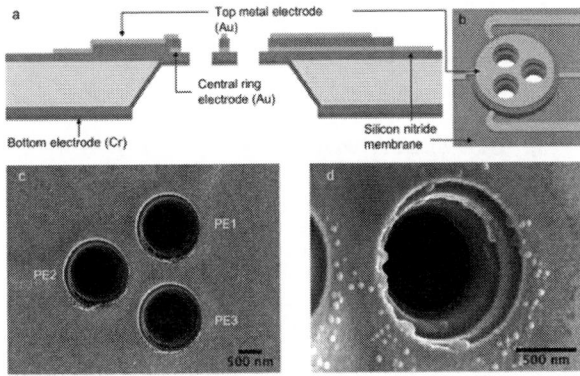

Fig. 2 (a) A cross section, and (b) 3d perspective view, show the schematic of five-layer Boersch phase shifter device with three elements along with three contact wires and a contact wire for grounding the top metal layer. (c) SEM image of three Boersch phase shifting elements. (d) A high magnification SEM image of the sidewall of one of the elements imaged at an angle of 20°.

Schematic cross section and perspective view of the fabricated five-layer phase shifter element is shown in Fig.2. The SEM image of the elements [Fig. 2c] shows that the lift-off process was clean and the alignment accuracy was lower than 50 nm. The inner sidewall of the element imaged at an angle of 20° [Fig. 2d] reveals that all three metal layers are electrically insulated from each other.

III. RESULTS

The fabricated device is tested in a transmission electron microscope with a cold field emission gun at a beam energy of 200 keV. The microscope was operated in diffraction mode. Fig. 3a and 3b show the hexagonal three-beam interference patterns: Fig. 3a shows the case when three beams were in-phase, and Fig. 3b shows the case when the electron beam traversing through PE1 was phase-shifted by π by applying a voltage bias of 1.54 V to it. Most importantly, the comparison of the relative position of the two interference patterns (see the shift with respect to the black circle drawn in the centre) demonstrates that beam deflection caused by the

voltage bias to PE1 for π shift, that was observed in the three-layer device previously, is suppressed in the present five-layer Boersch phase shifter device. Note that the offset voltages were applied to compensate the default phase shift due to yet-to-be-identified causes but with the amount of only a few volts. The observed phase shift per voltage agrees well with theory based on the 3-dimensional finite-element simulation (not shown, see [3]) within 20%.

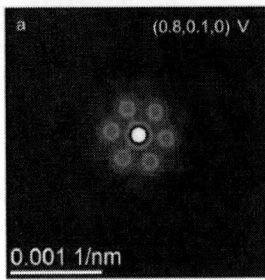 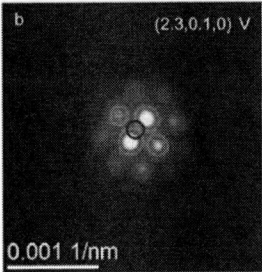

Fig. 3 Voltage-controlled three beam interference of electrons for five-layer Boersch phase shifter elements. The voltage applied to PE1 was equal to (a) 0.8 V for zero-phase condition and (b) 2.3 V for π phase shift respectively. PE2 is biased at 0.1 V to compensate for offset voltage. The circle in the centre is used as a reference to evaluate the beam deflection.

IV. CONCLUSION

Multi-element Boersch phase shifter device with the five-layer structure fabricated with scalable methods is demonstrated. With the proposed MIMIM structure, the cross talks between elements by parasitic beam deflection is minimized. This is an important step forward to realize an array of programmable phase shifter with a large number of elements to arbitrary manipulate electron wavefront for aberration correction, massively parallel electron beam lithography, and novel imaging applications.

ACKNOWLEDGMENT

We would like to extend our sincere thanks to Prof. Giulio Pozzi (University of Bologna, Italy) for his valuable insights. We thank Laboratory of Micro- and Nanotechnology (LMN), Paul Scherrer Insitut for providing their excellent cleanroom infrastructure for device fabrication and Jana Lehmann (LMN) for her kind help with silicon nitride membrane fabrication. The project was partially funded by the Swiss Nanoscience Institute, University of Basel (Project P1505).

REFERENCES

[1] J. Verbeeck, A. Béché, K. Müller-Caspary, G. Guzzinati, M. A. Luong, and M. Den Hertog, "Demonstration of a 2x2 programmable phase plate for electrons," *Ultramicroscopy*, vol. 190, pp. 58–65, 2018.

[2] H. Boersch, "Über die Kontraste von Atomen im Elektronen-mikroskop," *Zeitschrift für Naturforschung A*, vol. 2. p. 615, 1947.

[3] P. Thakkar, V. A. Guzenko, P.-H. Lu, R. E. Dunin-Borkowski, J. P. Abrahams, and S. Tsujino, "Fabrication of low aspect ratio three-element Boersch phase shifters for voltage-controlled three electron beam interference," *J. Appl. Phys.*, vol. 128, no. 13, p. 134502, 2020.

A standing molecule as a coherent single-electron field emitter

Taner Esat[1,2], Marvin Knol[1,2,3], Philipp Leinen[1,2,3], Matthew F. B. Green[1,2,3], Malte Esders[4], Niklas Friedrich[1], Michael Maiworm[5], Nicola Ferri[6], Pawel Chmielniak[2,7,8], Sidra Sarwar[2,7,8], Torsten Deilmann[9], Peter Krüger[9], Hadi H. Arefi[1,2], Daniel Corken[10], James Gardner[10], Kristof T. Schütt[4], Jeff Rawson[2,7,8], Paul Kögerler[2,7,8], Michael Rohlfing[9], Rolf Findeisen[5], Alexandre Tkatchenko[6,11], Klaus-Robert Müller[4,12,13], Reinhard J. Maurer[10], Christian Wagner[1,2], Ruslan Temirov[1,2,14] & F. Stefan Tautz[1,2,3]

[1]Peter Grünberg Institut (PGI-3), Forschungszentrum Jülich, Jülich, Germany
[2]Jülich Aachen Research Alliance (JARA)-Fundamentals of Future Information Technology, Jülich, Germany
[3]Experimentalphysik IV A, RWTH Aachen University, Aachen, Germany
[4]Machine Learning Group, Technische Universität Berlin, Berlin, Germany
[5]Otto-von-Guericke-Universität Magdeburg, Laboratory for Systems Theory and Automatic Control, Magdeburg, Germany
[6]Fritz-Haber-Institut der Max-Planck-Gesellschaft, Berlin, Germany
[7]Institute of Inorganic Chemistry, RWTH Aachen University, Aachen, Germany
[8]Peter Grünberg Institut (PGI-6), Forschungszentrum Jülich, Jülich, Germany
[9]Institut für Festkörpertheorie, Westfälische Wilhelms-Universität Münster, Münster, Germany
[10]Department of Chemistry, University of Warwick, Coventry, United Kingdom
[11]Physics and Materials Science Research Unit, University of Luxembourg, Luxembourg, Luxembourg
[12]Max Planck Institute for Informatics, Saarbrücken, Germany
[13]Department of Brain and Cognitive Engineering, Korea University, Seoul, South Korea
[14]Institute of Physics II, University of Cologne, Cologne, Germany

*Contact: s.tautz@fz-juelich.de, phone +49-2461 61 4561

Abstract— **The assembly of single-molecule devices with the help of the manipulation capability of scanning probe microscopes offers many opportunities for quantum- and nanotechnology. A key challenge is fabricating device structures that can overcome their attraction to the underlying surface and thus protrude from the two-dimensional flatlands of the surface. In my talk, I will report the fabrication of such a structure: we use the tip of a scanning probe microscope to lift a large planar aromatic molecule into an upright, standing geometry on a pedestal of two metal adatoms. This atypical upright orientation of the single molecule, whose stability can be understood as the result of a fine balance between chemical and dispersion forces, enables the system to function as a quantum dot and an on-demand coherent single-electron field emitter. If attached to the tip of the microscope, the standing molecule can also be applied as a sensitive quantum dot sensor. We anticipate that other metastable adsorbate configurations might also be accessible, thereby opening up the third dimension for the design of functional nanostructures on surfaces. Finally, we have made first steps into the direction of an autonomous robotic nanofabrication of single-molecule devices.**

I. INTRODUCTION

During the last 200 years, starting with Michel Chevreul in 1816 and Friedrich Wöhler in 1826, chemists have learned to synthesize almost any conceivable molecule. And there are many: It has been estimated that there are 10^{60} possible organic compounds with up to 30 carbon, nitrogen, oxygen or sulfur atoms [1]. To put this in perspective: There are 10^{78} to 10^{82} atoms in the known, observable universe [2]. So, if we wanted to synthesize even a small macroscopic quantity of each of these 10^{60} molecules (on the order of one mol each), not even thinking about larger molecules, we would quickly run out of atoms in the universe!

Nature knows how to build with molecules. Biological organisms are a living testimony to this. While nature uses self-assembly to make functional structures from molecules, human construction in the macroscopic world is largely based on the manufacturing paradigm. Therefore, the question arises whether building with molecules "brick-by-brick", following the manufacturing paradigm, is at all possible. The benefit would be tremendous: Getting a grip on molecules, we could fully exploit the potential of the most versatile collection of functional components in the universe for a range of human technologies.

The scanning probe microscope (SPM) has brought the vision of molecular-scale fabrication closer to reality, since it offers the capability to rearrange atoms and molecules on surfaces, thereby allowing the creation of metastable structures which do not form spontaneously. The standing molecule [3,4] is an instructive example.

II. RESULTS

A. The fabrication of the standing molecule

A single 3,4,9,10 perylene-tetracarboxylic dianhydride (PTCDA) molecule can be brought into a standing configuration on the Ag(111) surface, if a pedestal of two Ag adatoms is first created [3,4]: Once the adatoms have been attached to one side of the flat-lying molecule by atomic manipulation, the SPM tip is moved into contact with a carboxylic oxygen atom at the opposite side. Next, the tip lifts PTCDA on a curved trajectory into the vertical. When the molecule stands upright, the tip is moved straight up, whence its bond to the molecule breaks and the molecule remains standing.

B. Structure and stability of the standing molecule

The standing molecule adsorbs with the silver atoms of its pedestal in hollow sites of the Ag(111) surface. Two types of standing molecules exist: adatoms in identical hollow sites (fcc-fcc or hcp-hcp), adatoms in non-identical hollow sites (fcc-hcp). They can be distinguished by their azimuthal orientation and their thermal stability. The latter also reveals the potential energy barrier that stabilizes the molecule in its standing configuration. Density functional theory (DFT) calculations show that this barrier results from a fine balance between covalent and van der Waals (vdW) interactions. Because the experimentally measured barriers are in the range of only 30meV, standing PTCDA serves as a highly sensitive benchmarking system for state-of-the-art ab initio theory. For example, we find that many-body screening effects are an essential element of the standing molecule's stability.

C. Field emission from the standing molecule

The standing molecule functions as a quantum dot [5], accepting electrons one-by-one. Because the quantum dot is very small, it does not fulfil the constant interaction approximation [6]. Furthermore, under a large negative bias the additional electron is field-emitted towards the tip of the SPM [3]. Scanning the tip in the xy plane over standing molecule, one observes an interference pattern that can be modelled by combining coherent emission from the lobes of the relevant molecular orbital ("double-slit" experiment, Fig. 1) and field-emission resonances of the trapezoidal barrier (Gundlach oscillations). The standing molecule withstands extreme current densities of 10^8 A m^{-2}.

D. The standing molecules as sensor

If the standing molecule is fabricated on the SPM tip, its quantum dot functionality and the joint electrostatic screening by tip and surface enable quantitative surface potential imaging across all relevant length scales down to single atoms (SQDM-Scanning Quantum Dot Microscopy)[5,7-9] .

E. Prospects of autonomous fabrication

In a proof-of-concept study, we have demonstrated that autonomous robotic nanofabrication with single molecules is in principle possible, using the approach of reinforcement learning [10]. We anticipate that this work opens the way toward autonomous agents for the robotic construction of functional supramolecular structures with speed, precision, and perseverance beyond our current capabilities.

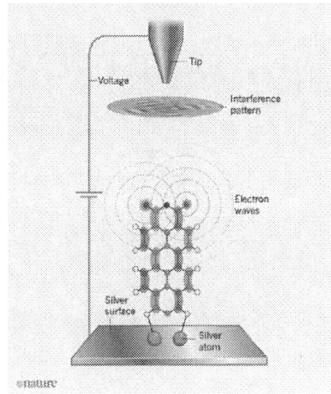

Fig. 1 A tiny „double-slit" experiment based on a standing PTCDA molecule, which serves as a coherent single-electron field emitter (From ref. [4]).

III. CONCLUSIONS

Metastable structures, apart from being more abundant than stable ones, tend to offer attractive functionalities, because their constituent building blocks can be arranged more freely and in particular in desired functional relationships. The metastable standing molecule exemplifies that the swift development of quantum- and nanotechnologies could be further advanced if we learned to freely design quantum matter by placing atoms and molecules in precisely the right places, not being constrained by availability and stability of exotic (quantum) materials.

REFERENCES

[1] R. S. Bohacek, C. MacMartin, W. C. Guida, The art and practice of structure-based drug design: A molecular modeling perspective. Medicinal Research Reviews, Vol. 16, No. 1, 3-50 (1996). https://doi.org/10.1002/(SICI)1098-1128(199601)16:1<3::AID-MED1>3.0.CO;2-6

[2] https://www.universetoday.com/36302/atoms-in-the-universe/, accessed July 19, 2021.

[3] T. Esat, N. Friedrich, F. S. Tautz & R. Temirov. A standing molecule as a single-electron field emitter. Nature 558, 573–576 (2018). https://doi.org/10.1038/s41586-018-0223-y

[4] T. Greber, Nature 558, 525-526 (2018). https://doi.org/10.1038/d41586-018-05502-5

[5] C. Wagner, M. F. B. Green, P. Leinen, T. Deilmann, P. Krüger, M. Rohlfing, R. Temirov & F. S. Tautz, Scanning Quantum Dot Microscopy. Physical Review Letters 2015, 115, 026101. https://doi.org/10.1103/PhysRevLett.115.026101

[6] R. Temirov, M. F. B. Green, N. Friedrich, P. Leinen, T. Esat, P. Chmielniak, S. Sarwar, J. Rawson, P. Kögerler, C. Wagner, M. Rohlfing & F.S. Tautz, Molecular Model of a Quantum Dot Beyond the Constant Interaction Approximation. Physical Review Letters 2018, 120, 206801. https://doi.org/10.1103/PhysRevLett.120.206801

[7] C. Wagner, M. F. B Green, M. Maiworm, P. Leinen, T. Esat, N. Ferri, N. Friedrich, R. Findeisen, A. Tkatchenko, R. Temirov & F. S. Tautz, Quantitative imaging of electric surface potentials with single-atom sensitivity. Nature Materials 2019, 18, 853, https://doi.org/10.1038/s41563-019-0382-8

[8] M. Persson, Electric potentials at the atomic scale. Nat. Mater. 18, 773–774 (2019). https://doi.org/10.1038/s41563-019-0383-7

[9] C. Wagner & F. S. Tautz, The theory of scanning quantum dot microscopy. Journal of Physics: Condensed Matter 2019, 31, 475901. https://doi.org/10.1088/1361-648X/ab2d09

[10] P. Leinen, M. Esders, K. T. Schütt, C. Wagner, K.-R. Müller & F. S. Tautz, Autonomous robotic nanofabrication with reinforcement learning. Science Advances Vol. 6, no. 36, eabb6987. https://doi.org/10.1126/sciadv.abb6987

Scanning Field Emission Microscopy with Spin and Energy Analysis

A-K Thamm[1]*, J. Wei[1], M. Demydenko[1], C.G.H. Walker[1,2], D. Pescia[1] and U. Ramsperger[1]
A. Pratt[2], S.P. Tear[2], M.M. El Gomati[3]

[1] Laboratory for Solid State Physics, ETH Zurich, 8093 Zurich, Switzerland.
[2] Departement of Physics, University of York, Heslington, York, YO10 5DD, U.;
[3] York Probe Sources Ltd, 7 Harwood Rd, York YO26 6QU, UK.
*Contact: athamm@ethz.ch, phone: +41 4463 32362

Abstract — **Scanning Field Emission Microscopy with Polarization Analysis was recently introduced to detect the spin polarization of electrons excited in the field emission regime of Scanning Tunnelling Microscopy. In this work, a miniature electron energy analyzer, called Bessel Box, is implemented into the Scanning Field Emission Microscope with Polarization Analysis setup. It is used to filter electrons according to their energy before they reach the spin detector. The Bessel Box allows, e.g., the spin polarization of elastically scattered electrons to be compared with the spin polarization obtained with the full energy spectrum. We use this technology to measure the local in-plane polarization signal as a function of the magnetic field B at room temperature for 10 monolayers Fe deposited on top of a W(011)- single crystal surface through a half mask (half of the surface is covered with Fe, the other half is uncovered). The spin polarization at the Fe-W crossing drops sharply from 9 % above Fe to 0 % above W(011) only if the *elastically* scattered electrons are selected for spin analysis. The mechanism of signal generation in Scanning Field Emission Microscope with Polarization Analysis including the formation of cascade of inelastically scattered electrons is discussed as an explanation for the different spin polarization profiles observed with and without Bessel Box (energy filtered).**

Keywords — *Electron Microscopy, Field Emission, Energy and Spin Analysis, Bessel Box*

I. INTRODUCTION

In Scanning Field Emission Microscopy with Polarization Analysis (SFEMPA) electrons are field emitted from a tip residing only few 10 nm away from the target (field emission regime of Scanning Tunneling Microscopy (STM)). Upon impinging onto the surface, the field emitted electrons excite secondary electrons off the surface, which can be analyzed in terms of their spin-polarization. The spatial resolution of SFEMPA is few nanometers [1], [2]. In references [1] and [2] all secondary electrons (SE) were collected and analysed in the spin detector. Here we add a miniature energy analyzer (Bessel-box (BB), [3], [4]) that allows to filter the secondary electrons according to their energy before entering the spin detector (Fig. 1A).

II. EXPERIMENTAL SETUP

The STM is converted into a SFEM(PA) by retracting the tungsten tip 5 to 200 nm distance from the surface. SFEM can be also regarded as a lensless low-energy Scanning Electron Microscope. At sufficiently high negative bias voltage applied to the tip (Fig. 1A, grey) electrons are field-emitted towards the sample. The primary electrons impact the sample surface with few 10 eV (typically 30 to 90 eV, Fig. 1A, red) exciting secondary electrons off the sample (Fig. 1A, green).

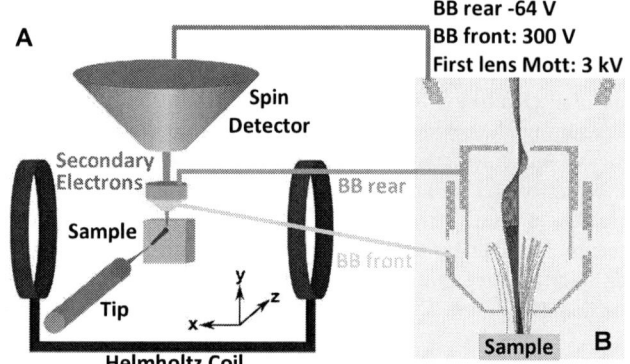

Fig. 1
A) SFEMPA setup with the BB (blue) between sample (yellow) and spin detector (grey). Primary electrons (red) are field emitted from the W-tip (grey) by applying a negative bias voltage (- 30 to - 90 V) and produce secondary electrons (green) at the sample surface. These electrons are accelerated towards the front cone of the BB (light blue). The BB-rear voltage (blue) filters the SE according to their energy and thus only energy selected SE are able to reach the spin detector (grey). The Helmholtz coil (dark blue) is used to apply a pulsed magnetic field for hysteresis loop measurements.
B) Simulation of the electron trajectories through the Bessel box (BB rear voltage = - 65 V, energy of electrons: red = 65 eV, blue = 30 eV, green = 2 eV).

Those electrons escaping the tip-sample junction are spin-analyzed (Fig. 1A, gray, Mott-type spin detector) after passing the miniature energy analyzer (Fig. 1A, cyan). The BB is placed between the sample (Fig. 1A, yellow) and the spin detector on a manipulator in order to optimize the position of the BB and to remove the BB for a measurement in the standard full spectrum mode. The two lenses of the BB can be adjusted in such a way that in the energy filtered mode all electrons with an energy lower than the primary energy are stopped in the BB (Fig. 1B). Only the elastically scattered SE are able to reach the spin detector. After leaving the BB the (elastic) SE are collected by the first spin detector lens and then spin analyzed.

III. MEASUREMENT AND RESULTS

The sample is produced in situ by Molecular Beam Epitaxy. 10 ML Fe are evaporated through a "half mask" on top of a clean W(011)-single crystal surface. The thickness of the Fe layer is determined by Auger Electron Spectroscopy as well as STM. The STM is also used to find the Fe-W crossing (Fig.2A). After finding the Fe-W crossing the STM is converted into SFEM.

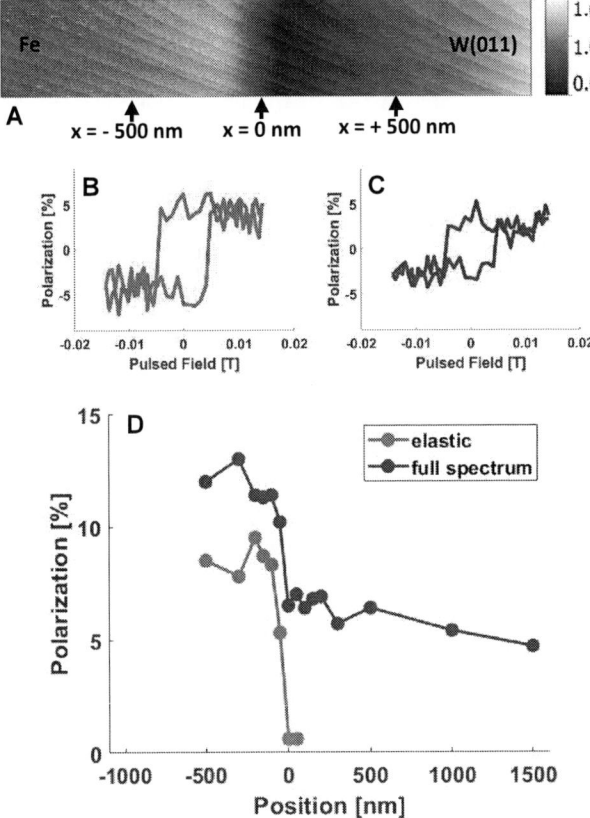

Fig. 2
A) (2×0.35) µm STM-image taken across the Fe-W crossing (V_{tip} = -0.2 V, I_{tunnel} = 0.1 nA). The Fe film is on the left (negative horizontal coordinates). The coordinate x=0 indicate the approximate position of the Fe-W crossing.
B) Hysteresis loop taken with the W-tip above the Fe, x = -500 nm in the energy filtered mode, V_{tip} = -60 V.
C) Hysteresis loop taken with the W-tip above pure W, x = + 1µm in the full spectrum mode, V_{tip} = -60 V.
D) Profile of the in-plane spin polarization as a function of position relative to the Fe-W crossing (x = 0 nm) obtained by hysteresis loop measurements. Energy filtered mode: red, full spectrum: blue.

The tip is moved to the starting position (e.g. x = − 500 nm) above Fe and a hysteresis loop is measured (Fig. 2B). The tip is then moved step by step from the initial position to the crossing

and up to x = + 1.5 µm above the clean W while hysteresis loops are taken at each position. The measurements are performed in the filtered mode and the full spectrum mode. The in-plane spin polarization signal drops sharply within 200 nm at the Fe-W crossing in the filtered mode (Fig. 2D, red) while a spin polarization signal (up to 6 %, Fig 2C) remains in the full energy spectrum mode when the tip is located on top of W (Fig. 2D, blue).

IV. DISCUSSION

When the tip is located above the W(011) (which is non-magnetic), a finite spin polarization is detected in the full energy spectrum mode of detection. A possible explanation for this is the mechanism of signal generation in SFEM. According to recent simulations [5], [6], the primary beam is the origin of a two-dimensional cascade of inelastic electrons that might extend up to mm away from the point of impact of the primary electrons. When the full energy spectrum is collected on top of W, electrons originating from Fe might contribute and produce the finite spin polarization detected on the W. If only the elastically SE are allowed to enter the spin detector (energy filtered mode), a local mapping of the surface magnetization might become possible, as a sizeable fraction of the elastic electrons [6] consists of those electrons that originate from underneath the tip. According to this scenario, the sharp boundary observed in energy filtered SFEMPA is produced by the elimination of the cascade electrons.

ACKNOWLEDGMENT

We would like to thank T. Michlmayr, U. Maier and T. Bähler for technical assistance. We thank the European Commission (SIMDALEE2: Marie Curie Initial Training Network (ITN), Grant number 606988 under FP7-PEOPLE-2013-ITN, the Swiss National Science Foundation (SNF grant number 20-134422) and the Commission for Technology and Innovation (CTI grant number 9860.1 PFNM-NM) for financial support.

REFERENCES

[1] L. De Pietro, G. Bertolini, Q. Peter, H. Cabrera, A. Vindigni, O. Gürlü, D. Pescia, U. Ramsperger, 'Spin-polarised electrons in a one-magnet-only Mott spin junction', Sci. Rep. 7, 13237 (2017).

[2] U. Ramsperger and D. Pescia, 'Vectorial, non-destructive magnetic imaging with scanning tunneling microscopy in the field emission regime', Appl. Phys. Lett. 115, 112402 (2019).

[3] A. Bellissimo et al., 'Improving the Detection and the Analysis of Energy Filtered and Spin Polarised Electrons with the implementation of a Miniature Energy Analyser', IEEE, DOI: 10.1109/IVNC 49440.2020. 9203493 (2020).

[4] A. Suri, 'Detection and Analysis of Low Energy Electrons in a Scanning Electron Microscope using a novel detector design based on the Bessel Box Energy Analyser', PhD-thesis, http://etheses.whiterose.ac.uk/26527/ (2020).

[5] W.S.M. Werner et al., 'Scanning tunneling microscopy in the field-emission regime: Formation of a two-dimensional electron cascade', Appl. Phys. Lett., vol. 115, 251604, (2019)

[6] C.G.H. Walker et al., 'Electron energy analysis in Scanning Field Emission Microscopy using a Bessel box energy analyzer', IVNC 2021

978-1-6654-2590-2/21 $31.00 © 2021 IEEE

A Plasmon-Mediated Cold-Cathode

Shaozhi Deng*, Yan Shen, Huanjun Chen, Ningsheng Xu

State Key Lab of Optoelectronic Materials and Technologies,
Guangdong Province Key Lab of Display Material and Technology,
School of Electronic and Information Technology, Sun Yat-sen University,
Guangzhou 510275, People's Republic of China
*Contact: stsdsz@mail.sysu.edu.cn, phone +86 20 84110916

Abstract— **Light-driven electron emission plays an important role in modern electronic and optoelectronic devices. However, such a process usually requires a light field either with a high intensity, or with a high frequency, which is not in favor of its applications. To solve these issues, we propose and realize a plasmon-mediated cold-cathode (PMCC), which combines plasmonic nanostructures with nano-electron-emitters of low work function. In a PMCC, hot-electrons generated by plasmon resonances upon light excitation will enter into adjacent emitter, and subsequently emit into vacuum. PMCCs greatly benefit from strong energy coupling of plasmonic effect, and work based on a plasmon-mediated electron emission process. Electron emission of high efficiency can be obtained with light field of moderate intensities and visible wavelengths. This electron emission is of interest for applications including cold-cathode electron sources, advanced photocathodes, and micro- and nano-electronic devices relying on free electrons. The above findings will be presented with a review of progresses in the study of such a light-driven electron emission.**

I. INTRODUCTION

Light-driven electron emission forms the basis of various applications including sensitive photodetection, surface analysis, electron beam source, lithography and microscope. Here, we present a review of developments in light-driven electron emission and report our results about a plasmon-mediated cold-cathode.

Light-driven electron emission was first concerned because of the photoemission effect, in which the photon energies are absorbed by electrons inside a solid, and gain enough energy to overcome the energy barrier at the solid surface and escaped into vacuum, explained by Albert Einstein in 1905. Along with the development of laser technology, the cathodes can be triggered by high-intensity continuous lasers or laser pulses to generate various laser-assisted electron emissions, such as photo-field emission (PFE), multiphoton photoemission (MPP), above-threshold photoemisison (ATP) and optical field emission (OFE) [1]. For example, in OFE process, the surface potential barrier may be decreased by strong optical fields, and the electrons can directly tunnel through such a barrier into vacuum. This process is similar to the field emission driven by static electric fields in physical behavior. Typically, the laser intensities driven on the order of 1 TW·cm⁻² are required for strong OFE. Recently, another type of light-driven electron emission process, *i.e.* the plasmon-photoemission has gained much attention. By the confined and enhanced electromagnetic fields, delicately engineered plasmonic nanostructures as nano-electron-emitters can be excited by laser pulses with energy density over 1 GW·cm⁻² to generate energetic hot electrons. However, for practical device application, electron emission with high current and high current density under moderate light intensity excitation is still preferred [2].

In our study, we propose to employ the plasmonic nanostructures to mediate electron emission, *i.e.* the plasmon-mediated electron emission (PMEE) process, and develop the corresponding plasmon-mediated cold cathode (PMCC) [3]. The plasmonic nanostructure acts as nanoantenna to efficiently adsorb incidence light field. The decay of plasmon resonances generates hot electrons, which then transit into a nano-electron-emitter that the plasmonic nanostructure is adhered to. The hot electrons then can be released into vacuum under a small static electric field applied to the emitter, which is shown to exhibit outstanding performance.

II. METHODS

One key to realize the PMEE process is that the selected nano-electron-emitter should be with high aspect ratio and low work function. Another key is that the interface between the plasmonic nanostructures and emitters should be as perfect as possible so that the generated hot electrons can pass through without any hindrance.

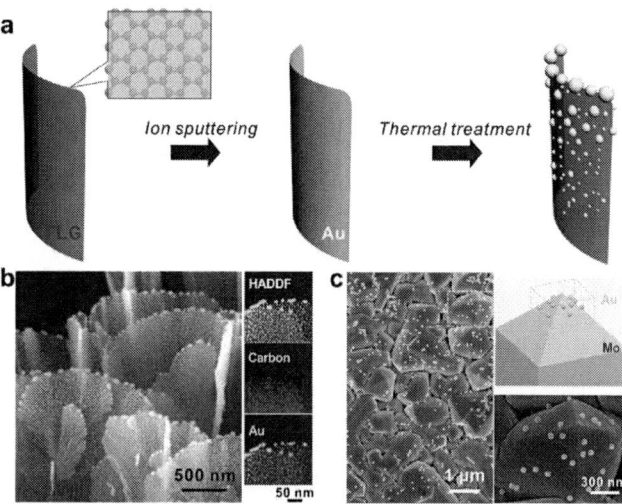

Fig. 1 (a) Schematic illustration of synthesis of the PMCC (Au-on-Gr nanostructures). (b) and (c) are typical morphologies of the Au-on-Gr and Au-on-Mo PMCC emitters, respectively.

We have demonstrated our proposed design using gold-on-graphene and gold-on-molybdenum nanostructures based on conducting substrates. For example, as illustrated in Fig. 1(a), the vFLG is employed as substrate for deposition of ultra-thin gold film *via* an ion sputtering process, and then the gold nanoparticles are formed onto the surface of the vFLG after a thermal annealing treatment. Fig. 1(b) and (c) show the typical morphologies of the Au-on-Gr and Au-on-Mo PMCC emitters, respectively. Both of the main bodies of vFLG flake and Mo nanopyramid are of micrometer scale in length and width. The gold nanoparticles distribute uniformly onto the surface of the emitters, with the diameter about 50 nm along their top edges, which show plasmonic resonance in range of 520~530 nm.

To initiate the light-driven electron emission, a laser excitation in combination with a small pulsed electrostatic field is applied to the samples in vacuum. Focused by a super-continuum white-light laser, the emission current is firstly recorded under various laser and electric field excitations, and then the incident laser beam is divided into different excitation bands for the frequency-dependence measurement. Moreover, picosecond and femtosecond lasers are used to demonstrate the possible ultrafast and ultrashort pulsed electron emission through the PMEE process.

III. RESULTS AND DISCUSSIONS

The effect of the super-continuum white-light laser intensity (I_{laser}) on the PMCCs driven by a small pulsed electrostatic field (E) can be seen in Fig. 2. Laser intensity as low as 0.073 W·cm^{-2} can already give rise to the PMEE. A maximum emission current density over 155.6 mA·cm^{-2} can be obtained with low laser intensity of around 5 W·cm^{-2}, without use of light source of intensities as high as GW·cm^{-2} or TW·cm^{-2}. Electron emission yield as high as 40.16% is observed, higher than most of photocathodes can provide. Such enhanced electron emission is strongly dependent on the plasmon resonance wavelength of the gold nanostructures decorated on the PMCCs.

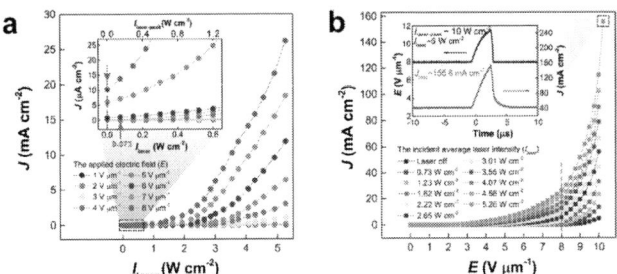

Fig. 2 PMEE properties of the Au-on-Gr nanostructures. (a) Dependence of the current density on the I_{laser} under various electric field. Inset: J-I_{laser} behavior at low I_{laser} region. (b) J-E curves under different incident I_{laser}. Inset: waveforms of input pulsed electric field and maximum current density.

Furthermore, the PMEE pulses excited by picosecond and femtosecond lasers are also obtained. In a demo structure, as shown in Fig. 3(a), we successfully capture the periodic pulses of PMEE under low-power picosecond light excitation, and an ultrashort pulsed electron beam source is provided.

Physical mechanism is further studied. By using an *in situ* micro-zoned analysis, we firstly find that the PMEE process occurs at low sample temperature, thus thermal electron emission is negligible. Fig. 3(b) shows the potential emission

mechanisms: The hot electrons with high energy directly tunnel into vacuum through the surface barrier of the gold nanoparticles under the electric field (I), or inject into the graphene by going through the barrier at the interface between graphene and gold nanoparticle, which may subsequently tunnel into the vacuum from the graphene's edges (II). The plasmon-enhanced near-field is strong enough to reduce the barrier width experienced by the electrons within the graphene next to the gold nanoparticles (III). Fourthly, the enhanced near-field induced by the plasmon resonances accelerate the optical absorption of the graphene, which consequently increase the electron interband transitions of the graphene. These addition electrons tunnel through the graphene edges under the static electric field (IV). We believe that the mechanism responsible for the PMEE should be mainly associated with the high-energetic hot electrons transferring from the plasmonic nanoparticles (Au) into the graphene (Fig.3(c)). These additional hot electrons can increase the supply of electrons which experience smaller surface potential width for emission with high efficiency. Further analyses of the delay time, rising and falling edges for the picosecond PMEE pulses give more detailed process.

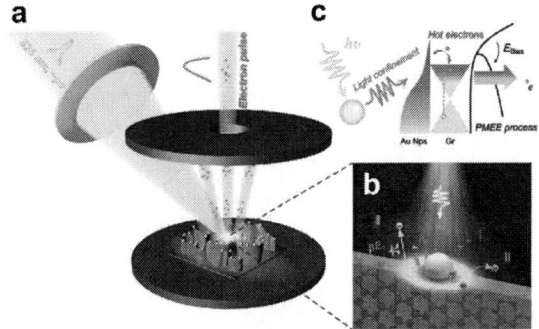

Fig. 3 (a) PMEE pulsed electron beam source under picosecond laser excitation. (b) Schematic illustration of possible light-driven electron emission mechanisms and (c) the corresponding energy level diagram of PMEE.

IV. CONCLUSIONS

Our study demonstrate the potential of proposed PMCC in high efficient, portable and integrated high-time-resolution vacuum electron device applications.

ACKNOWLEDGMENT

This work was financially supported by the National Key Basic Research Program of China (Grant Nos. 2019YFA0210201, 2019YFA0210203), the National Natural Science Foundation of China (Grant No. 52072416), the Fundamental Research Funds for Central Universities, and the Science and Technology Department of Guangdong Province.

REFERENCES

[1] S. H. Zhou, K. Chen, M. T. Cole, Z. J. Li, J. Chen, C. Li, and Q. Dai, "Ultrafast field-emission electron sources based on nanomaterials," *Adv. Mater.*, vol. 31, pp. 1805845, 2019.

[2] E. Forati, T. J. Dill, A. R. Tao, and D. Sievenpiper, "Photoemission-Based Microelectronic Devices," *Nat. Commun.*, vol. 7, pp. 13399, 2016.

[3] Y. Shen, H. J. Chen, N. S. Xu, Y. Xing, H. Wang, R. Z. Zhan, L. Gong, J. X. Wen, C. Zhuang, X. X. Chen, X. M. Wang, Y. Zhang, F. Liu, J. Chen, J. C. She, and S. Z. Deng, "A plasmon-mediated electron emission process," *ACS Nano*, vol. 13, pp. 1977–1989, 2019.

Enhancement of thermionic emission and conversion characteristics using polarization- and band-engineered n-type AlGaN cathodes

Shigeya Kimura[1]*, Hisashi Yoshida[1], Hisao Miyazaki[1], Takuya Fujimoto[2], and Akihisa Ogino[2]

[1] *Corporate Research & Development Center, Toshiba Corporation, Kawasaki, Japan*
[2] *Graduate School of Integrated Science and Technology, Shizuoka University, Hamamatsu, Japan*
*Contact: shigeya.kimura@toshiba.co.jp, phone +81-44-549-2138

Abstract— **We observed enhancement of thermionic emission (TE) and conversion (TC) characteristics by controlling spontaneous and piezo polarization and the band diagram of n-type AlGaN/GaN thermionic cathodes. The obtained TE current from N-polarity n-AlGaN films grown on an n-GaN substrate was 0.29 mA at 500 °C in a Cs gas atmosphere in the vacuum gap between the cathode and a stainless steel anode. This TE current was 5.0 times and 1.6 times higher than that from the surface of Ga-polarity n-GaN substrate and that of the Ga-polarity n-AlGaN film on the substrate, respectively.**

I. INTRODUCTION

Wide-bandgap semiconductors, such as hydrogen-terminated diamond and Cs-adsorbed GaN, are expected to become used as electron emitters owing to their low or negative electron affinity [1,2]. We have focused on GaN-based semiconductors for application in high-efficiency thermionic converters through band engineering using heterostructures [3]. These GaN-based semiconductors have two inequivalent surfaces perpendicular to the c-axis, namely, the Ga-face and the N-face, which are polarized in opposite directions. Recently, we reported that thermionic emission (TE) from N-polarity n-GaN was higher than that of Ga-polarity [4]. In the present study, we report on the TE and thermionic conversion (TC) characteristics from polarization- and band-engineered AlGaN/GaN thermionic cathodes.

II. SAMPLE PREPARATION AND STRUCTURAL PROPERTIES

We prepared four types of thermionic cathode samples as shown in Fig. 1. Fig. 1(a) shows a 280-μm-thick n-type GaN substrate with Ga-polarity (the front surface is referred to as the Ga-face) that was subjected to chemical mechanical polishing (CMP) and chemical etching with HCl on the (0001) surface. Fig. 1(c) shows a 20-nm-thick Si-doped $Al_{0.25}Ga_{0.75}N$ film grown by metal organic chemical vapor deposition (MOCVD) on the Ga-polarity n-type GaN (0001) substrate. The Si concentration is 1.0×10^{18}/cm^3 in the AlGaN film. Fig. 1(b) shows an n-type GaN substrate with N-polarity (the front surface is referred to as the N-face) that was similarly processed on the (000-1) surface. Fig. 1(d) shows a 20-nm-thick Si-doped $Al_{0.25}Ga_{0.75}N$ film grown by MOCVD under the same conditions as in Fig. 1(c) on the N-polarity n-type GaN (000-1) substrate. Tungsten-based ohmic metal contacts were formed on the side opposite to the prepared front surface before chemical etching.

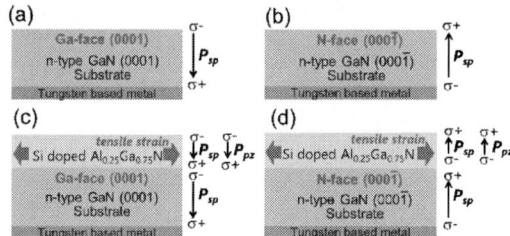

Fig. 1 Schematics of the nitride-based cathode structures examined in this study.

X-ray reciprocal space mapping found that the Si-doped AlGaN films shown in Fig. 1(c) and (d) were epitaxially and coherently grown on the n-type GaN substrates (0001) and (000-1), respectively. This means that these films were under tensile strain. Fig. 2(a) and (b) show cross sectional transmission electron microscopy (TEM) images of the samples shown in Fig. 1(c) and (d), respectively. The insets in both figures show enlarged views by high-angle annular dark-field (HAADF) scanning in the AlGaN films. TEM analysis found that the Si-doped AlGaN film shown in Fig. 1(c) had Al- or Ga-polarity, while that shown in Fig. 1(d) had N-polarity. From these structural analyses, we confirmed that the AlGaN films shown in Fig. 1(c) and (d) had spontaneous and piezo polarization in opposite directions.

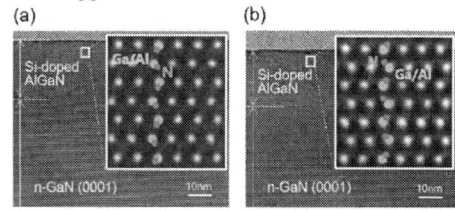

Fig. 2 Cross sectional TEM images of the samples shown in Fig. 1(c) and (d).

III. SURFACE PROPERTIES AND BAND DIAGRAM

Fig. 3 shows the results of ultra-photoemission spectroscopy (UPS) analysis of the sample surfaces. The estimated work functions of the cleaned Ga-polarity surface (Fig. 1(a)) and N-polarity surface (Fig. 1(b)) of n-GaN substrates were 3.5 and 3.3 eV, respectively. The work functions of the cleaned Al- or Ga-polarity surface (Ga- or Al-face, shown in Fig. 1(c)) of AlGaN film was 3.2 eV, and that of the N-polarity surface (N-face, shown in Fig. 1(d)) was 2.9 eV. The UPS analysis suggests the band diagrams shown in Fig. 4 (a)-(d), which correspond to Fig. 1(a)-(d). That is, the surface work functions and affinities could be reduced in AlGaN films not only by the increased the band gap but also by inversion of the spontaneous- and piezo-polarization direction.

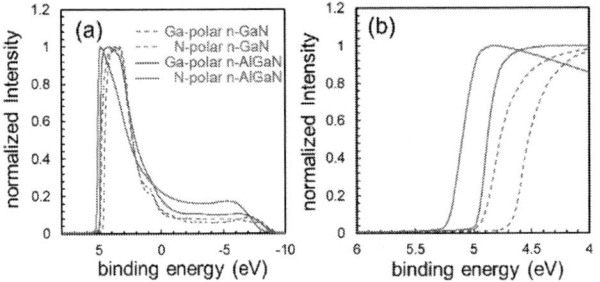

Fig. 3 (a) UPS spectra of the samples in this study. (b) Enlarged view of (a).

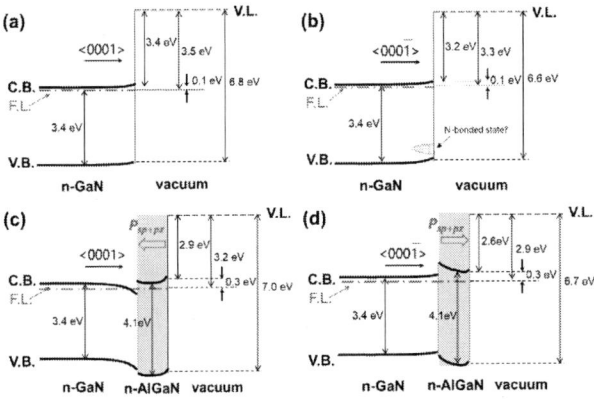

Fig. 4 Band diagrams corresponding to Fig. 1(a)-(d).

IV. THERMIONIC EMISSION AND CONVERSION CHARACTERISTICS

Fig. 5(a) shows the configuration used for measuring TE/TC in this study. The HCl-etched front surface faced the anode through a vacuum gap. During measurements, Cs gas controlled by the Cs dispenser current was supplied to the vacuum gap area as shown in Fig. 5(a). The vacuum gap width was maintained at 0.5 mm, and the cathode temperature was controlled by a heater stage on the backside of the samples. Fig. 5(b) shows TE current versus temperature for the samples shown in Fig. 1(a)-(d). The applied anode bias was maintained at +30V during the measurements. As shown in Fig. 5(b), TE current increased with increasing temperatures up to 500 °C, but decreased at higher temperature. This decrease was caused by Cs atom desorption from the cathode sample surfaces by heating. We note the following two points in Fig. 5(b): the TE current from the cathode samples with AlGaN film was higher than that without the film, and the TE current from the samples with N-polarity was higher than that from the samples with Ga-polarity. The latter effect was observed more clearly in the sample without AlGaN film than in the sample with the film. Fig. 6

shows the TE current as a function of applied voltage for each cathode sample. As shown in Fig. 6, TE current was detected at 0 V, meaning that our measurement configuration using nitride-based cathodes had TC characteristics. The short circuit current of the polarization- and band-engineered cathode samples (Fig. 1(b)-(d)) were larger than that of the Ga-polar n-GaN substrate.

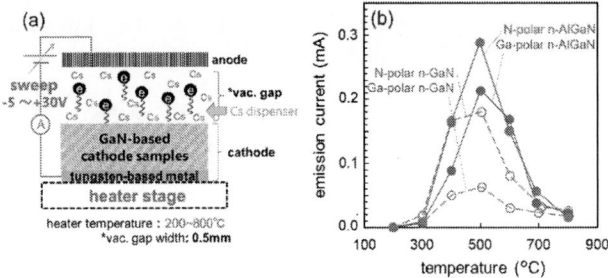

Fig. 5 (a) Configuration for measuring TE/TC characteristics in a vacuum chamber. (b) TE current versus temperature for the cathode samples.

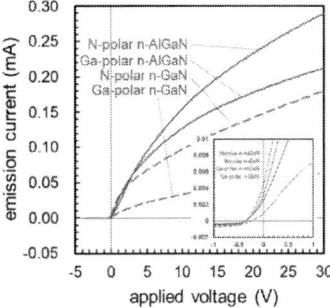

Fig. 6 TE current versus applied voltage for the cathode samples. The heater stage temperature was 500 °C.

V. CONCLUSION

The TE and TC characteristics of Ga-polar and N-polar Si-doped n-type AlGaN/GaN based samples with Cs atom adsorption were investigated. We confirmed that polarization- and band-engineering of nitride-based cathodes together with Cs dipole formation on the surface were effective for enhancing the performance of TE and TC characteristics.

REFERENCES

[1] T. Sun, F. A. M. Koeck, C. Zhu, and R. J. Nemanich, "Combined visible light photo-emission and low temperature thermionic emission from nitrogen doped diamond films," Appl. Phys Lett., vol. 98, pp. 202101, Novemver 2011.

[2] CI Wu and A Kahn, "Negative electron affinity and electron emission at cesiated GaN and AlN surfaces," Appl. Surface Science, vol. 162–163, pp. 250-255, August 2000.

[3] S. Kimura, H. Yoshida, S. Uchida, and A. Ogino, "Thermionic emission and conversion properties of n-type AlGaN thin film cathodes grown on 6H–SiC substrates," Jpn. J. Appl Phys, vol. 59, pp. SGGF01, February 2020.

[4] S. Kimura, H. Yoshida, H. Miyazaki, T. Fujimoto, and A. Ogino,, "Surface polarity dependence of thermionic emission and conversion characteristics of n-type GaN cathodes," J. Vac. Sci. & Technol. B, vol. 39, pp. 014201, January 2021.

978-1-6654-2590-2/21 $31.00 © 2021 IEEE

Planar type electron emission device using atomic layered materials and it applications

Katsuhisa Murakami[1,2,*], Naoyuki Matsumoto[3], Yukino Kameda[4], Yoshinori Takao[5], Yoichiro Neo[4], Yoichi Yamada[2], Kazutaka Mitsuishi[6], Masahiro Sasaki[2], Hidenori Mimura[4], and Masayoshi Nagao[1]

[1]National Institute of Advanced Industrial Science and Technology (AIST), Tsukuba, Japan
[2]Faculty of Pure and Applied Sciences, University of Tsukuba, Tsukuba, Japan
[3] Department of Mechanical Engineering, Materials Science, and Ocean Engineering,
Yokohama National University, Yokohama, Japan
[4]Research Institute of Electronics, Shizuoka University, Hamamatsu, Japan
[5]Division of Systems Research, Yokohama National University, Yokohama, Japan
[6]National Institute for Materials Science, Tsukuba, Japan

*E-mail: murakami.k@aist.go.jp

Abstract—The planar type electron emission devices using atomic layered materials of graphene and hexagonal boron nitride (h-BN) were developed to suppress inelastic electron scattering within the device structure. High emission efficiency of more than 40 % and high emission current density of more than 100 mA/cm^2 were achieved by the suppression of the inelastic electron scattering within the topmost gate electrode using graphene,. In addition, highly monochromatic electron emission with an energy spread of 0.18 eV in the full width at half maximum were realized by the suppression of the inelastic electron scattering within the topmost gate electrode and insulating layer using the graphene/h-BN heterostructure. These results would lead to several practical applications of planar type electron emission devices.

Keywords—Graphene; Hexagonal Boron Nitride; Metal-Oxide-Semiconductor (MOS) structure, electron emission

I. INTRODUCTION

A planar type electron emission device[1] has several advantages, such as small divergence angle of electron beams emitted from the planar surface, low operation voltage less than 20 V, operation in a low vacuum condition[2], in an atmospheric pressure gas environment[3], and in a liquid[4]. These advantages would lead to several practical applications, such as a low-cost, high-resolution, and low-vacuum scanning electron microscope (SEM), an ion neutralizer for space propulsion, and the reforming of liquid and gas materials by a irradiation of the low energy electron beams. However, the very low electron emission efficiency, the low electron emission current density and the wider energy spread of the electron beams are the critical issues to overcome for the realization of the practical applications of the planar type electron emission devices. These issues are caused by the electron inelastic scattering within the gate electrode and insulating layer of the planar type electron emission devices. The device structure of the planar type electron emission devices is a metal/insulator/metal or metal/insulator/semiconductor stack. The thickness of the topmost metal electrode and insulator layer are nanometer scale,

respectively. The potential barrier of the insulating layer is thinned by the application of voltage to the top electrode, and then, electrons of the lower conductive substrate penetrate the potential barrier due to the tunneling effect. The electrons that tunnel through the potential barrier travel through the insulating layer by the electric field, and then, the electrons having the higher energy than the work function of the top metal electrode are emitted into the vacuum. However, the energy of the most of electrons lose by the electron inelastic scattering within the insulating layer and topmost electrode and become lower than the work function of the topmost gate electrode. Therefore, the electron emission efficiency is usually less than 1 %, and 99 % of the electrons are collected by the top metal electrode. In addition, the energy distributions of the emitted electrons spread by the inelastic scattering within the device. Therefore, the suppression of the inelastic scattering within the planar type electron emission devices solves their low electron emission efficiency and wider energy spreads of the electron beams and lead to their practical applications. The inelastic scattering cross section of an electron within a material is generally proportional to the atomic number of the material and inversely proportional to the energy of the incident electron. Since the driving voltage of the planar electron-emission device is as low as 10 V, the inelastic scattering cross section within the material is very large. It is, therefore, important to search for materials with small atomic number to suppress the inelastic scattering within the devices. Graphene is an atomic layered conductor composed of carbon atoms, which is a lighter element compared to the conventional gate metals such as Al and Au. Hexagonal boron nitride (h-BN) is also an atomic layered insulator composed of lighter element of N and B compared to the conventional insulator materials of SiO$_2$ and Al$_2$O$_3$. Therefore, the electron inelastic scattering within the device structure is expected to be suppressed using two-dimensional atomic layered materials of graphene and h-BN, which would lead to the planar type electron emission device with high-emission efficiency, high emission current density, and highly monochromatic electron beams. In this paper, planar type electron emission device using

978-1-6654-2590-2/21 $31.00 © 2021 IEEE

atomic layered materials of graphene and h-BN, and its possible applications will be discussed.

II. RESULTS AND DISCUSSION

A. Improvement of electron emission efficiency using graphene electrodes

First, we tried to suppress the electron inelastic scattering within the topmost electrode by using graphene as the top electrode[5-9]. The Graphene/Oxide/Semiconductor (GOS) type electron emission devices were fabricated by a catalyst free direct synthesis manner of graphene on an insulating substrate using the originally developed plasma enhanced chemical vapor deposition (PECVD). As a result, an electron emission efficiency of up to 48.5% and an emission current density of over 100 mA/cm^2 were achieved, which is an improvement of up to 10,000 times compared to conventional devices using topmost metal electrode[9]. From the energy analysis of the emitted electrons, it was found that the energy distribution of the emitted electrons has a low energy edge at about 2.5 eV higher than the work function of graphene. This suggests that all electrons tunneling through the insulating layer reach the electrodes while retaining an energy higher than the work function of graphene. In other words, it is found that the reason for the low electron emission efficiency of the conventional planar electron emission devices is not the electron inelastic scattering within the insulating layer but the energy loss due to the electron inelastic scattering within the topmost electrode. Based on the electron transmission of single-layer graphene, the maximum electron emission efficiency of the GOS device was found to be about 70%[9]. This is a remarkable phenomenon that about 70% of the electrons flowing in a Si substrate can be extracted from its surface by applying a voltage of only about 10 V to the graphene electrode.

B. Monochromatic electron emission from graphene/h-BN heterostructure

Although the graphene electrode achieves high electron emission efficiency, the energy distribution of the emitted electrons spreads over 1 eV in the full width at half maximum (FWHM) due to the electron inelastic scattering within the SiO$_2$ layer. Next, by using h-BN as the insulating layer, we attempted to monochromatize the energy of emitted electrons by suppressing electron inelastic scattering within the insulating layer[10-11]. We formed an h-BN insulating layer on a Si substrate via the wet-transfer process of a multilayer h-BN on cupper foil synthesized by CVD. The graphene electrodes were deposited by the same PECVD method as for GOS-type electron-emitting devices. Since the h-BN insulating layer was formed by wet transfer, wrinkles and cracks appeared in the h-BN insulating layer. However, even by using such wet-transferred h-BN as the insulating layer, we were able to achieve monochromatic electron emission from the graphene/h-BN/n-Si heterostructure. The narrowest energy spread of 0.18 eV in the FWHM were achieved, which is superior to the energy spread of a tungsten field emitter of about 0.3 eV[11]. In addition, the shape of the energy spectrum tails toward the high-energy side, which is very different from the shape of the energy spectrum when SiO$_2$ is used as the insulating layer. This suggests that the shape of energy spectrum reflects the electron distribution in the conduction band of n-Si by the suppression of electron inelastic scattering within the h-BN insulating layer and graphene electrode.

III. SUMMARY

The planar type electron emission device with high emission efficiency of more than 40 %, high current density of more than 100 mA/cm^2, and highly monochromatic electron emission with an energy spread of 0.18 eV at FWHM were demonstrated using the atomic layered materials of graphene and h-BN. These results solve the critical issues of the planar type electron emission devices and would lead to the several practical applications, such as a low-cost, high-resolution, and low-vacuum SEM, an ion neutralizer for space propulsion, and the reforming of liquid and gas materials by a irradiation of the low energy electron beams.

ACKNOWLEDGMENT

This work was partly supported by JSPS KAKENHI Grant Numbers JP18H01505, JP18K18910, 19K04516, and JP21H01401. Support was also received from the AIST Nanofabrication Platform in the Nanotechnology Platform Project sponsored by MEXT, Japan.

REFERENCES

[1] K. Yokoo, H. Tanaka, S. Sato, J. Murota, S. Ono, *J. Vac. Sci. Technol. B* **11**, 429-432 (1993).

[2] T. Ichihara, T. Hatai, N. Koshida, *J. Vac. Sci. Technol. B* **27**, 772-774 (2009).

[3] T. Ohta, A. Kojima, H. Hirakawa, T. Iwamatsu, N. Koshida, *J. Vac. Sci. Technol. B* **23**, 2336-2338 (2005).

[4] K. Koshida, T. Ohta, B. Gelloz, *Appl. Phys. Lett.* **90**, 163505 (2007).

[5] K. Murakami, S. Tanaka, A. Miyashita, M. Nagao, Y. Nemoto, M. Takeguchi, J. Fujita, *Appl. Phys. Lett.* **108**, 083506 (2016).

[6] K. Murakami, S. Tanaka, T. Iijima, M. Nagao, Y. Nemoto, M. Takeguchi, Y. Yamada, M. Sasaki, *J. Vac. Sci. Technol.* **36**, 02C110 (2018).

[7] K. Murakami, J. Miyaji, R. Furuya, M. Adachi, M. Nagao, Y. Neo, Y. Takao, Y. Yamada, M. Sasaki, H. Mimura, *Appl. Phys. Lett.* **114**, No. 213501 (2019).

[8] R. Furuya, Y. Takao, M. Nagao, K. Murakami, *Acta Astronaut.* **174**, 48-54 (2020).

[9] K. Murakami, M. Adachi, J. Miyaji, R. Furuya, M. Nagao, Y. Yamada, Y. Neo, Y. Takao, M. Sasaki, and H. Mimura, *ACS Appl. Electron. Mater.* **2**, 2265-2273 (2020).

[10] K. Murakami, T. Igari, K. Mitsuishi, M. Nagao, M. Sasaki, Y. Yamada, *ACS Appl. Mater. Interfaces* **12**, 4061-4067 (2020).

[11] T. Igari, M. Nagao, K. Mitsuishi, M. Sasaki, Y. Yamada, and K. Murakami, *Phys. Rev. Applied* **15**, 014044 (2021).

Gap in pagination due to withheld paper.

Pages 62-63

Oxygen Resistance Investigation of Graphene-Oxide-Semiconductor Planar-Type Electron Sources for Low Earth Orbit Applications

Naoyuki Matsumoto[1,2,+], Yoshinori Takao[3], Masayoshi Nagao[2], and Katsuhisa Murakami[2]

[1]Department of Mechanical Engineering, Materials Science, and Ocean Engineering, Yokohama National University, Yokohama, Japan
[2] Device Technology Research Institute, National Institute of Advanced Industrial Science and Technology (AIST), Tsukuba, Japan
[3]Division of Systems Research, Yokohama National University, Yokohama, Japan

+Corresponding author: matsumoto-naoyuki-hv@ynu.jp

Abstract—A graphene-oxide-semiconductor (GOS) planar-type electron source is coated with a hexagonal boron nitride (h-BN) film to improve its oxidation resistance. The h-BN film is selected as a coating material owing to its high oxidation resistance and electron transmissivity. To evaluate oxygen resistance, the electron emission density versus applied gate voltage is measured before and after an atomic oxygen (AO) exposure for 4 minutes using an oxygen plasma asher. As a result, the device covered by h-BN emits electrons even after the AO exposure, indicating that the graphene electrode under the h-BN remains.

Keywords— electric propulsion, electron sources, low Earth orbit, graphene, hexagonal boron nitride, oxygen resistance

I. Introduction

Advanced space missions on low Earth orbit that require orbit transfer and station keeping become feasible at low cost with the advent of micro/nanosatellites equipped with miniature propulsion systems. One of such propulsion systems is a miniature ion thruster, which generates the thrust by extracting and accelerating positive ions from the plasma source [1]. To neutralize the ion beam and keep the satellite electrically neutral, it is also required to emit electrons from a neutralizer. Conventional electron sources, e.g., hollow cathodes, consume the propellant. However, small satellites need to reduce propellant consumption because of their capacity restriction. Here, a graphene-oxide-semiconductor (GOS) planar-type electron source is attractive as a propellantless neutralizer [2]. The GOS device is based on a graphene/SiO2/n-Si structure utilizing graphene as electron transparent electrodes [3][4]. By applying the gate voltage between the graphene electrode and the n-Si substrate, electrons in the substrate pass through the SiO2 insulating layer by tunneling effect, and then those which surmount the work function of the graphene electrode are emitted into a vacuum. The GOS device provides high electron emission densities of 1–100 mA/cm² at low applied voltages of 10–20 V without propellant consumption and is promising as neutralizers of miniature ion thrusters.

On low Earth orbit where many small satellites fly, oxygen molecules dissociate into atomic oxygen (AO) by ultraviolet rays, and AO is the dominant gas component at altitudes of 200–600 km [5]. Because AO is chemically reactive, it is concerned that the graphene electrode might be oxidized with AO on the low Earth orbit, resulting in the loss of the electron emission capability. To use the GOS device as neutralizers of small satellites, it is required to protect the graphene electrode from AO without performance degradation of the electron emission. In this study, the GOS electron source is coated with a hexagonal boron nitride (h-BN) film to improve its oxidation resistance. The electron emission properties and oxygen resistance are discussed.

II. Experimental Methods

A. Fabrication of the GOS with h-BN

Fig. 1 shows the fabrication procedure of the GOS with the h-BN protection layer. (a) A n-type Si substrate with SiO2 of 300 nm thickness was used. (b) The electron emission area of 10 μm square was formed by photolithography and wet etching. (c) Then, thin SiO2 dielectric layer was grown on Si by thermal oxidation. (d) The graphene electrode was synthesized by chemical vapor deposition (CVD). (e) The Ni/Ti contact electrode was deposited by electron-beam evaporation and lift-off process. (f) Monolayer h-BN synthesized on the Cu foil (EMJapan Co., Ltd. Cat No. G-54) was used as an oxidation protection layer of the GOS device. The h-BN was mechanically transferred onto the device from the Cu foil [6].

B. Oxygen Resistance Evaluation of the GOS Devices

The GOS devices with/without monolayer h-BN were fabricated. The electron emission densities versus applied gate voltage were measured before and after AO exposure for 4 minutes by oxygen plasma asher (ashing). The oxygen resistance of the device was evaluated by comparing them. Fig.

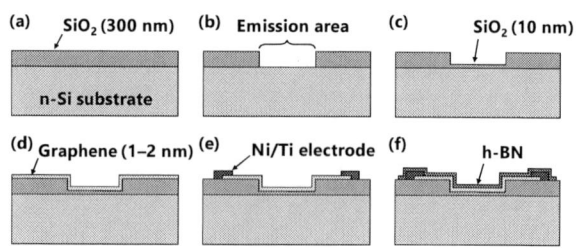

Fig. 1. The fabrication procedure of the graphene-oxide-semiconductor electron source with the h-BN layer.

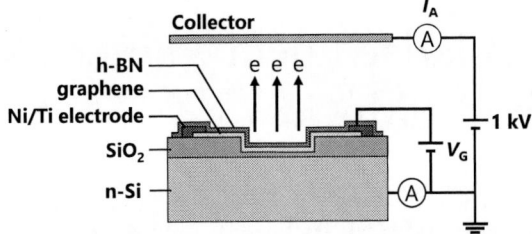

Fig. 2. A schematic diagram of the electron emission measurement setup.

2 shows a schematic diagram of the electron emission measurement setup. By applying the gate voltage V_G, electrons are emitted into a vacuum. These electrons encounter the 1 kV applied collector, and electron emission current I_A was measured. For evaluation, emission current density J_A ($= I_A$ / emission area) was used.

III. RESULTS AND DISCUSSION

Fig. 3 shows the J_A–V_G curves of the GOS devices with/without monolayer h-BN. Before the ashing for 4 minutes, the emission current density with monolayer h-BN reached over 1 mA/cm^2 at $V_G > 8.4$ V as shown in Fig. 3(b). Even though the device is covered by monolayer h-BN, it could emit nearly the same amount of electrons as the device without the h-BN layer shown in Fig. 3(a). This result indicates that the h-BN layer less affects the device's electron emission even though it exists on the graphene electrode because of less electron scattering within it. After the ashing for 4 minutes, the device without h-BN lost the electron emission capability, as indicated in the red line in Fig. 3(a). This result means that most areas of the graphene electrode were oxidized and disappeared by AO, and the device could not apply a sufficient electric field at the insulating layer. On the other hand, the GOS device with monolayer h-BN emits electrons even after the ashing for 4 minutes, which indicates that the graphene electrode underneath the h-BN remained under the AO exposure. However, in Fig. 3(b), the maximum emission current density at $V_G = 9$ V decreased from 1.8 mA/cm^2 before the ashing to 3.2×10^{-2} mA/cm^2 after the ashing for 4 minutes even though the device was protected by h-BN. In Fig. 3(b), the slope of the J_A–V_G curve after the ashing becomes more gradual than that before the ashing, resulting in the lower emission current. This degradation would be due to the drop in the effective voltage applied to the emission area caused by the following steps: (1) AO intruded under the h-BN layer from the cracks of transferred h-BN during the ashing. (2) The graphene electrode was partially oxidized by AO and lost its thickness or completely disappeared. (3) Electrical resistance of the graphene electrode increased. (4) During the electron emission, a voltage drop occurred within the graphene electrode, resulting in the drop in effective voltage applied to the emission area. However, this problem would be prevented by depositing the h-BN layer on the graphene electrode utilizing CVD because this method can provide the h-BN layer with fewer cracks. The crack-less h-BN layer by CVD prevents AO intrusion under the h-BN layer and further improves the oxygen resistance of the GOS device.

IV. SUMMARY

The GOS planar-type electron source was coated with the h-BN film. The oxygen resistance was evaluated from the electron emission properties measured before and after the AO exposure for 4 minutes using the oxygen plasma asher. The

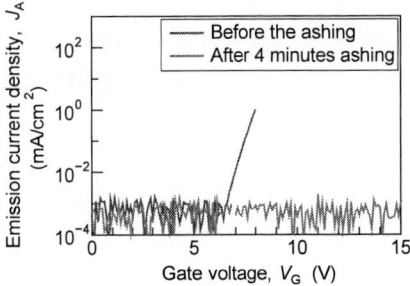

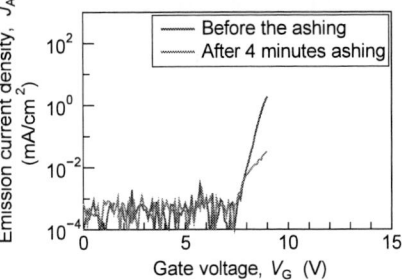

Fig. 3. The J_A–V_G curves of the graphene-oxide-semiconductor electron sources: (a) without h-BN and (b) with monolayer h-BN.

oxygen resistance was improved by the h-BN protection layer. Less electron emission disturbance of h-BN was also indicated. It is indicated that the h-BN protection layer can improve oxygen resistance of the GOS electron source while avoiding the deterioration of emission performance due to h-BN itself. In the future, to prevent performance degradation caused by AO intrudes from cracks of transferred h-BN, the h-BN will be deposited on GOS by CVD.

ACKNOWLEDGMENT

This work was partly supported by JSPS KAKENHI (Grant Numbers JP18H01505, 18K18910, and 21H01401), and the AIST Nanofabrication Platform in the Nanotechnology Platform Project sponsored by MEXT, Japan.

REFERENCES

[1] H. Koizumi, K. Komurasaki, J. Aoyama, and K. Yamaguchi, "Development and flight operation of a miniature ion propulsion system," J. Propul. Power, vol. 34, pp. 960–968, December 2017.

[2] R. Furuya, Y. Takao, M. Nagao, and K. Murakami, "Low-power-consumption, high-current-density, and propellantless cathode using graphene-oxide-semiconductor structure array," Acta Astronaut., vol. 174, pp. 48–54, September 2020.

[3] K. Murakami, S. Tanaka, A. Miyashita, M. Nagao, Y. Nemoto, M. Takeguchi, and J. Fujita, "Graphene-oxide-semiconductor planar-type electron emission device," Appl. Phys. Lett., vol. 108, 083506, February 2016.

[4] K. Murakami, M. Adachi, J. Miyaji, R. Furuya, M. Nagao, Y. Yamada, Y. Neo, Y. Takao, M. Sasaki, and H. Mimura, "Mechanism of highly efficient electron emission from a graphene/oxide/semiconductor structure," ACS Appl. Electron. Mater., vol. 2, pp. 2265–2273, July 2020.

[5] Y. Kimoto, E. Miyazaki, J. Ishizawa, and H. Shimamura, "Atomic oxygen effects on space materials in low earth orbit and its ground evaluation," J. Vac. Soc. Jpn., vol. 52, pp. 475–483, October 2009.

[6] K. Murakami, T. Igari, K. Mitsuishi, M. Nagao, M. Sasaki, and Y. Yamada, "Highly monochromatic electron emission from graphene/hexagonal boron nitride/Si heterostructure," ACS Appl. Mater. Interfaces, vol. 12, pp. 4061–4067, December 2019.

978-1-6654-2590-2/21 $31.00 © 2021 IEEE

Development of highly spin-polarized field emitter using Heusler alloy Co_2MnGa

Shigekazu Nagai
Mie University, Japan
*Contact: nagai@elec.mie-u.ac.jp, phone +81-59-231-9769

Abstract— **Spin-polarized field emitter using Heusler alloy Co2MnGa was fabricated by using a focused ion beam technique. To obtain a stable and highly spin-polarized electron beam, the surface of the Co₂MnGa emitter was cleaned by hydrogen-promoted field-evaporation and characterized by field ion microscopy. A spin polarization of field emitted electrons from clean Co₂MnGa(100) surface is up to 66% under conditions that the emitter temperature and emission current are 44 K and 1 nA, respectively. Although the magnitude of spin polarization decreased with elapsed time due to adsorption of residual gases such as hydrogen, the magnitude can be reversed to initial one by the field evaporation in UHV.**

I. INTRODUCTION

In the field of materials analysis, it has been increasingly important to clarify phenomena involving electron spins, in addition to structural and electronic properties. Currently, GaAs-based photocathodes are used as probes for detecting electron spins at surfaces, and have been applied to analytical techniques such as spin-polarized low-energy electron microscopy[1] and spin-resolved inverse photoemission spectroscopy[2] which has been remarkable achievement for surface magnetism. However, the source size of the photocathode is restricted by the focusing diameter of the laser beam for electron excitation, which limits the spatial resolution of analysis. Therefore, both high spin polarization and the extremely small source size are necessary to improve the spatial resolution of these analytical methods. To improve the spin polarization of field emitters, which are expected to have high brightness properties, we have developed a field emitter based on a half-metallic ferromagnet Co_2MnGa[3, 4], which is theoretically predicted to be perfectly spin-polarized at the Fermi level.

In this paper, author will discuss the cleanness of the emitter surface and its effect on the spin polarization of emitted electrons for realization of a highly spin-polarized field emitter.

II. EXPERIMENTAL

In an UHV sputtering system, 100 nm thick Co_2MnGa thin films was grown on MgO(001) substrate with a 50 nm thick Cr buffer layer. Then, the prepared thin film was annealed at 1023 K to promote atomic ordering to an $L2_1$ structure. The Co_2MnGa film was fixed on the apex of a electrochemically etched tungsten tip and was milled by the focused Ga ion beam to form a needle with a radius of curvature of about 50 nm in order to generate enough high field strength.

The fabricated emitters were introduced into a load-locked system consisting of a field ion microscope (FIM) and a field emission microscope (FEM) with a Mott-type spin detector[5] in the second stage. The base pressure in each chamber is of the order of 10^{-8} Pa, and H_2 and Ne can be introduced as imaging gases of the FIM observation.

The spin polarization of electrons passing through a probe hole in the center of a fluorescent screen, where the FEM image is projected, is measured by a retarding type Mott spin detector. The operating voltage of the Mott spin detector is 25 kV, and the effective Sherman function is 0.14 calibrated with field-emitted electrons from Fe/W surface[6].

III. RESULTS AND DISCUSSION

Since the evaporation field strength of the constituent elements of Co_2MnGa is different, the elements with lower evaporation fields are preferentially evaporated. To avoid preferential field evaporation, H2 gas was introduced when the cleaning of the surface by the field evaporation was carried out. Figure 1 shows the FIM and FEM images of Co_2MnGa after the field evaporation treatment. In the centre of the FIM shown in Fig. 1(a), a ring structure indicating the atomic ordering was observed. In Fig. 1(b), the spin polarization of field-emitted electrons from the (001) surface was measured, because this surface has a low work function and preferentially emits electrons and is expected to be highly polarized.

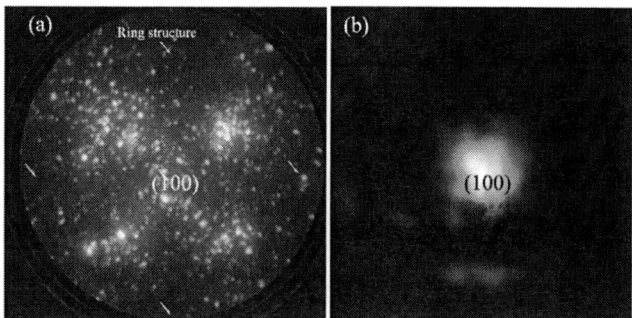

Fig. 1 (a) FIM and (b) FEM image of Co_2MnGa emitter.[4]

Table I shows the spin polarization of field-emitted electrons from the (100) plane under the conditions of an emission current of 44 nA and an emitter temperature of 46 K. The x- and y-directions are the horizon and vertical direction in the fluorescent screen plane, and the z-direction is the component parallel to the emitter axis. The magnitude of the spin polarization is calculated as the vector sum of the components in each direction. The atomic layers along the (100) orientation of the $L2_1$ structure with half-metal properties are alternating Co-Co and Mn-Ga layers. The spin polarization at the Fermi level of each surface is 67% and 62% according to first-principles calculations, and the effect of surface composition is relatively small. The measured value of 66% in this study is in good agreement with the value predicted by the theoretical calculation. At an ejection current of 1 nA, the spin polarization of the emitted electrons decreases to 30% in one hour, but it can be restored to 60% by cleaning the surface by field evaporation. This result indicates that the adsorption of residual gases, mainly hydrogen, on the surface reduces the spin polarization of the emitted electrons. Therefore, to operate Co2MnGa emitters stably at high polarization for a long time, an extremely high vacuum of less than 10^{-8} Pa may be required.

TABLE II

EACH COMPONENT OF SPIN POLARIZATION OF FIELD EMITTED ELECTRONS FROM CO$_2$MNGA(100) SURFACE AT 46 K

Component	x	y	z
Spin polarization [%]	9.5	59	-7.0

IV. CONCLUSIONS

In this study, the spin polarization of field-emitted electrons from the <100>-oriented Co$_2$MnGa emitter fabricated by focused ion beam technology was measured. The spin polarization of the emitted electrons from clean Co2MnGa(100) surface is up to 66% at 46 K. The measured magnitude of the spin polarization is in good agreement with theoretical one. The decrease in the spin polarization due to the adsorption of residual gas such as hydrogen was observed. Therefore, it is important to operate under an extremely high vacuum for highly and stable spin-polarized field emitters.

ACKNOWLEDGMENT

Author thanks the Murata Science Foundation for the presentation of this study.

REFERENCES

[1] T. Duden, and E. Bauer, A compact electron-spin-polarization manipulator. Review of Scientific Instruments **66**, pp. 2861-2864, 2009.

[2] M. Cantoni, and R. Bertacco, High efficiency apparatus for spin polarized inverse photoemission. Review of Scientific Instruments **75**, pp. 2387-2392, 2004.

[3] T. Ishikawa, S. Nagai, K. Oh-ishi, H. Ikemizu, and K. Hata, Fabrication and characterization of spin-polarized Co$_2$MnGa field-emission electron source. Applied Physics Express 12, 075004, 2019.

[4] S. Nagai, H. Ikemizu, K. Hata, and T. Ishikawa, Highly spin-polarized field emission from ⟨100⟩-oriented Co$_2$MnGa tips. J Appl Phys **126**, 135302, 2019.

[5] G. C. Burnett, T. J. Monroe, and F. B. Dunning, High-efficiency retarding-potential Mott polarization analyzer. Rev Sci Instrum **65**, pp. 1893-1896, 1994.

[6] T. Irisawa, T. K. Yamada, and T. Mizoguchi, Spin polarization vectors of field emitted electrons from Fe/W tips. New Journal of Physics **11**, 113031, 2009.

A HfC nanowire field emission point electron source

Shuai Tang[1*], Jie Tang[1,2*], Ta-Wei Chiu[1,2], Wataru Hayami[1], Lu-Chang Qin[3]

[1]*National Institute for Materials Science, Tsukuba, Ibaraki 305-0047, Japan*

[2]*Graduate School of Pure and Applied Science, University of Tsukuba, Tsukuba, Ibaraki 305-8857, Japan*

[3]*Department of Physics and Astronomy, University of North Carolina at Chapel Hill, Chapel Hill, NC 27599-3255, USA*

*Corresponding author: tang.shuai@nims.go.jp; tang.jie@nims.go.jp

Abstract— **Searching for viable field-emission structures with a stable emission current, high emission brightness, and low energy spread has always been highly demanded for the development of high-performance electron-optical instrumentation. In this work, a single-crystalline HfC nanowire emitter with an oxycarbide surface is investigated. This new emitter shows an excellent current stability with a fluctuation less than 1%/h, as well as a narrow energy spread and high emission brightness. It exhibits great potentials to be a practical electron source for experimental trials in electron microscopes.**

I. INTRODUCTION

Room-temperature (cold) field-emission electron sources are used widely in applications. Developed since 1960's, single-crystalline tungsten (W) filament is still the only operational cold field-emission point electron source for contemporary electron microscopes. However, it still suffers great limitations including (i) rapid decays in emission current even in ultrahigh vacuum and (ii) a high work function leading to a large energy spread in the emitted electrons [1-2].

Many materials, including nanotips, nanowires, and nanotubes have been investigated for applications as an field emission electron source [3-5]. Hafnium carbide (HfC) is a promising field-emission material due to its high melting point, excellent thermal and electrical conductivity, and outstanding mechanical strength. However, until now, the field emission stability has been unsatisfactory for practical applications.

In this work, a density function theory (DFT) calculation has been performed and showed that an oxycarbide surface can reduce the work function of HfC effectively. A HfC nanowire with an oxyarbide surface structure is therefore selected for designing and fabrication of a field-emission point electron source. A single-crystalline HfC nanowire was grown, picked-up, and fixed onto a tungsten hairpin structure to obtain a field emitter. A post-treatment involving field evaporation and oxidation was applied to optimize the surface morphology and composition. Current stability, energy spread, and emission brightness are also measured to characterize the electron optical performance of the HfC nanowire emitter.

II. EXPERIMENTAL

The single crystalline HfC nanowires were grown using a chemical vapor deposition (CVD) method [6]. The preparation of single HfC nanowire emitter includes three steps (Figure 1):

(i) A W(310) needle was cut by focused ion beam (FIB) to produce a flat platform; (ii) a HfC nanowire was picked up and placed onto the tungsten tip; and (iii) the HfC nanowire was fixed with a carbon pad using e-beam deposition [7-8].

In the post-treatment, an electric field was applied to the nanowire to sharpen the tip and oxygen was let into the chamber for the formation of an oxycarbide on the surface of the HfC nanowire.

The field-emission properties were measured in a high vacuum chamber at 10^{-7} Pa. The energy spread and reduced brightness were calculated using the established methods [9-10].

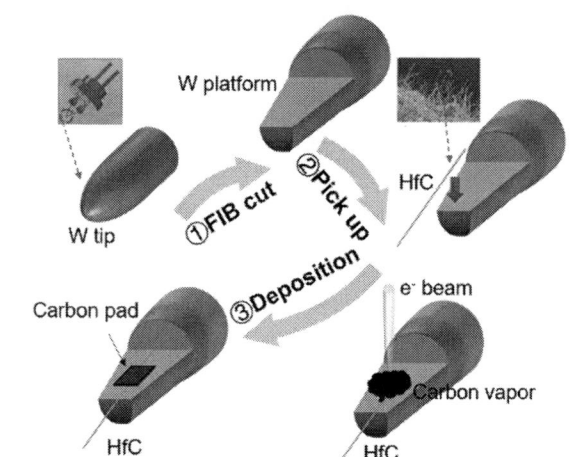

Figure 1 Illustration of preparation of single HfC nanowire emitter.

III. RESULTS AND DISCUSSION

A. Field emission I-V and F-N curve

Figure 2a is the field emission current vs. voltage (I-V) curve and Figure 2b is its corresponding linearized Fowler-Nordheim (F-N) curve. It shows a good linearity with a slope of −3912. By using "1/5r" to estimate the field enhancement factor, the work function is calculated as 2.5 eV, which is about 1.1 eV lower than the HfC crystal. However, it matches well with the DFT calculations – the oxycarbide at the surface would reduce the work function of the HfC nanowire by 1.2 eV because the O atoms contribute more valence electrons to the structure than the C atoms [7, 11].

978-1-6654-2590-2/21 $31.00 © 2021 IEEE

The corresponding field emission pattern (inset of Figure 2a) shows a converged single spot which is good for obtaining a high emission brightness. The reduced brightness and energy spread are obtained as greater than 10^9 A m^{-2} sr^{-1} V^{-1} and small than 0.3 eV, respectively.

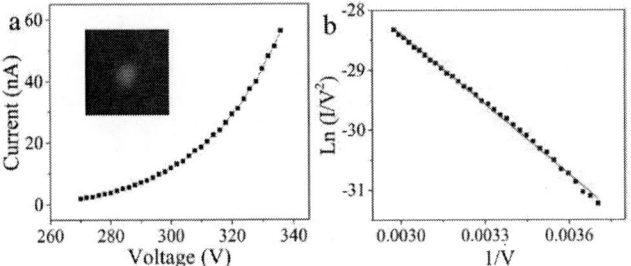

Figure 2 (a) Field emission I-V curve. Inset is the field emission spot. (b) Corresponding linearized F-N curve.

B. Long-term current stability

The long-term current stability is the most crucial criterion for applications as an electron source. The long-term current stability of the HfC nanowire emitter at 41 nA is measured in a high vacuum chamber at 10^{-7} Pa (Figure 3). The fluctuation, defined as $<\Delta I^2>^{1/2}/I$, was 2.9% for 8 hours continuous emission without decay, which is much better than the W(310) field emission electron source with fluctuation and decay of 10% and 13% in 8 hours, respectively, even in an extreme high vacuum (EHV) of 10^{-9} Pa [2]. The recovery and reproducibility of stable emission current after exposure to air were also verified experimentally. The experimental trials in a scanning electron microscope (SEM) are underway.

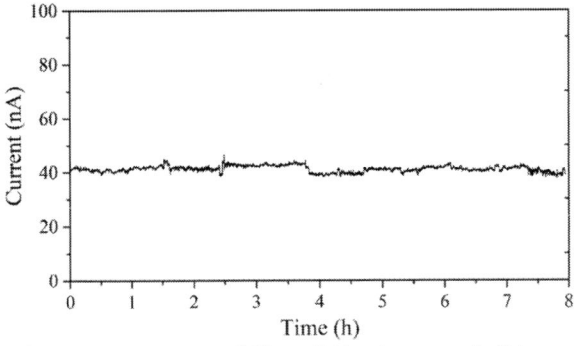

Figure 3 Long-term stability of 41 nA current in 8 hours

C. Large current stability

A high signal-to-noise ratio is of great importance for high-resolution imaging and a high signal-to-noise demands a large emission current. Figure 4a shows an emission stability with a current of 177 nA for 1 hour with a fluctuation of 1.6%/h. Figure 4b shows the emission stability at an even larger current of 700 nA with a fluctuation is 1.4% in 5 mins. These results show that the HfC nanowire emitter can emit electrons stably at current as large as 700 nA.

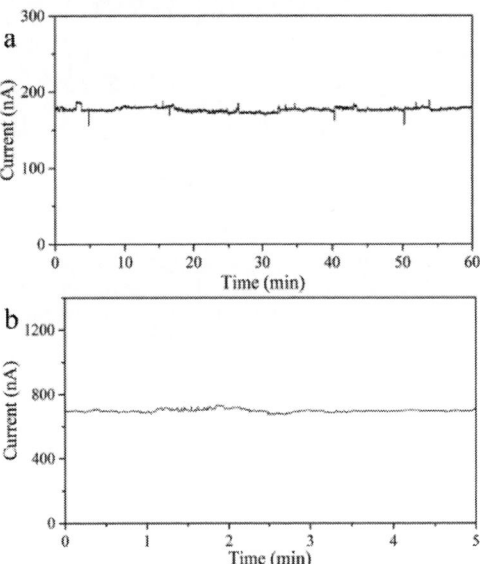

Figure 4 (a) Emission stability at current of 177 nA in 60 mins; and (b) emission stability at 700 nA in 5 mins.

IV. Conclusions

A single HfC nanowire with oxycarbide on surface has been fabricated and characterized for applications as a point electron source. It shows a high emission stability, high emission brightness, and narrow energy spread. It also shows great potential for large current emission from a point electron source.

Acknowledgment

This work was supported partially by the NIMS-DENKA Centre of Excellence for Next Generation Materials. A part of this work was also supported by NIMS Microstructural Characterization Platform as a program of "Nanotechnology Platform" of the Ministry of Education, Culture, Sports, Science and Technology (MEXT), Japan.

References

[1] H. Sawada, N. Shimura, K. Satoh, et al., JEOL News 49, 51-58 (2014).

[2] L. W. Swanson and L. C. Crouser, Phys. Rev. 163, 622-641 (1967).

[3] S. Tang, J. Tang, J. Uzuhashi, T. Ohkubo, W. Hayami, J. S. Yuan, M. Takeguchi, M. Mitome and L.-C. Qin, Nanoscale Adv. 3, 2787-2792 (2021).

[4] H. Zhang, J. Tang, J. S. Yuan, Y. Yamauchi, T. T. Suzuki, N. Shinya, K. Nakajima and L.-C. Qin, Nat. Nanotechnol. 11, 273-280 (2016).

[5] M. Irita, S. Yamazaki, H. Nakahara and Y. Saito, IOP Conf. Ser.: Mater. Sci. Eng. 304, 012006 (2018).

[6] J. Yuan, H. Zhang, J. Tang, N. Shinya, K. Nakajima and L.-C. Qin, J. Am. Ceram. Soc. 95, 2352-2356 (2012).

[7] S. Tang, J. Tang, T.-W. Chiu, W. Hayami, J. Uzuhashi, T. Ohkubo, F. Uesugi, M. Takeguchi, M. Mitome and L.-C. Qin, Nano Res. 13, 1620-1626 (2020).

[8] S. Tang, J. Tang, T.-W. Chiu, J. Uzuhashi, D.-M. Tang, T. Ohkubo, M. Mitome, F. Uesugi, M. Takeguchi and L.-C. Qin, Nanoscale 12, 16770-16774 (2020).

[9] R. Young, Phys. Rev. 113, 110-114 (1959).

[10] M. Bronsgeest, J. Barth, L. Swanson and P. Kruit, J. Vac. Sci. Technol. B 26, 949-955 (2008).

[11] W. Hayami, S. Tang, T.-W. Chiu and J. Tang, ACS Omega 2021, in press doi.org/10.1021/acsomega.1c0167.

Field Emission from Genuine Graphene: An Experimental Study

Philippe Poncharal[*+], Anthony Ayari[*], Pascal Vincent[*], Sorin Perisanu[*], Stephen T. Purcell[*]

[*]Univ Lyon, Univ Claude Bernard Lyon 1, CNRS, Institut Lumière Matière, F-69622, VILLEURBANNE, France.
[+]Corresponding author: philippe.poncharal@univ-lyon1.fr

Abstract —**In this study, we expose our methodology to prepare and characterize a single Graphene sheet to observe Field Emission from a clean edge. The total energy distribution was measured and found to be extremely narrow.**

Keywords—field emission, graphene

I. INTRODUCTION

Although considerable work has been published on graphene Field Emission [1-7] there is a dearth of measurements on well-characterized, individual single sheet graphene layers that can guide the community towards future developments. The available work is primarily focused on multi-emitter films or individual multilayer emitters with no clear proofs of single layer edge emission. To our knowledge, only one case of some ultra-high vacuum cleaning is reported [5]. On the theoretical side a few papers study FE from edge states [8-12], but none predict the energy distributions of single layer edges.

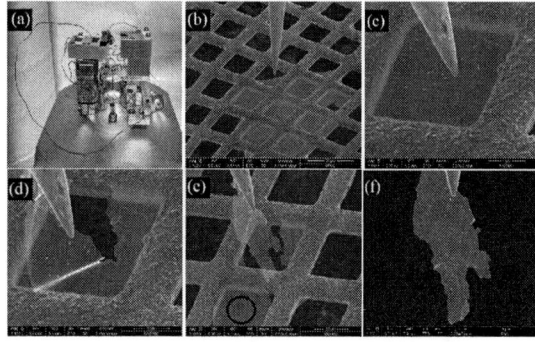

Fig. 2. Graphene sample preparation. (a) 3-axis nanomanipulator installed in our SEM. (b) to (e) pick-up sequence of a graphene flake, starting from CVD grown graphene on a TEM grid to finish with a graphene flake on a tip. The black circle shows the location of the TEM imaging analysis (see [13]). (f) The graphene sample studied.

The total energy distributions (TEDs) of the emitted electrons are of particular interest because they give access to the density of states near the Fermi level and thus of the graphene edge and corner states. However, the sole TEDs of emitted electrons published were from an uncleaned surface covered with multilayer graphene [6].

II. SAMPLE PREPARATION

Graphene was produced by Chemical Vapor deposition (CVD) with a strict control of hydrogen etching to achieve a coverage with mostly single layer graphene with some bi-layer islands [14] and transferred to transmission electron microscopy (TEM) grids. Graphene flakes were torn from the layers and picked-up at the apex of etched tungsten tips (see figure 2).

The sample was confirmed to be single layer (although poly-crystalline with small single crystal grain size) via complementary Raman and TEM diffraction studies [13]. These tips were then mounted in a polyvalent field emission system with a base pressure of 10^{-10} Torr, sample heating by a W loop, imaging by microchannel plate for FE microscopy and an electron energy analyzer.

We first study graphene field emission without any cleaning. The first FEM images consisted of one spot (figure 1a) presumably from the dominant nanostructure. A mechanical vibration signature test is performed: while our

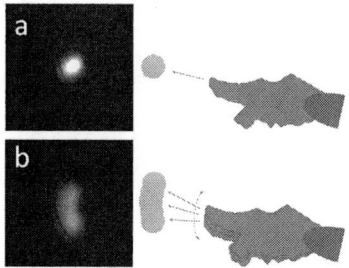

Fig. 1. (a) FE pattern observed when graphene sheet is out of mechanical resonance exhibiting a single emission spot. (b) When the graphene sheet enter in mechanical resonance, the pattern is spread vertically as the tip vibrates. As we expect the ''easy vibration'' first mode for a 2D object to be perpendicular to its plane, we can conclude than the graphene flake must be typically horizontal in our set-up.

sample is emitting, a small AC voltage is scanned in frequency until an electro mechanical resonance occurs, which is observable by an elongation of the FE spot (figure 2b). As the emission pattern shape during mechanical vibration is vertically elongated, we infer that the graphene flake is almost horizontal within our system. The resonance frequencies were in the MHz range.

The cleaning procedure consists in gradually heating up the sample to 1000°C under a FE extraction field. After thermal cleaning, the FEM images still consisted of one spot (figure 3c) but with a clear elliptical shape as expected for an edge emitter. To better eliminate these nanostructure, we used field-induced desorption techniques which is more local than thermal cleaning [15]. Field desorption eliminates local asperities where the electric field is the highest. As the first principal emitters are removed, new emission sites appear at higher voltage (see figure 3e). Each emitter is elliptically shaped in the same direction as expected, although they are not perfectly aligned.

This could be a consequence of local crinkling of the graphene sheet which will tilt the average trajectory of each emitter. It is worth noting that these kind of emission patterns exhibiting two symmetric lobes with a darker central band

978-1-6654-2590-2/21 $31.00 © 2021 IEEE

have already been observed by Yokoyama *et al.* on graphene edges [5] and were attributed to edge state symmetry.

The TED spectra of respectively pristine, heat treated and desorbed graphene are compared on figure 4. All these TED

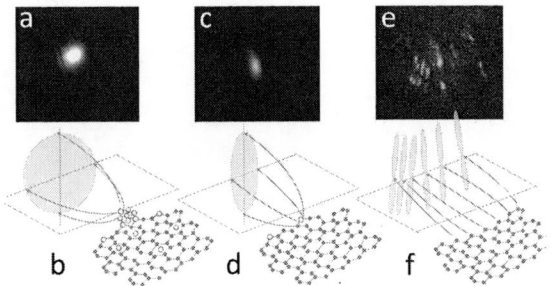

Fig. 3. (a) FE pattern from unprocessed dirty graphene dominated by a single emitter (nanoprotrusion). The nanoprotusion consist usually of bunch of atoms or molecules (white spheres on the sketch) with a cone-like local environment which produces a symmetrical pattern as described in the sketch (b). (c) FE pattern form thermally cleaned graphene. Thermal treatment removed foreign physiosorbed molecules but leaves graphene irregularities and strongly chemisorbed atoms. The emission is still dominated by a single emitter as there will always be a place with a local higher β. However, the local environment is no-longer symmetric: the pattern is focused along the graphene edge direction, but not on the perpendicular direction. As a consequence, the pattern is elliptical with main axis perpendicular to the graphene sheet (d). (e) Field ion microscopy (FIM) pattern from desorbed graphene using oxygen. As field desorption eliminates local asperities until the field is even, emission is no longer dominated by a single spot but multiple emitters are now visible. Note that the pattern of each emitter show vertically elongated shapes with two wings separated by a darker center. The set of projected patterns are not perfectly aligned but follow the graphene edge local bending and crinkling as depicted on (f).

were recorded around 10 pA of emission at room temperature. The multiple peaks feature produced by adsorbates visible on untreated sample disappears after thermal curing.

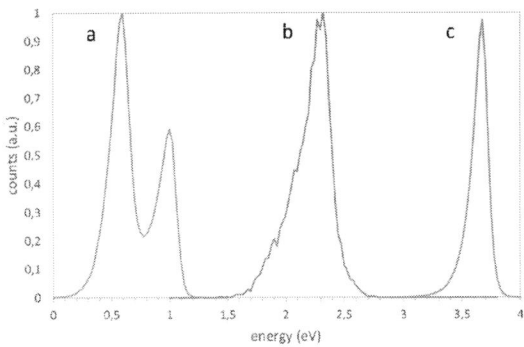

Fig. 4. Total energy distribution (TED) curves of (a) untreated, (b) heated and (c) desorbed sample for comparable emission current (respectively 7.5, 12 and 8 pA). FWHM are respectively 0.20 (highest peak), 0.26 and 0.15 eV. Curves have been arbitrarily shifted for clarity.

While there is an adsorbate, the TED peaks are rather narrow (0.2 eV) consistent with localized state in the adsorbate. Once adsorbates are thermally removed, the electrons are emitted through a triangular barrier and the TED becomes slightly wider (0.26 eV) with broadening on low energy side, signature of the field effect. Oddly enough, on the desorbed sample, the TED peaks FWHM shrinks down to 0.15 eV. Note that the main reduction resides on the low energy side of the TED peak which exact shape is controlled by the triangular tunnel barrier details.

We can speculate that the triangular tunnel barrier model might not be valid for a single graphene layer, either due to local states or because of field penetration.

III. CONCLUSIONS

We explore the preparation process of clean graphene sample for field emission studies. Cleaning procedure, ultra-high vacuum environment and mechanical vibration control strategy confirms that the measured object is indeed a carefully desorbed single layer graphene. Clean graphene TED peak is extremely narrow (0.15 eV) at room temperature.

ACKNOWLEDGMENT

The authors would like to thank the Plateforme Nanofils et Nanotubes Lyonnaise of the University Lyon1.

REFERENCES

[1] Z.-S Wu et al. "Field Emission of Single-Layer Graphene Films Prepared by Electrophoretic Deposition", Advanced Materials 21, 1756-1760 (2009).

[2] J. Meng et al. "Damage to few layer sheets during electron field emission" Carbon, 125, 370-374 (2017).

[3] X. Guo et al. "Vertical graphene nanosheets synthesized by thermal chemical vapor deposition and the field emission properties" J. Phys.D:Appl. Phys. 48, 385301-6 (2016).

[4] S. Tang et al. "In situ study of graphene crystallinity effect on field emission characteristics" J. Vac. Sci. Technol. B, 35, 02C107-5 (2017).

[5] N. Yokoyama et al. "Field emission patterns showing symmetry of electronic states in graphene edges", Surf. Interface, Anal. 48, 1217-1220 (2016).

[6] G.N. Fursey et al. "Peculiarities of the total energy distribution of field emission electrons from graphene-like structures", Journal of Communication Technology and Electronics, 61, 72-75 (2016).

[7] J. Tsai et al. "Field Emission from individual freestanding graphene edges", Small, 8, 3739-3745 (2012).

[8] M. Luo et al. "Multi-Field electron emission pattern of 2D emitter: illustrated with graphene", J. Appl. Phys, 120,204304-5 (2016)

[9] V.L. Katkov et al. "Energy distribution of emitted electrons from few layer graphene sheets with AB and ABC stacking" Physics of Partcle and Nucleis, 41, 1027-1030 (2010)

[10] Z. Li et al. "Coherent field emission images of graphene predicted with microscopic theory" Phys. Rev. B, 85, 115427-115435 (2012)

[11] W. Wang et al. "Field electron emission characteristics of graphene » J. Appl. Phys. 109, 044304-7 (2011)

[12] Y. Gao et al. "Field emission properties of edge functionalized graphene" Carbon, 142, 190-195 (2019)

[13] R. Diehl et al. "Narrow energy distributions of electrons emitted from clean graphene edges", Phys.Rev .B. 102, 035416 (2020).

[14] S. Choubak et al. "Graphene CVD: Interplay between growth and etching on morphology and stacking by hydrogen and oxidizing impurities", J. Phys. Chem. C, 118, 21532-40 (2014).

[15] E.W. Muller et al. "Field Ion Microscopy, principles and Applications" Elsevier, New York (1969).

Combined effect of single-electron charging and quantum confinement on field electron emission from heterostructured nanotips

Victor I. Kleshch

Department of Physics, Lomonosov Moscow State University
Moscow, Russia
klesch@polly.phys.msu.ru

Abstract — Numerical simulations of a heterostructured field emitter consisting of a quantum dot attached to a nanotip are presented. The simulations take into account the combined effect of single-electron charging and quantum confinement, both of which could be observed under certain conditions. The model is based on the theory of Coulomb blockade in quantum dots in combination with the theory of field emission. The current-voltage characteristics are calculated for various parameters of the heterostructured field emitter. As a result, the values of the model parameters corresponding to different regimes of charge transport were determined.

Keywords—field electron emission, quantum confinement, Coulomb blockade, quantum dots

I. Introduction

Theoretical consideration of field electron emission from nanostructures weakly coupled to the cathode shows that under certain conditions charge transport in such a system may be controlled by the Coulomb blockade effect, leading to a stepwise increase of the emission current with an applied voltage (Coulomb staircase) due to single-electron charging of the nanostructure [1]. Several experimental observations of the Coulomb staircase in field emission current-voltage characteristics of various nanostructures [2-7] have been reported over the last decade. In experiments with heterostructured carbon nanotips, we recently observed an additional short-period modulation of the field emission current amplitude in the Coulomb staircase, which was explained by the presence of discrete states in the emitter [7]. Here we perform numerical simulation of electron transport in a system with such a heterostructured emitter, which takes into account the combined effect of single-electron charging and quantum confinement.

II. Model

The system under study is shown in Fig. 1a. It consists of a small conducting object (nanoparticle) attached to a metal tip and placed at a macroscopic distance from an anode. The nanoparticle is electrically isolated from the tip by a thin dielectric layer. The corresponding energy diagram is presented in Fig. 1b. Electron transport through such double-barrier structure has been considered previously either for the case of strong quantum confinement (see, e.g. [8-10]), or for the case of Coulomb blockade [1,2]. Here, we assume that both of these effects influence the transport properties of the system.

To determine the field emission current, I, at a certain voltage, V, it is necessary to calculate the probability distribution P_N of finding the nanoparticle in a state with N electrons on it, so that:

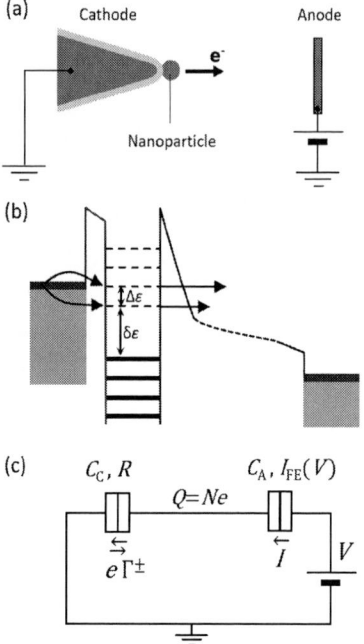

Fig. 1. (a) Schematics of the system and (b) corresponding energy diagram, where $\delta\varepsilon$ is the Coulomb-blockade charging energy and $\Delta\varepsilon$ is the energy level separation in the nanoparticle. (c) Equivalent circuit of the system.

$$I = \sum_N P_N I_N, \qquad (1)$$

where I_N is the partial field emission current from the nanoparticle in the N-electron state. P_N can be calculated by solving the master equation given by [1,11]:

$$\frac{\partial P_N}{\partial t} = Q_{N+1} - Q_N, \qquad (2)$$

where

$$Q_N = P_N(\Gamma_N^- + I_N/e) - P_{N-1}\Gamma_N^+.$$

Here $\Gamma_N^{\pm}$ are the rates of tunneling between the nanoparticle in N-electron state and the cathode, e is the magnitude of the electron charge. In the stationary case, (2) is reduced to $Q_N=0$. In the case of strong quantum confinement, when the separation of the energy levels in the nanoparticle exceeds the characteristic energy of thermal fluctuations k_BT, the tunneling rates and partial currents are given by the following equations [11]:

978-1-6654-2590-2/21 $31.00 © 2021 IEEE

$$\Gamma_N^- = \sum_k \Gamma_k \big(1 - f(\varepsilon_k + \Delta U_N)\big) g_N(\varepsilon_k),$$

$$\Gamma_N^+ = \sum_k \Gamma_k \big(1 - g_N(\varepsilon_k)\big) f(\varepsilon_k + \Delta U_N), \qquad (3)$$

$$I_N = \sum_k I_{FE}(\varepsilon_k, F_N),$$

where ε_k are the nanoparticle discrete energy levels, $f(\varepsilon)$ is the Fermi distribution, $g_N(\varepsilon)$ is the distribution function in the nanoparticle, F_N is the electric field at the surface of the nanoparticle and ΔU_N is the difference in the Coulomb energies in states with N and N-1 electrons, which is given by:

$$\Delta U_N = U_N - U_{N-1} = \delta\varepsilon(N - 1/2 - C_A V/e), \qquad (4)$$

where $\delta\varepsilon = e^2/C$ is Coulomb-blockade charging energy, C_A is the capacitance of the nanoparticle with respect to the anode electrode and C is the total capacitance $C = C_A + C_C$ (see the equivalent circuit in Fig. 1c).

III. Results and Discussion

Here we consider the simplest case of equidistant energy levels $\varepsilon_k = k\Delta\varepsilon$ occupied according to the Fermi distribution, i.e. $g(\varepsilon) = f(\varepsilon) = 1/(1+\exp(\varepsilon/k_B T))$. The partial field emission currents associated with discrete states ε_k are considered to be proportional to the classical total energy distribution $j_{FE}(\varepsilon)$, so that $I_{FE}(\varepsilon_k, F_N) = k_j \Delta\varepsilon\, j_{FE}(\varepsilon_k, F_N)$, where k_j is the pre-exponential factor. The field at the surface of the nanoparticle F_N is proportional to the applied voltage V and electron number N, i.e. $F_N = \alpha_V V + \alpha_N N$, where α_V and α_N are parameters estimated either by electrostatic simulations or by fitting of experimental data [1,2,4]. The result of calculation is shown in Fig. 2 for the following parameters values $C_A = 6.3 \times 10^{-21}$ F, $C = 2 \times 10^{-19}$ F, $\Gamma_k = \Gamma_0 = 4.8 \times 10^9\,\mathrm{s}^{-1}$, $k_j = 1.3 \times 10^{-19}\,\mathrm{cm}^2$, $\alpha_V = 4.5 \times 10^7\,\mathrm{m}^{-1}$, $\alpha_N = 8.2 \times 10^8$ V/m, $T = 300$ K. At $\delta\varepsilon = 30\,k_B T$ and $\Delta\varepsilon = 1\,k_B T$ (curve 2, Fig. 2a) the Coulomb staircase is observed. With an increase in the separation energy to $\Delta\varepsilon = 10\,k_B T$ (curve 3), an additional current modulation is observed due to the fact that the number of levels, n_t, available for tunneling to the nanoparticle increases with voltage (for example, in Fig. 1b $n_t = 2$). At $\delta\varepsilon = \Delta\varepsilon = 1\,k_B T$ (curve 1) the step-like features in $I(V)$ curve are washed out.

Figure 2b shows the $I(V)$ curve, the derivative dI/dV and the probability distribution map $P_N(N, V)$ obtained at $\delta\varepsilon = 30\,k_B T$ and $\Delta\varepsilon = 10\,k_B T$. Three regions can be distinguished. Region 1 corresponds to the pure Coulomb blockade regime, because transport is mainly determined by the outer triangular barrier. In region 2, additional peaks in the derivative dI/dV appear, since the role of the internal tunneling barrier increases. In region 3, the oscillations in dI/dV are washed out because states with different N become equally probable.

IV. Conclusions

It is shown that the combined effect of single-electron charging and quantum confinement leads to the appearance of periodic modulations in field emission current-voltage characteristics, similar to those observed in solid-state systems with quantum dots in the Coulomb blockade regime [11]. The results obtained can be used in the studies of nanotips functionalized with quantum dots, which are widely explored at present [12,13].

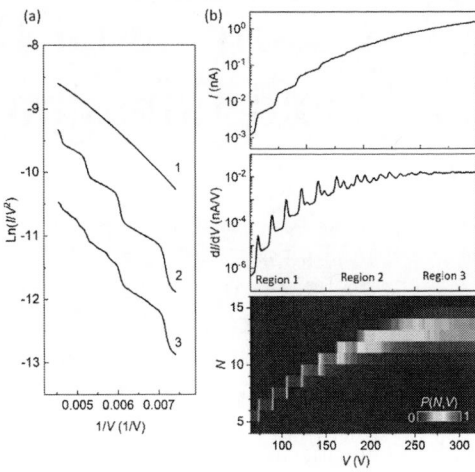

Fig. 2. (a) $I(V)$ curves in Fowler-Nordheim coordinates simulated at different values of charging energy $\delta\varepsilon$ and levels separation $\Delta\varepsilon$. $1 - \delta\varepsilon$, $\Delta\varepsilon \sim k_B T$, $2 - \delta\varepsilon \gg k_B T$, $\Delta\varepsilon \sim k_B T$, $3 - \delta\varepsilon$, $\Delta\varepsilon \gg k_B T$. (b) $I(V)$ curve, the derivative dI/dV and the probability distribution map $P_N(N, V)$ obtained at $\delta\varepsilon = 30\,k_B T$ and $\Delta\varepsilon = 10\,k_B T$.

Acknowledgment

The work was supported by Russian Science Foundation (Project No. 19-72-10067).

References

[1] O. E. Raichev, Coulomb blockade of field emission from nanoscale conductors, Phys. Rev. B **73**, 195328 (2006).

[2] A. Pascale-Hamri, S. Perisanu, A. Derouet, C. Journet, P. Vincent, A. Ayari, and S. T. Purcell, Ultrashort single-wall carbon nanotubes reveal field-emission Coulomb blockade and highest electron-source brightness, Phys. Rev. Lett. **112**, 126805 (2014).

[3] C. Kim, H. S. Kim, H. Qin, and R. H. Blick, Coulomb-controlled single electron field emission via a freely suspended metallic island, Nano Lett. **10**, 615 (2010).

[4] V. I. Kleshch, V. Porshyn, A. S. Orekhov, A. S. Orekhov, D. Lützenkirchen-Hecht, and A. N. Obraztsov, Carbon single-electron point source controlled by Coulomb blockade, Carbon **171**, 154 (2021).

[5] V. I. Kleshch et al., Coulomb blockade in field electron emission from carbon nanotubes, Appl. Phys. Lett. **118**, 053101 (2021).

[6] V. M. Lobanov and E. P. Sheshin, Periodic deviations of the field emission current of a carbon nanotube from the Fowler-Nordheim law, Tech. Phys. Lett. **33**, 365 (2007).

[7] V. I. Kleshch, V. Porshyn, D. Lützenkirchen-Hecht, and A. N. Obraztsov, Coulomb blockade and quantum confinement in field electron emission from heterostructured nanotips, Phys. Rev. B **102**, 235437 (2020).

[8] J. W. Gadzuk, Resonance-tunneling spectroscopy of atoms adsorbed on metal surfaces: theory, Phys. Rev. B **1**, 2110 (1970).

[9] L. D. Filip, M. Palumbo, J. David Carey, and S. R. P. Silva, Two-step electron tunneling from confined electronic states in a nanoparticle, Phys. Rev. B **79**, 245429 (2009).

[10] V. Filip and H. Wong, Comparative study of resonant and sequential features in electron field emission from composite surfaces, Thin Solid Films **608**, 26 (2016).

[11] D. V. Averin, A. N. Korotkov, and K. K. Likharev, Theory of single-electron charging of quantum-wells and dots, Phys. Rev. B **44**, 6199 (1991).

[12] M. Duchet et al., Femtosecond laser induced resonant tunneling in an individual quantum dot attached to a nanotip, ACS Photonics **8**, 505 (2021).

[13] C. Wagner et al., Scanning quantum dot microscopy, Phys. Rev. Lett. **115**, 026101 (2015).

Negative Differential Resistance in Laser-Assisted Field Emission from Si Nanowires

M. Choueib
Département de chimie
Université de Montréal
Montréal, Québec H3T 1J4, Canada
LPMCN
Université Lyon 1, CNRS, UMR 5586
F-69622 Villeurbanne, Cedex, France

A. Derouet, P. Vincent, A. Ayari, P. Poncharal
LPMCN
Université Lyon 1
F-69622 Villeurbanne, Cedex, France

C. S. Cojocaru
LPICM
Ecole Polytechnique
UMR 7647, F-91128 Palaiseau, France

R. Martel
Département de chimie
Université de Montréal
Montréal, Québec H3T 1J4, Canada
r.martel@umontreal.ca, 0000-0002-9021-4656

S.T. Purcell
LPMCN
Université Lyon 1, CNRS, UMR 5586
F-69622 Villeurbanne, Cedex, France

Abstract—**Semiconducting nanowires (NWs) are studied in field-emission (FE) for expanding electron gun performances and functionality in terms of stability, brightness and pulsed emission. Here we report on a pronounced double negative differential resistance (NDR) in the FE IV characteristics measured during photo-assisted field emission experiments on highly crystalline p-type silicon nanowires (NWs). The main feature is a double NDR in the current saturation regime, which can be modulated by both temperature and light intensity. These results contrast with previous FE studies in which only barely discernable single NDR was reported. Several mechanisms for the physical explanation of the NDR are currently under consideration : photo-generated carrier instabilities in the depletion region, which give raise to a pulsed space-charge current build-ups in the nanowire or the presence of a double quantum well formed by confinement near the Si nanowire apex. Because NDRs are signatures of pulsed currents, these results suggest new functionalities for which pulsed electron sources can potentially be achieved at high repetition rates.**

Index Terms—**Negative Differential Resistance; field emission; silicon nanowire; photo-induced diode; space-charge effects**

I. INTRODUCTION

There is a growing interest in developing time modulated and ultrashort pulsed electron sources for both fundamental studies of the emission processes, particularly when induced by ultrafast lasers [1], [2], and applications such as dynamic studies in electron microscopes, diffractometers, radio frequency amplifiers and X-ray imaging systems. The schemes require, however, an important infrastructure, *e.g.* short wavelengths and pulsed laser sources, and are not well adapted to many instruments requiring compact sources.

This work was carried out within the framework of the "Plateforme nanofils et nanotubes lyonnaise". R.M. acknowledges support from Canada Research Chair and NSERC under grants RGPIN-2019-06545 and RGPAS-2019-00050 and Canada Research Chair.

In contrast, electron emitters based on field-emission provide well-defined point sources having high brightness, low energy distribution and stable emission. However, there is currently no-known field-emitting source providing ultra-fast oscillating electron emission that can be powered cheaply by a *dc* energy source, namely a self-oscillating field emitters. [3] Here, we test new ideas targeting the promotion of oscillatory (pulsed) emission during field-emission (FE) of low-doped semiconducting nanowires (NWs) under light excitation. We report on a double N-shape NDR in the photo-induced field-emission I-V characteristics of individual highly crystalline Si nanowires (NWs).

II. EXPERIMENTS

Photo-induced FE measurements were carried out on two individual silicon nanowires (NW1: diameter $d = 92$nm, lengths $L = 11.5\mu$m and mechanical oscillation frequency $f = 119$kHz and NW2: $d = 100$ nm, $L = 20\mu$m and $f = 107$ kHz) attached by conductive carbon glue to a tungsten tip. Grown by a vapor-liquid-solid (VLS) process, these Si NWs are unintentionally p-doped ($N_a =\sim 2 \times 10^{15}/cm^{-3}$) and present a highly crystalline core covered with a 10 nm oxide passivation layer all along the sidewall. The apex of NW1 (Figure 1b) was sharpened using *in-situ* field-induced atomic desorption in a Transmission Electron Microscope (TEM) and the apex of NW2 was used "as is" and consists of a rounded silicon tip containing the gold nanoparticle catalyst. The rounded apex showed for both NWs a good directionality to the electron beam and a high stability on the MCP detector (Figure 1b, inset). Prior to FE measurements, the NWs were annealed a few times in UHV at 500 °C during 5 min to remove contaminants and exposed at 400 °C to an atomic hydrogen beam for 5 min. We have previously reported on the FE properties of Si NWs

978-1-6654-2590-2/21 $31.00 © 2021 IEEE

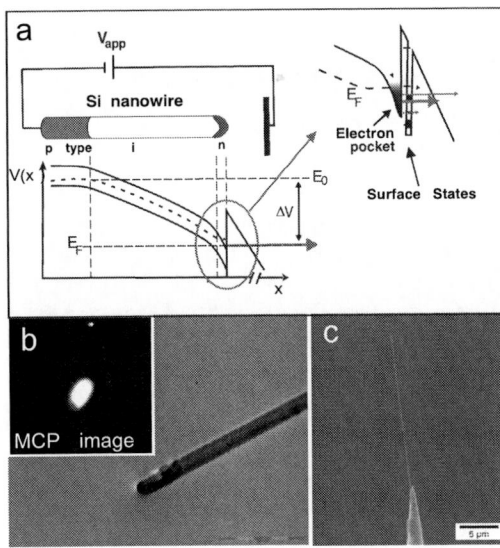

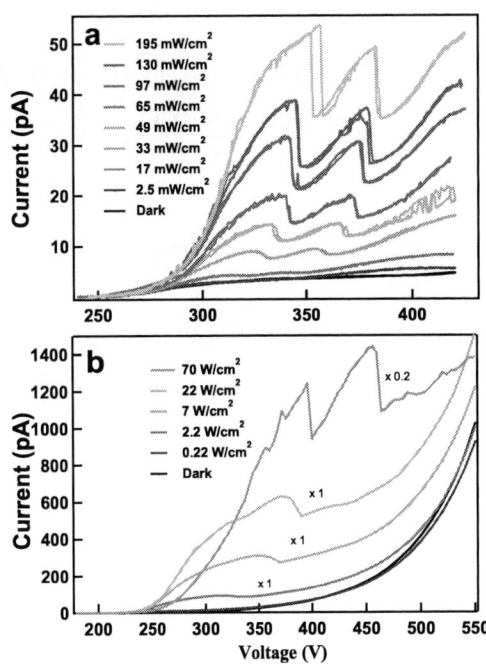

Fig. 1. *a,* Schematic for photo-assisted FE from a p-type Si Nanowire (top) and a plot of the electrical potential model for FE (middle). Zoom on the emission area with a proposed double quantum dot termination. *b,* Transmission Electron Micropscopy (TEM) image of NW1 zoomed on the rounded tip obtained after field-induce atom desorption for 5 min. The scale is 500 nm. Inset: Typical emission image of NWs with rounded apex acquired by the microchannel plate (MCP) *c,* Scanning electron microscopy of Si NW2 of diameter d=100nm and length L=20μm attached to a tungsten tip.

Fig. 2. (a) Current-Voltage plots with increasing laser intensity for NW1 (a) and NW2 (b): In saturation region, the current is highly sensitive to laser intensity as predicted by theory for field emission from semiconductors and features a two negative differential resistance peaks

from the same batches and extensively discussed the nonlinear IVs obtained after passivation, which show a characteristic saturation current in the sub nA range. [4] Although no dopant were included in the synthesis, the IVs obtained with NW1 and NW2 are consistent with what is expected for p-doped NWs and behave as predicted by the FE theory of p-doped semiconductors. [5]

The room temperature IV plot of NW1 and NW 2 under continuous illumination (λ=514nm) are shown in Figure 2a and 2b, respectively. Of central interest is the response of the FE current in the saturation region with increasing laser power. For both NWs, the IV curves in the saturation region show a high sensitivity to laser light : Increasing the laser intensity boosts the saturated current and shifts the onset of saturation toward higher voltages. Surprisingly, the plots of both NWs exhibit at the high intensity two sharp peaks, the so-called N-shaped NDRs. They are highly stable and present almost no hysteresis loop between forward and backward bias scans.

An important observation is a linear shift in voltage and a strong increase in amplitude with light intensity. For example, increasing the intensity on NW1 from 49 to 195 mW/cm^2 induces a 18V shift to the first NDR and increases its amplitude from 3 pA to 18 pA. The second NDR follows overall similar trends and both NDRs are located in the saturation region of the IV curve where the emission is limited by generation-recombination current (I_{GR}).

Several mechanisms are under consideration to explain the double NDR features presented here. On one hand, the analysis of the results with temperature and laser intensity, which

are not shown here, indicates that we can safely eliminate thermal effects. Two other mechanisms are being investigated at the moment: 1) NDRs due to photo-generated carrier instabilities in the depletion region, which can give raise to space-charge current build-ups in the nanowire. 2) A double quantum well formed by confinement near the Si nanowire apex (see Figure 1a bottom). The latter would essentially depend on the distribution of the surface states of the Si NW under strong electric fields. The former would be due to a transient state caused by an accumulation of excess positive charges in the depletion region, which can cause a transient electric field causing variations of the width of the depletion region and hence oscillating currents. These original results suggest that new functionalities and experiments in vacuum nano-electronics can be designed to explore pulsed electron sources. This work can potentially be expanded to design pulsed electron sources working at high repetition rates.

References

[1] M. Kruger, M. Schenk and P. Hommelhoff, "Attosecond control of electrons emitted from a nanoscale metal tip," Nature, 475, pp. 78, 2011.

[2] M. F. Ciappina *et al.*, "Attosecond physics at the nanoscale, Rep. Prog. Phys., vol. 80, pp. 054401-1-50, 2017.

[3] A. Ayari *et al.*, "Self-oscillations in field emission nanowire mechanical resonators: A nanometric dc-ac conversion," Nano Letters, vol. 7, pp. 2252-2257, 2007.

[4] M. Choueib, R. Martel, C.S. Cojocaru, A. Ayari, P. Vincent, P and S. T. Purcell, "Current saturation in field emission from H-passivated Si nanowires," ACS Nano, 6, pp. 7463-7471, 2012.

[5] L. M. Baskin, O. I. Lvov and G. N. Fursey, "General Features of Field Emission from Semiconductors," Phys. Status Solidi B, vol. 47, pp. 49-62, 1971.

Vertical Si Nano Vacuum Channel Transistors: Building Blocks for Empty State Electronics

Akintunde I. Akinwande*, Girish Rughoobur, Nedeljko Karaulac, Winston Chern and Olusoji O. Ilori.

Microsystems Technology Laboratories, Massachusetts Institute of Technology, Cambridge, MA, USA
*Contact: akinwand@mtl.mit.edu, phone +1-617 258 7974

Abstract— **Vacuum nano channel transistors (VNCT) could potentially have superior performance compared to solid state devices of equivalent channel length owing to ballistic transport of electrons, shorter transit time and higher intrinsic breakdown voltage. Furthermore, they are expected to have very high breakdown voltage. Hence, NVCT have promise for very high Johnson Figure of Merit that is as high as 10^{14} V/s.**

I. INTRODUCTION

Electronic systems based on silicon integrated circuits (ICs) have enabled many transformative consumer and industrial devices that have impacted the society in ways never before imagined. Exponential growth of the number of devices on IC chips was predicted by Gordon Moore in his 1965 paper[1]. Dennard provided a framework for scaling devices to smaller dimensions enabling the very high level of device integration and performance improvement with IC generation[2]. However, transistors are reaching the limits of the performance that could be attained by scaling. This has necessitated a number of significant changes to the design of nanodevices and nano-systems. These changes range from (1) limitation of threshold voltage scaling because of leakage currents due the 60 mV/decade rule, (2) limitation of the maximum supply voltages due to breakdown, and (3) limitation of the clock speed due to dynamic thermal dissipation. All these limitations can be attributed to materials properties associated with electron transport in semiconductors. Several approaches have been used at the system level to circumvent these limitations such as multi-core processors, system on a chip (SoC) and system in a package (SiP). At the device level there are also several approaches that are being used to address performance limitation such as the use of III-V semiconductors, 2D materials, carbon nanotubes and 3D integration. These approaches do not address the fundamental limitations of using a semiconductor channel – velocity saturation and breakdown which ultimately result in a limited Johnson Figure of Merit.

Solid state transistors are inherently limited by two materials properties related to motion of carriers in the device channel – velocity saturation and breakdown. Saturation velocity is related to inelastic scattering with optical phonons which in most semiconductors range from 30 meV$<E_{phonon}<$160 meV. Saturations velocities are estimated to range from 1×10^7 cm/s $< v_{sat} < 2.5 \times 10^7$ cm/s. On the other hand, semiconductor breakdown is related to the strength of the semiconductor bond

and hence the bandgap. The critical breakdown field is related to high electrostatic field carrier transport in the semiconductor which results in impact ionization and avalanche multiplication. The first major consequence is that the material breakdown field which depends on the bandgap is limited to 3×10^5 V/cm in silicon and 10^7 V/cm in diamond. Scaling the channel length limits the voltage that could be applied to the drain (or collector) due to the limited breakdown field. This fundamental limitation is captured by the Johnson Figure of Merit (J_{FoM}) which provides a comparison of the relative performance of a device based on their ability to deliver high power at high frequency in RF/THz devices or their ability to deliver high switching speed with adequate noise margins in digital systems. The Johnson Figure of Merit is given by:

$$J_{FoM} = \frac{E_{BD} \times v_{sat}}{2\pi}$$

where E_{BD} is the breakdown field of the semiconductor channel and v_{sat} is the saturation velocity in the semiconductor channel. The Johnson Figure of Merit for Si, GaN and Diamond are 4.8×10^{11} V/s, 7.4×10^{12} V/s, and 2.4×10^{13} V/s, respectively.

II. NANO VACUUM CHANNEL TRANSISTORS (NVCT)

An obvious way to overcome the limitations of Si transistors is to use vacuum as the device transport channel. There is no scattering in the vacuum channel (ballistic transport) and there is no impact ionization in a vacuum channel (no channel breakdown). Work at NASA is exploring the nano vacuum channel transistor (NVCT) in which the transport channel for a MOSFET is replaced by vacuum as long as carriers are efficiently injected into the channel [3]. Carrier injection into the channel is by field emission. There are several questions that are being posed by device physicists / engineers.

The NVCT and the Si transistor are both based on density modulation of electrons by a gate; however, it is instructive to compare their applications to electronics. A key distinction between the MOSFET and the NVCT is the method of electron injection into the channel. In the MOSFET, electron injection occurs through the lowering of the n$^+$-source/p-channel barrier through the application of a voltage to the gate of the MOS capacitor formed with the p-channel. This barrier is of the order of the bandgap of silicon ($E_G = 1.12$ eV). In contrast, for the NVCT, electrons are injected into the channel by tunneling

978-1-6654-2590-2/21 $31.00 © 2021 IEEE

through the barrier as the barrier is "narrowed" by applying a voltage between the gate and emitter tip (Fig. 1). This barrier height is of the order of the electron affinity of Si (χ = 4.05 eV).

Vacuum devices and by extension NVCTs have excellent output characteristics such as low output conductance, high voltage and high-power handling capabilities because output electrode can withstand a high voltage. However, their input characteristics are relatively poor because they have low current capabilities, low transconductance, high modulation / turn-on voltage and poor noise characteristics. The last 30 years witnessed tremendous research efforts to improve the input characteristics of vacuum microelectronic devices for various applications by increasing the transmission through this barrier.

By scaling down all the critical device dimensions by a factor of 10, very high density (10^8 tips/cm^2), self-aligned gate field emitter arrays with low turn-on voltage (8.5 V), low operating voltage (20 V), high current density (150 A/cm^2) and long lifetime (>300 hours) have been demonstrated [4].

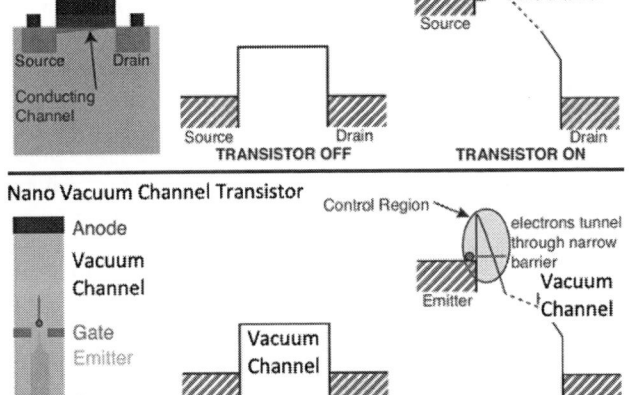

Fig. 1. Comparison of the nanoscale vacuum channel transistor (NVCT) to a metal oxide semiconductor field effect transistor (MOSFET).

III. SCALING OF NVCTs FOR HIGHER PERFORMANCE

The Johnson Figure of Merit (J_{FoM}) for the NVCT is 4.8 $\times$ 10^{14} V/s which is 10x as large as the J_{FoM} for the widest bandgap semiconductor. The breakdown field of vacuum channel is very high and the electron velocity does not saturate. This makes the NVCT as a candidate for a practical high frequency and high power device. A key building block for electronic application is the switch. In this work we shall explore two switches that are based on NVCTs.

Chern et al. reported a vacuum transistor fabricated based upon Si operating in excess of 38 kV applied bias. They demonstrated a high-performance vacuum transistor with the potential of having a semiconductor-like footprint. Electrons are emitted from the gated field emission array into vacuum through tunneling, travel through a vacuum channel and are collected at the anode. The vacuum determines the properties of transport and the high voltage isolation. Important figures of merit show the benefits of using a vacuum drift channel, especially for applications which require both high frequency

and power. The origin of the improved performance comes from the high critical field of vacuum and most importantly the unbounded nature of the electron velocity due to a lack of scattering.

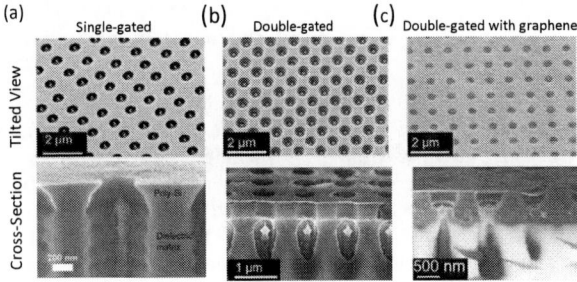

Fig. 2. (a) Single-gated device, (b) Double-gated device with two self-aligned gates, and (c) Double-gated device encapsulated by two-layers of graphene[6].

We also present an electronic switch which focuses on low voltage operation and hence lowering the turn-on voltage on NVCTs. Rughoobur et al.[6] recently reported three different architectures of Si field emitter devices with self-aligned gates (single-gated, double-gated and graphene encapsulated double-gated), shown in Fig. 2, for high field factors, β, and low turn-on voltages, V_{ON}. The lowest V_{ON} achieved was < 8 V using the graphene encapsulated double-gated device. All the device architectures demonstrated linear correlation between V_{ON} and the magnitude of the slope of the Fowler-Nordheim plot, b_{FN}. They explored the effects of tip radius, aperture diameters and tip position relative to the gate on β, in order to reduce V_{ON}.

IV. CONCLUSIONS

NVCTs could potentially have superior performance compared to solid state devices of equivalent channel length owing to ballistic transport of electrons, shorter transit time and higher intrinsic breakdown voltage; however, significant technology development is required to attain this potential.

ACKNOWLEDGMENT

Research sponsored by (a) AFOSR/MURI grant # FA9550-18-1-0436 and (b) IARPA/AFRL contract # FA8650-17-C-9113.

REFERENCES

[1] G. Moore, "Cramming More Components onto Integrated Circuits," *Electronics*, vol. 38, no. 8, 1965, pp. 114-117.

[2] R. Dennard et al., "Design of Ion-Implanted MOSFETs with Very Small Physical Dimensions," *IEEE J. Solid State Circuits*, vol. 9, no. 5, 1974, pp. 256-268.

[3] Jin-Woo Han, Dong-Il Moon, and M. Meyyappan, "Nanoscale Vacuum Channel Transistor," *Nano Lett.*, vol. 17, pp. 2146–2151, 2017.

[4] S. A. Guerrera et al., "Nanofabrication of arrays of silicon field emitters with vertical silicon nanowire current limiters and self-aligned gates," *Nanotechnology*, vol. 27, no. 29, p. 295302 (2016).

[5] W. Chern et al., "Demonstration of a ~40 kV Si Vacuum Transistor as a Practical High Frequency and Power Device," *Technical Digest of the 2020 International Electron Device Meeting*, December 2020.

[6] G. Rughoobur, et al., "Nanoscale Vacuum Channel Electron Sources," *Technical Digest of the 33rd International Vacuum Nanoelectronics Conference*, July 2020.

Investigation on the Emission Behaviour of p-doped Silicon Field Emission Arrays with Individually Controllable Single Tips

Philipp Buchner[1+], Vitali Bomke[1], Matthias Hausladen[1], Simon Edler[2], Michael Bachmann[2] and Rupert Schreiner[1]

[1]Faculty of Applied Natural Sciences and Cultural Studies, OTH Regensburg, D-93053 Regensburg, Germany

[2]Ketek GmbH, D-82737 Munich, Germany

[+]Corresponding author: philipp.buchner@oth-regensburg.de, phone +49-941-943-1257

Abstract—**Four individually controllable emission tips consisting of <111> p-Type silicon, were structured on a glass substrate by laser ablation. A matching extraction grid was manufactured in the same manner and aligned with the emitters. The resulting samples were characterized in ultra-high vacuum. As expected, the individual currents show a strong saturation and in the saturation region a considerably lower current fluctuation than n-type silicon due to charge carrier depletion. The individual tips behave completely independent behaviour from each other and the overall emission can be deduced from the sum of the currents through the individual tips.**

Keywords — field emission, field emitter array, p-Type silicon, current fluctuation

I. INTRODUCTION

P-doped silicon exhibits significant current saturation in the current-voltage characteristics, which was observed in various field emission arrays (FEAs). [1, 2] It is expected that the geometry of the individual tips as well as the distribution of their geometrical parameters (tip radii, tip heights) have a great influence on the integral current saturation characteristics of an FEA. A geometric variation of the individual emission spots was proposed to explain this saturation behaviour for different arrays. [3] However, for the arrays used so far, there was no possibility to measure the saturation currents of the individual emitters of an array and the integral saturation current of the complete FEA simultaneously. To study these phenomena in detail, we designed an electron source that allows controlling and simultaneously characterizing both the individual emitter currents and the total current of the complete array.

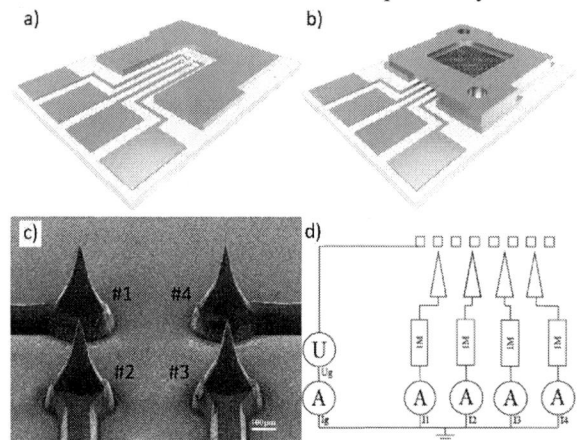

Fig 1: Renderings of the emitters (a), extraction grid (b), SEM image of sample F (c) and schematic for the measuring setup (d).

The research work was funded by the Bavarian Research Foundation under project-number AZ-139619.

II. FABRICATION

The fabrication process is similar to that described in [4]. Initially, the p-Si substrate was anodically bonded to an insulating Borofloat substrate. Afterwards, the emitters and extraction grids were fabricated by laser micromachining and wet etching (Fig. 1a-c). [4, 5]

Three samples (E, F and G) of p-Type (Boron, DSP, <111>, 1-20 Ωcm) silicon and one sample (C) of n-type (Phosphor, DSP, <111>, 1300 - 2700 Ωcm) silicon were manufactured. The extraction grids were manufactured from an n-Type silicon wafer (Arsenic, DSP, <100>, < 5 mΩcm). The grid was fixed to the emitter with vacuum compatible, conducting silver adhesive and simultaneously aligned to the emission tips under an optical microscope (Fig 1a-c).

III. CHARACTERIZATION

In order to measure the emission characteristics, the samples were attached to insulating ceramic substrates and then clamped into a milled metal sample holder. The sample was inserted into an ultra-high vacuum chamber (residual pressure $< 10^{-7}$ Pa). Tip 2 and the extraction grid were connected by manipulators with tungsten needles. Tips 1, 3 and 4 by metallic spring contacts.

All data were measured in a diode configuration with the emission tips connected to ground and a positive voltage applied to the extraction grid (Fig. 1d). The extraction voltage was provided by a Keithley KE6517B. The four emission tips were individually connected to a series resistor of 1 MΩ and a current measurement unit (I_1: Keithley KE6485; I_2: Keithley KE6487; I_3: Keithley KE6487; I_4: Keithley DMM7510). The grid current I_g was measured with a second Keithley DMM7510. For each data point, the intended voltage was applied and after a stabilisation time of two seconds the six currents were measured with an integration time of 100 ms. All voltages were incremented in 5V steps.

Each sample was conditioned by applying a voltage of 1 kV to the grid for six hours before measuring the characteristics. Then the whole array was characterized up to 500 V with various connected emitters. Lastly, a longterm measurement was performed for 1 h at 1 kV for sample C and F to gather information on the current fluctuations.

IV. DATA EVALUATION

To prepare the data, for each point of measurement the voltage drop through the current limiting resistor was calculated and subtracted from the corresponding measured voltage. For each curve, the corresponding measurements from Up- and Down-Sweep were averaged. The resulting IV-characteristic of

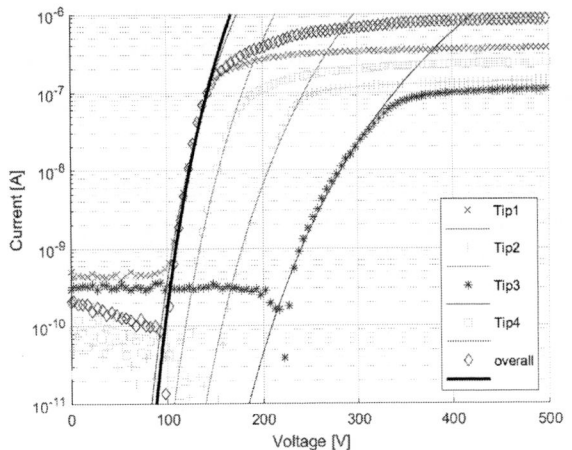

Fig 2: IV-characteristics for the simultaneous measurement of all five currents (individual tip currents and integral current of the array) of sample F.

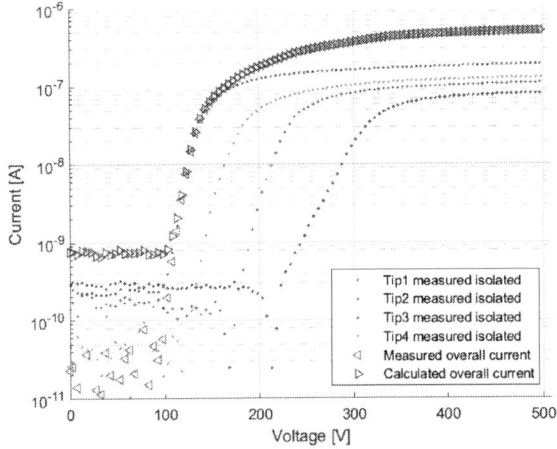

Fig 3: IV-characteristics of the four individually measured tip currents (grey) and the calculated overall current (blue) of sample F. This is in good agreement with the integral current measurement of the complete array (red).

all five currents shows, that the total current into the grid accurately reflects the behaviour of the individual currents out of the tips (Fig. 2). In addition, the calculated sum of four individually measured tip currents matches the integral measured grid current of the array, too (Fig. 3). Sample G and sample E show a similar behaviour. Fig. 4 shows the 1h constant voltage behaviour of sample F (p-type) and sample C (n-type) at 1 kV.

For all individual tip currents and the overall current the average over the hold time, as well as the current fluctuations were calculated as described in [4]. For both n-type emitters the fluctuations are comparable for the individual tips as well as for the integral array current (Tab. I). The p-type emitter shows, as expected much lower noise due to depletion of charge carriers. The overall fluctuations of the p-type sample are almost one order of magnitude smaller than its n-type counterpart for each individual tip as well as for the whole array.

V. CONCLUSION

Further investigations will explore ways to draw conclusions about the emission behaviour of the individual emitters in the array from the integral IV-characteristics of an array of p-type silicon tips. For this purpose, the results of the experiments presented here will be combined with the theoretical predictions from [3].

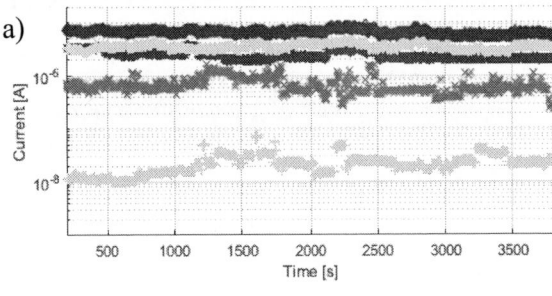

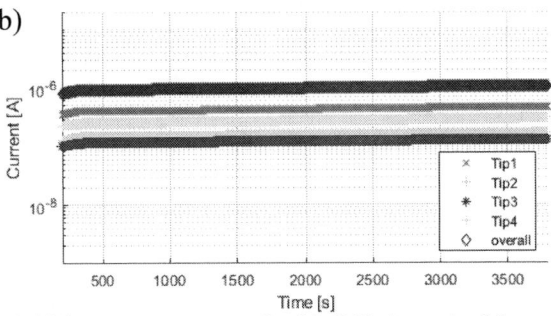

Fig 4: 1 h longterm measurement of each individual current and the overall current of sample C (a) and sample F (b) at a voltage of 1 kV.

TABLE I
CURRENT FLUCTUATIONS OF SAMPLE F (P-TYPE) AND SAMPLE C (N-TYPE)

	n-type (C)		p-type (F)	
	I_{avg} [A]	$\Delta I(10/90)$ [%±]	I_{avg} [A]	$\Delta I(10/90)$ [±%]
I_1	$6{,}52*10^{-7}$	40,59	$4{,}54*10^{-7}$	9,67
I_2	$2{,}26*10^{-8}$	50,83	$1{,}61*10^{-7}$	8,40
I_3	$2{,}40*10^{-6}$	24,82	$1{,}27*10^{-7}$	7,69
I_4	$3{,}43*10^{-6}$	17,15	$2{,}91*10^{-7}$	10,56
I_g	$6{,}50*10^{-6}$	14,09	$1{,}03*10^{-6}$	9,47

REFERENCES

[1] F. Dams et al., "Homogeneous field emission cathodes with precisely adjustable geometry fabricated by silicon technology", IEEE TRANSACTIONS ON ELECTRON DEVICES, VOL. 59, NO. 10, OCTOBER 2012

[2] C. Langer et al., „Field emission properties of p-type black silicon on pillar structures", Journal of Vacuum Science & Technology B **34**, 02G107 (2016); doi: 10.1116/1.4943919

[3] J. Breuer *et al.*, "Extraction of the current distribution out of saturated integral measurement data of p-type silicon field emitter arrays," *Journal of Vacuum Science & Technology B*, vol. 36, no. 5, p. 051805, Sep. 2018, doi: 10.1116/1.5035189

[4] R. Lawrowski et. al., "Silicon Field Emission Electron Source with Individually Controllable Single Emitters", Transactions on Electron Devices, paprer in review

[5] C. Langer et al., "Silicon chip field emission electron source fabricated by laser micromachining"; Journal of Vacuum Science & Technology B **38**, 013202 (2020); doi: 10.1116/1.51

Failure Mode of Si Field Emission Arrays based on Emission Pattern Analysis

Reza Farsad Asadi[1], Tao Zheng[1], Jaime da Silva[1], Girish Rughoobur[2], Akintunde I Akinwande[2], Bruce Gnade[1]*

[1]School of Space Electrical Engineering Department, Southern Methodist University, Dallas, TX 75205 USA

[2] Microsystems Technology Laboratories, Massachusetts Institute of Technology, Cambridge, MA 02139 USA

*Contact: bgnade@smu.edu, phone +1-214-768-1717

Abstract— **Large arrays of nanoscale silicon field emitters are tested in a UHV system equipped with a phosphor screen. The anode current is measured and the emission pattern is captured at the same time. The emission pattern shows the whole Field Emission Arrays (FEAs) are emitting from the beginning of the test. But as the test progresses, the regions at the edge of the array are more vulnerable and tend to be damaged/shorted first. A few shorted emitters along the edges could reduce the effective gate bias applied to the central region of the array, causing the emission current of the whole array to drop dramatically, and prompt the FEA to fail, while the physical integrity of majority emitters in the array are intact.**

I. INTRODUCTION

The field emitter with proximity gate has the advantage of low turn on voltage and high emission current [1]. When built into large Field Emission Arrays (FEAs), they can provide high emission current density [2]. It is important to conduct long term test and explore the failure mechanism of the FEAs before applying them in critical applications. In earlier study [3], we proposed the possible mechanism of shorted emitters close to the edge cause the failure of the whole array. However, at that moment there exists a possibility that in the array, only the emitters on the edge were emitting from the very beginning of the test. To answer the question and further explore the failure mechanism, we conduct this study.

II. EXPERIMENTS

Nano Si field emitter arrays are manufactured as in reference [4]. Then the 6-inch wafer is cut into individual die of 1cm × 1cm resulting in several FEAs on each die. Later, the die is packed and wired bonded, then loaded into a custom-built vacuum testing system. Inside the system, FEAs are facing a P22 phosphor screen placed 4-5 mm away. The phosphor screen is used as anode during the test to collect the emitted electrons. The excited emission pattern on the phosphor screen is captured by a Basler Aca1920 CMOS camera equipped with a Pentax C60636KP lens through a glass vacuum viewport, while the camera is placed outside the vacuum system. The DIE and phosphor screen are approximately parallel to each other with less than 5-degree error. The base pressure of the system is 5×10^{-9} Torr. Two Keithley source measurement units (SMUs) are used to apply the constant DC gate voltage and anode voltage, while measuring the gate and anode current. A LabVIEW program is used to control the SMUs and the camera so that the current measurement and the pattern capture are synchronized. The data is collected at a rate of 500 ms.

III. RESULTS AND DISCUSSION

The stability test results of one 1000×1000 FEA are shown in Figure 1. When the gate bias was 35V, the emission pattern has a square shape mirroring the FEA geometry, see inset in Figure 1. The brightness of the square pattern is relatively uniform over the most area. The observed square pattern confirms the whole array are emitting. However, the emission pattern was lost when the emission current suddenly dropped from 180nA to 6.4nA at 594th second, corresponding to a 96% drop. This loss of the majority of the emission current can be defined as the failure of the FEA. Post-testing optical and SEM images revealed that there are damaged regions in the array active area. The dark/damaged areas only occupied about 1% of the total emitter array area. This observation implies the failure mechanism cannot be a linear relation between the burned area and the emission current.

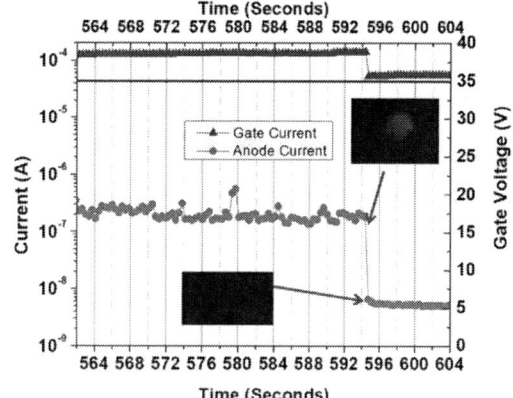

Fig. 1: Gate current and Anode current vs. time for one 1000×1000 FEA. The insets show the corresponding emission pattern right before and after the sudden drop in the emission current. The gate voltage is held at 35V, and the anode voltage is 700V.

Assuming the dark areas were damaged emitters shorting to the substrate which is held at 0V during the test, an electrostatic simulation using COMSOL Multiphysics software shows the effective gate potential over the array gate region, with the presence of the shorting areas, shown in Figure 2. In the simulation, the array size is set as 50×50 due to the availability of computation resources, while other physical dimension in the array is kept the same as the original 1000×1000 array. The damaged/dark area in the original 1000×1000 array is proportionally scaled onto the 50×50 array. The simulation results show that in the central region of the FEA, the effective gate bias seen by the emitters have dropped far below 35V, the nominal applied gate bias.

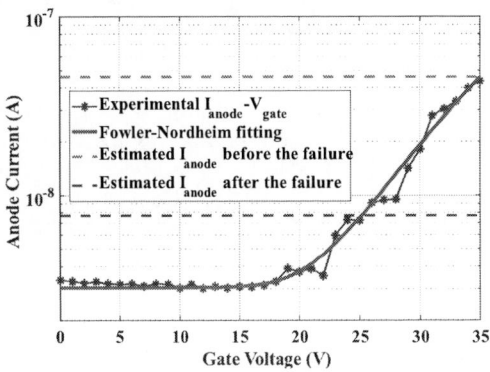

Fig. 3 Last known I_{anode} vs V_{gate} curve of the 1000x1000 array in Figure 1 before the reliability test. The fitted I_{anode} vs V_{gate} curve using Fowler-Nordheim equation is shown in green. And the estimated total emission current before and after the failure is also labelled.

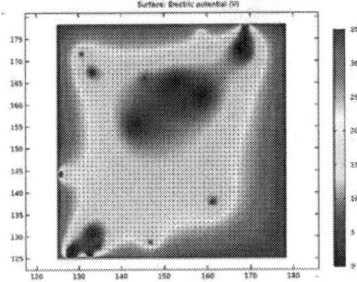

Fig. 2 (a) Post-testing optical images of the 1000x1000 array tested in Figure1. (b) COMSOL simulations results shows the effective gate bias over the example 50x50 array assuming the dark regions in (a) are shored to the substrate/ground.

Based on the last-known I_{anode} vs V_{gate} curve of the 1000x1000 array taken before the stability test, as shown in Figure 3, the anode current after 594th second is calculated using the simulated results in Figure 2. The original I-V curve of the array is fitted to a curve in the format of $aV^2 e^{-\frac{b}{V}} + c$ where a, b, c are constants. Then the fitted curve is divided by 10^6 to create the I-V behaviour of a single emitter. The gate bias around each individual emitter in the 50×50 array in Figure 2 is determined by averaging over 4 points around the emitter, then the emission current of that specific emitter is extracted from the fitted curve. Summing up the emission current over each individual emitter generates the total current of the 50×50 array, then the total current is scaled up to generate the value corresponding to the 1000×1000 array. The final calculated anode current is shown in Figure 3, which is 17% of the original value, or an 83% drop in total anode current. The calculated anode current changes are in good agreement to the experimental observation.

IV. CONCLUSIONS

Large nano-scale Si FEAs are tested to study the failure mechanisms. It is observed that the majority of the emitters were turned on at the beginning of the test but later were suddenly turned off. This sudden change is likely due to a few shorted emitters around the edge of the array which lowers the effective gate bias applied to the central/undamaged region of the array.

ACKNOWLEDGMENT

The authors would like to acknowledge the funding support by the Air Force Office of Scientific Research through MURI 2018 program -- AFOSR Grant No. FA9550-18-1-0436.

REFERENCES

[1] M. Ding, G. Sha, and A. I. Akinwande, "Silicon field emission arrays with atomically sharp tips: turn-on voltage and the effect of tip radius distribution," in IEEE Transactions on Electron Devices, vol. 49, no. 12, pp. 2333-2342, Dec. 2002

[2] S. A. Guerrera and A. I. Akinwande, "Silicon Field Emitter Arrays With Current Densities Exceeding 100 A/cm2 at Gate Voltages Below 75 V," in IEEE Electron Device Letters, vol. 37, no. 1, pp. 96-99, Jan. 2016

[3] T.Zheng, R. Asadi1, J. Silva, G. Rughoobur, A. I Akinwande, B. Gnade, " Failure Mechanism of Large Scale Field Emission Array", 2021 IEEE International Vacuum Electronics Conference (IVEC), 2021

[4] G. Rughoobur, Á. Sahagún, O. O. Ilori and A. I. Akinwande, "Nanofabricated Low-Voltage Gated Si Field-Ionization Arrays," in IEEE Transactions on Electron Devices, vol. 67, no. 8, pp. 3378-3384, Aug. 2020.

Field Emission Arrays from Graphite Fabricated by Laser Micromachining

Robert Ławrowski[1], Michael Bachmann[2] and Rupert Schreiner[1]

[1]*Faculty of Applied Natural Sciences and Cultural Studies, OTH Regensburg, D-93053 Regensburg, Germany*
[2]*Ketek GmbH, D-81737 Munich, Germany*
*Contact: robert.lawrowski@oth-regensburg.de

Abstract— **Arrays of 4x4 conical shaped emitters were directly structured by laser micromaching. Assembled as an electron source with a spacer and Si-grid, they were investigated by means of IV-measurement in ultra-high vacuum. The IV-measurements of the sample show integral emission currents up to 10 µA at a voltage of 1000 V (11 MV/m). Initially, an ideal FN-behaviour is noticeable. The deviation for higher voltage values in the FN-plot is caused by the change of the work function, which was proven by numerical calculations.**

Keywords —field emission array, graphite, laser micromachining

I. INTRODUCTION

UV laser microstructuring is a promising approach for the realization of silicon field emission electron sources [1]. By this maskless direct-writing process, silicon structures with different geometries can be realized on one chip. However, this fabrication process is not limited to silicon only. In principle, any material that absorbs UV light can be patterned with it. Due to its physical and chemical properties, graphite appears to be an interesting material for the realization of field emission cathodes.

II. FABRICATION

Field emitter arrays containing 4x4 emitters were fabricated on two different graphite substrates, one with a smoother and the other one with a rougher surface (Fig. 1). For this purpose, the laser beam was scanned in parallel lines across the substrate. The lines were interrupted in certain areas. Due to the

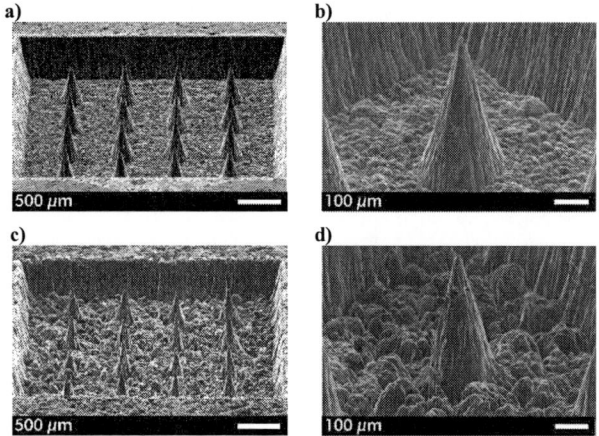

Fig. 1. SEM images of a structured graphite with a smooth surface (a, b) and a rough surface (c, d) including an array containing 4x4 emitters and corresponding detailed view of an emitter.

The research work was funded by the Bavarian Research Foundation under project-number AZ-139619.

beam waist, less graphite was ablated in these areas and conical structures were formed. The arrangement was rotated by 120° and repeated, until the emitter height was reached. Afterwards, no further treatment was performed.

III. CHARACTERIZAZION

Integral FE-measurements were performed in a diode configuration with a 50 µm mica spacer and a Si-grid in a vacuum chamber at pressures of about 10-9 mbar [2]. At an applied voltage of 1000 V the measurements show an integral emission current greater than several microamperes (Fig. 2a). The onset voltage for a current of 1 nA is at 450 and 550 V, respectively. Initially, for lower voltages the corresponding Fowler-Nordheim (FN) plot shows a linear behaviour for both samples (Fig. 2b). This results in a field enhancement factor β of about 275 and 450, respectively (Tab. I). The work function of the graphite was assumed to be 4.8 eV, since the plasma of the laser treatment produces amorphous carbon, which is greater than that of the native graphite (4.4 eV) [3]. A deviation from the ideal behaviour in the FN-plot is noticeable. A change

a)

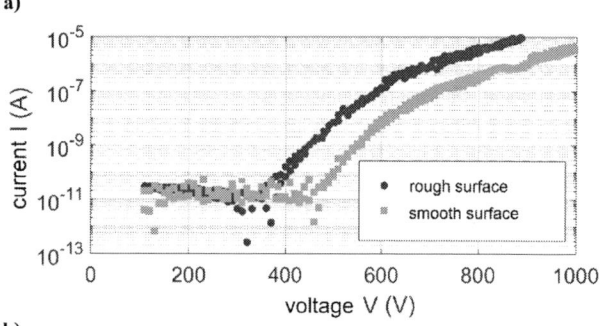

b)

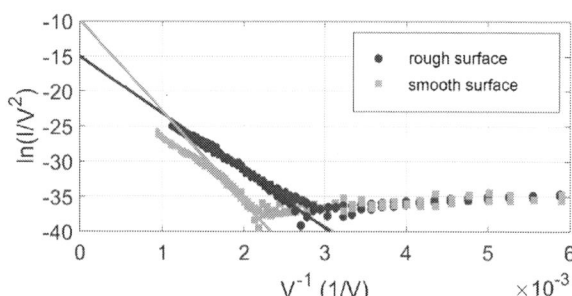

Fig. 2. IV-plot (a) and corresponding FN-plot (b) of the emission characteristics for both graphite arrays with corresponding regression.

Tab. I COMPARISON OF EMISSION BEHAVIOURS

emitter type	E_{on} [MV/m]	~β	γ [cm^{-1}]
rough surface	9.0	450	$8.81 \cdot 10^4$
smooth surface	11.0	275	$5.54 \cdot 10^4$

emitter type	a_{FN}	b_{FN}	S [cm^2]
rough surface	$3.22 \cdot 10^{-7}$	$8.16 \cdot 10^3$	$1.45 \cdot 10^{-12}$
smooth surface	$5.78 \cdot 10^{-5}$	$1.30 \cdot 10^4$	$6.60 \cdot 10^{-10}$

Tab. II REGRESSION PARAMETERS AND CALCULATED VALUES

emitter type	m_1	m_2	$(m_1/m_2)^{2/3}$	$P_{1/2}$ [V]
rough surface	$8.16 \cdot 10^3$	$3.74 \cdot 10^3$	1.68	650
smooth surface	$1.30 \cdot 10^4$	$7.82 \cdot 10^3$	1.33	625

emitter type	$\Delta\Phi$ [V]	F [GV/m]	V [V]	~β
rough surface	3.81	10.1	$5.05 \cdot 10^5$	775
smooth surface	3.19	7.08	$3.54 \cdot 10^5$	550

of the slope in the FN plot can be caused by a change of the enhancement or the work function, which takes place with a power of 1.5. Assuming a constant field enhancement, the quotient of the slopes (m_1/m_2) from the two regimes allows a statement about the ratios of the work function (Fig. 3). Since high electric fields lead to a lowering of the work function, it have to be considered in the quotient:

$$\left(\frac{m_1}{m_2}\right)^{2/3} = \frac{4.8\ eV - \Delta\phi}{4.4\ eV - \Delta\phi}$$

This equation can be solved numerically. Afterwards, the microscopic electric field at the transition of the two slopes in the FN-graph can be calculated:

$$\Delta\phi = \sqrt{\frac{e^3 \cdot F}{4 \cdot \pi \cdot \varepsilon_0}} = 3.79 \cdot 10^{-5} \cdot \sqrt{F}$$

Finally, the macroscopic field strength at the transition can be used to determine the field enhancement factor and compare it with that from the regression line. However, larger values are obtained for the field enhancement factor (Tab. II). This is, due to the underestimation of the distance between cathode and anode (50 µm), since the emitters are lowered compared to the chip edge due to laser process. Already an adjustment of the distance to 90 µm leads to the determined values. Thus, the deviation in the FN-behaviour is caused by the change of the work function.

a)

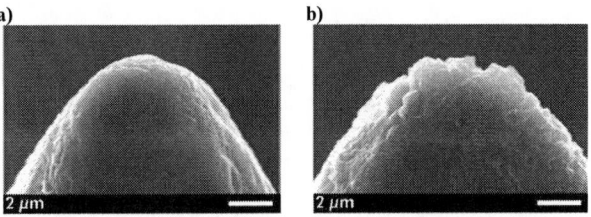

Fig. 4. Constant voltage measurement for both graphite arrays.

Tab. III SUMMARY OF THE LONGTERM MEASUREMENTS

emitter type	V [V]	I [A]	ΔI [%]	I_{drift} [A/h]
rough surface	750	$4.15 \cdot 10^{-6}$	24	$1.48 \cdot 10^{-6}$
rough surface	1000	$1.01 \cdot 10^{-5}$	8	$9.36 \cdot 10^{-7}$
smooth surface	1000	$2.33 \cdot 10^{-6}$	30	$3.78 \cdot 10^{-7}$
smooth surface	1250	$6.18 \cdot 10^{-6}$	24	$1.13 \cdot 10^{-6}$

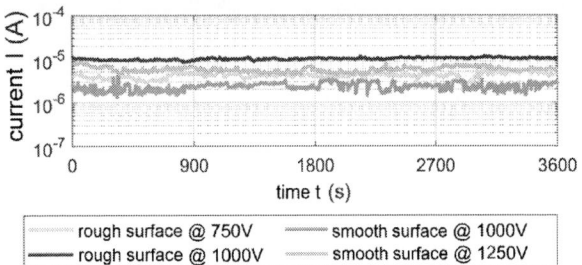

Fig. 5. SEM images illustrating a unchanged (a) and strongly changed (b) emitter tip surface after the constant voltage measurements.

In addition, constant voltage measurements over a period of 60 min with a sampling rate of 0.5 Hz and an integration time of 100 ms were performed (Fig. 4). At several micro amps the current fluctuations ΔI were in the range of 24 to 30%. For higher currents lower flcutuations are noticeable (8%). The current drift I_{drift} were for all measurements positive and up to 1.5 µA/h. This can be explained by the forming of additional emission spots on the surface of the emitters (Fig. 5).

The FE behaviour of such graphite arrays looks promising and is comparable with our previous Si emitters [1], despite the much simpler fabrication process.

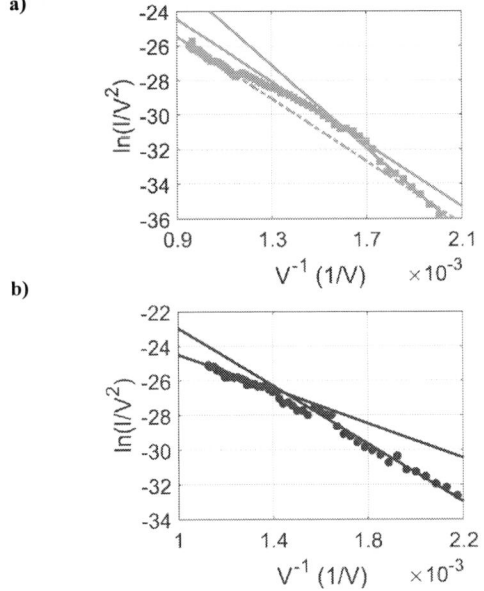

Fig. 3. Detailed view of FN-plots for smooth (a) and rough (b) surface of the graphite arrays with corresponding regression in both regimes.

REFERENCES

[1] C. Langer et al., "Silicon chip field emission electron source fabricated by laser micromachining," *J. Vac. Sci. Technol. B*, vol. 38, no. 1, p. 013202, Jan. 2020, doi: 10.1116/1.5134872.

[2] R. Schreiner et al., "Semiconductor field emission electron sources using a modular system concept for application in sensors and x-ray-sources," in *2015 28th International Vacuum Nanoelectronics Conference (IVNC)*, Jul. 2015, pp. 178–179. doi: 10.1109/IVNC.2015.7225572.

[3] H. Ago et al., "Work Functions and Surface Functional Groups of Multiwall Carbon Nanotubes," *J. Phys. Chem. B*, vol. 103, no. 38, pp. 8116–8121, Sep. 1999, doi: 10.1021/jp991659y.

EFFECTS OF ULTRA VIOLET LIGHT EXPOSURE ON GATED SILLICON FIELD EMITTER ARRAYS

Ranajoy Bhattacharya[1], Mason Canon[1], Nedeljko Karaulac[2], Girish Rughoobur[2],
Winston Chern[2], Akintunde I. Akinwande[2] and Jim Browning[1]

[1]Boise State University, Boise, ID, 83725 USA
[2]Massachusetts Institute of Technology, Cambridge, MA, 02139 USA
*Contact: jimbrowning@boisestate.edu, phone +1-208-426-2347

Abstract— **Ultra violet (UV) light assisted residual gas desorption was performed on silicon gated field emitter array (Si-GFEAs) with 1000×1000 tips. These GFEAs can be used as vacuum nano-transistors. Here, Si-GFEAs with tips are studied experimentally in normal condition as well as under ultra-violet (UV) light exposure for a period of 100 minutes with a time step of 20 minutes. After the experiment, it was found that UV exposure enhanced the field emission current by tenfold and decreased the gate to emitter leakage by greater than ten times. This enhancement can be attributed to the residual gas desorption (water) stimulated by UV exposure, thus reducing the gate surface leakage as well as the work function of emitter tips.**

I. INTRODUCTION

A non-invasive, non-complex water desorption system using UV light can be a simple solution to remove residual gasses trapped inside of modern-day vacuum tube nano structures. These modern-day nano structures can be applied to several electronic devices where high gain and low noise is required along with immunity to harsh environment. Current state-of-the-art approaches such as those reported in [1, 2] do not simultaneously exhibit all these attributes, resulting in significant compromises in the projected system performance. In particular, the field emission (FE) performance of cold cathodes is severely impacted by adsorbates. The primary objective of this work is to improve the FE of gated Si vacuum nano-electronic cold cathodes [1] by means of reduction in residual gas adsorbates. In this work, silicon GFEAs with 1000×1000 tips fabricated as described in [1] have been characterized before and after ultra violet (UV) light exposure. The devices investigated consist of arrays of silicon emitter tips each surrounded by a self-aligned gate aperture.

II. EXPERIMENT SETUP

Experiments were carried out in a stainless-steel high vacuum chamber equipped with a 3-axis manipulator probe arm, electrical and thermocouple feedthroughs, a residual gas analyzer, and a UV light source. The chamber is pumped using a turbomolecular pump backed by a rotary pump achieving a high vacuum of $\approx 5 \times 10^{-8}$ Torr. Measurements were carried out a Keysight B2902A source measurement unit (SMU). An in-house developed test jig consisting of a Low Temperature Co-Fired (LTCC) was used to mount the test die. A molybdenum

micromanipulator pin, mounted on the probe arm was used to connect to the gate. A fixed 100 V DC and 0-40 V DC sweep voltage were applied to collector and gate, respectively, while the substrate is grounded. The test fixture mounted in the LTCC test jig is shown in Figure 1(a). Figure 1(b) shows the magnified view where the probe pin can be seen in contact with the emitter array gate pad.

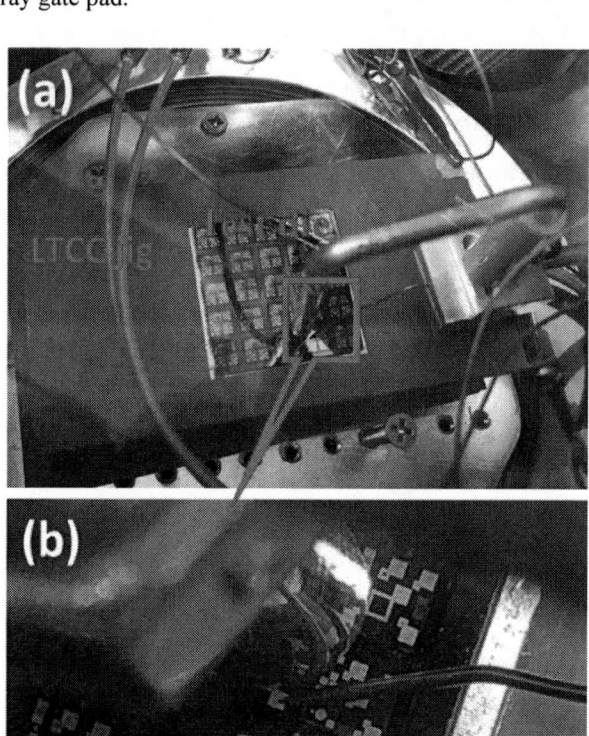

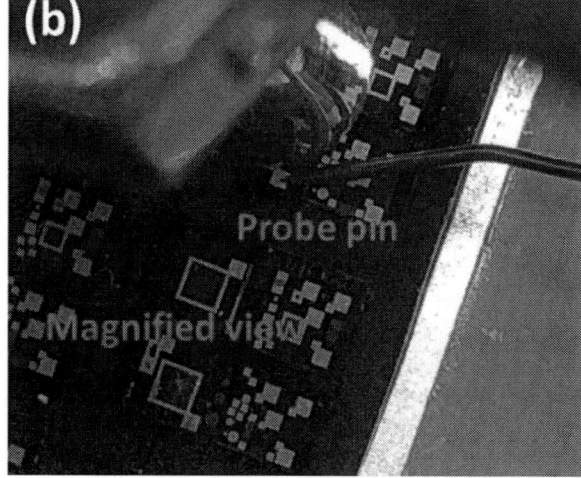

Fig.1 (a) Test die is mounted in the LTCC test jig. (b) Magnified view of the test jig. The probe can be seen connected to the gate pad.

Figure 2(a) shows the UV lamp power supply attached to the test chamber; whereas Figure 2(b) shows the operating UV lamp inside the chamber. The RBD Instruments miniZ UV light source has a 185 nm wavelength and power density of 350 $\mu W/cm^2$.

III. EXPERIMENTAL RESULTS

I-V curves (Figure 3) show that for a gate voltage of 40 V, the collector current was $\approx$ 74 μA while a large gate leakage current of $\approx$ 6 mA was observed. This low collector and large gate current are partially because of the large collector to array gap (>2 mm); however, the low emission current and high gate current are also the result of gas adsorbates. To study the effect of UV light on the performance, the GFEAs were first exposed to UV, and then in-situ I-V characterizations were carried out over 100 min in 20 min intervals.

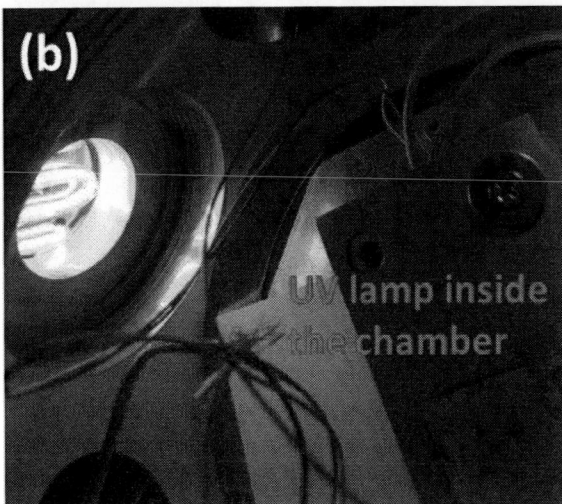

Fig.2 (a) UV lamp power supply attached to the chamber. (b) UV lamp inside the chamber.

During the study, it was observed that after 60 min of UV exposure the collector current increased by >10x (Figure 3) while the gate leakage current decreased by >10x, and even after turning off the UV after 100 minutes, this result remained unchanged. Residual Gas Analyzer (RGA) measurements taken during the UV exposure confirmed water vapor desorption [3]. This improvement is identical to that observed during previous high temperature (400 °C) experiments [4]. To confirm this phenomenon, the GFEA die were taken out of the vacuum chamber, kept in room air for 3 days, and then tested again. After each test cycle, the high gate current and low emission current returned until being exposed to UV light again. These results demonstrate that UV-assisted desorption can be used to clean emitter tips and reduce surface leakage.

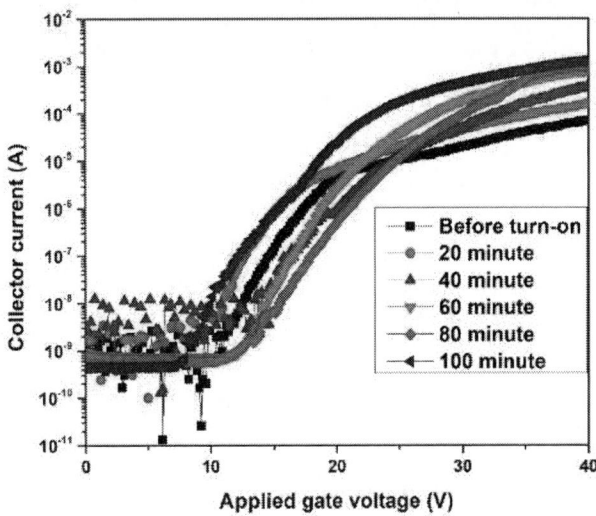

Fig.3 I-V characterization data showing the collector current with respect to different UV exposure times.

IV. CONCLUSIONS

I-V characterization of Si-GFEA with 1000 × 1000 tip arrays were carried out before and after UV light exposure. It was found that the gate current reduced by more than ten times and the emission current increased by more than ten times after 60 min of UV exposure. These results were repeated for several arrays. Using an RGA during UV illumination showed very clear water desorption during the hour-long exposure. If the devices were kept in atmosphere for three days, the collector and gate currents went back to their previous values, but subsequent UV exposure recovers the gate and collector currents to their enhanced levels.

ACKNOWLEDGMENT

Material support for this work is provided by the Air Force Office of Scientific Research under grant FA9550-18-1-0436.

REFERENCES

[1] N. Karaulac, G. Rughoobur and A. I. Akinwande, J. Vac. Sci. Technol. B **38**, 023201 (2020)

[2] Jin-Woo Han, Dong-Il Moon, M. Meyyappan, Nano Lett. 2017, 17, 4, 2146-2151

[3] R. Bhattacharya et. al., "Ultra-Violet Light Stimulated Water Desorption Effect on Emission Performance of Gated Field Emitter Array", Journal of Vacuum Science & Technology B **39**, 033201 (2021).

[4] R. Bhattacharya, et. al., "Temperature effects on gated silicon field emission array performance" J. Vac. Sci. Technol. B **39**, 023201 (2021).

Emission Behavior of Planar Nano-Vacuum Field Emitters

Marco Turchetti[1], Yujia Yang[1], Mina R. Bionta[1], Alberto Nardi[1], Luca Daniel[1], Karl K. Berggren[1], Philip D. Keathley[1]*

[1] *Research Laboratory of Electronics, MIT, 50 Vassar Street, Cambridge, MA 02139, USA*
Contact: pdkeat2@mit.edu

Abstract— **In this work, we analyze the behavior of planar nano-vacuum channel emitters. We developed a fabrication process to reliably pattern metallic (Au) and refractory (TiN) nanoemitters with sub-10nm features. We then investigated the field emission properties of the devices highlighting the transition between Schottky to Fowler-Nordheim field emission, and finally to the Child-Langmuir saturation regime.**

I. INTRODUCTION

Nano-vacuum channel (NVC) electronics promise fast switching times, and low power-delay product due to reduced turn-on voltages and high velocities. This is achieved thanks to ballistic transport across the channel free of phonon and charged impurity scattering. Also, since free electrons are effectively insensitive to ionizing radiation and temperature fluctuations, these devices are attractive for applications in harsh environments such as space technology [1,2]. Field-emitter devices commonly use vertical geometries making them difficult to integrate with traditional electronics. Alternatively, planar NVC (pNVC) field-emitters could be easily incorporated into integrated circuits on a large scale. However, the dominant electron emission mechanisms for pNVC devices is not well understood. For instance, to fit recent experimental data from pNVCs using traditional Fowler-Nordheim (FN) tunnelling models, >100x field enhancements (FE) have to be assumed [3,4], while electromagnetic models predict ~10x FE.

In this work, we fabricated metallic (Au) and refractory (TiN) pNVC bow-tie diodes having ~10-20 nm vacuum gaps (see Fig. 1a,b), and we analyzed their I-V response to investigate the underlying emission mechanisms. We observed three distinct emission regimes: Schottky, Fowler-Nordheim field emission, and saturation that we attribute to Child-Langmuir space charge effects. This work will inform the further development of pNVC devices for integrated electronics applications.

II. FABRICATION

The devices were fabricated on a SiO2 on Si substrate. The Au devices were patterned using a PMMA based electron beam lithography (EBL) and a subsequent e-beam evaporation and lift-off. Then the pads for electrical contact were laid out with a standard photolithography process. Initial testing indicated that oxide charging in the gap region reduced electron emission by orders of magnitude and caused hysteretic IV behavior. To resolve this issue and provide a truly free transport path from emitter to collector, we implemented a subsequent etching process to release the tips from the substrate by creating an undercut. This was performed in two steps: first a trench was formed with CF4 reactive ion etching (RIE), and then the undercut was formed with buffered HF wet etching. Fig. 1 illustrates the different steps involved in the fabrication, and a typical structure obtained with the process.

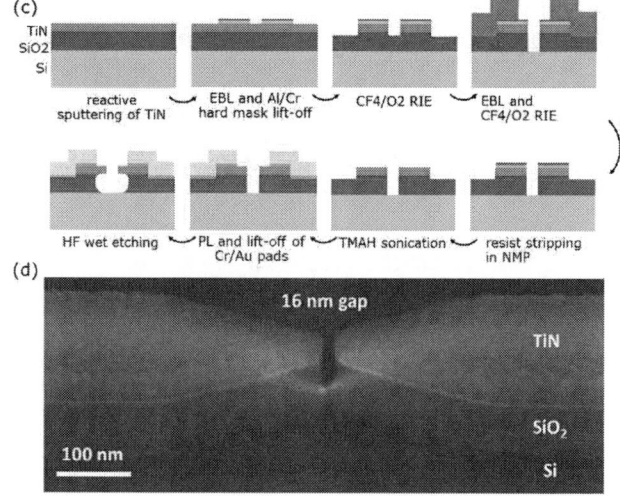

Fig. 1 (a,b) Fabrication procedure (a) and SEM micrograph of a typical resulting structure (b) for Au devices. (c,d) Fabrication procedure (c) and SEM micrograph of a typical resulting structure (d) for TiN devices.

978-1-6654-2590-2/21 $31.00 © 2021 IEEE

For the refractory devices, TiN was first reactively sputtered on the substrate, then a bilayer hard mask of Al/Cr was patterned using EBL and lift-off. This mask showed the best characteristics in terms of feature size, sidewall steepness and removability [5]. Subsequently, the TiN devices were defined with a CF4/O2 RIE process and the hard mask was removed through sonication in TMAH. After laying down the pads a process similar to that used for Au was employed for creating an undercut. Fig 2 illustrates the process and the obtained results.

III. RESULTS

After the fabrication, we tested the IV characteristics of the devices in a vacuum chamber at 10^{-6} mbar. Fig. 2 shows the IV response from a typical Au device. We can see that the device have a turn-on voltage of approximately 5V which is consistent with the Au work function of 5.1eV. The device then reaches a current of 5nA at 8V. For these devices we cannot increase the potential further because it causes a breakdown of the oxide. If we plot the data in a ln(I) vs $V^{0.5}$ plot we notice a linear behavior, which indicates that Schottky emission dominates at these low voltages. A ~50 FE would have to be assumed to fit this I-V response using typical a FN tunneling model.

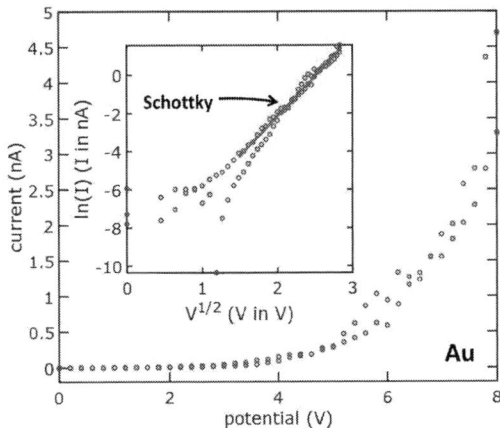

Fig. 2 I-V characteristics of a Au device with inset of the same data in a "Schottky" plot (ln(I) vs V0.5). A linear response in these "Schottky" plots indicates Schottky emission behavior.

We then proceeded with a similar characterization for the refractory emitters. Thanks to their more resilient nature and a thicker oxide, we were able to run the TiN devices at higher potentials. Fig. 3 shows a typical IV characteristic for these devices. We observed different regimes that emerge at higher bias potentials. In particular, if we look at the "Schottky" plot (ln(I) vs $V^{0.5}$) we notice that at low voltages the characteristics looks consistent with Schottky emission, similarly to the Au devices, but at approximately 10V, it becomes superlinear. This is the transition from Schottky to Fowler-Nordheim tunneling. The strong exponential component of FN emission brings the current to the μA level. We validated this result by fitting this superlinear part of the curve with a FN model which predicts a FE of 10x. We then performed an electromagnetic simulation,

which predict a FE of ~7x, which is comparable with the experimental data. After approximately 13 V we can see that the current reaches a saturation regime, which appears to be due to Child-Langmuir space-charge limitations.

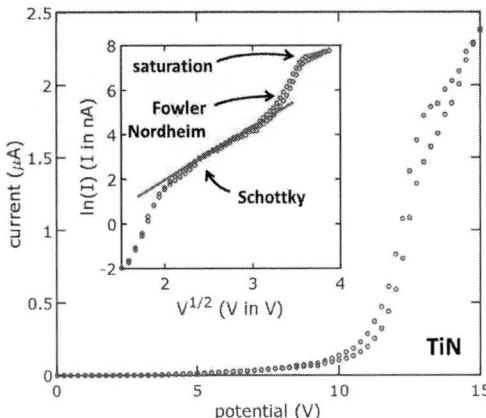

Fig. 3 I-V characteristics of a TiN device with inset of the same data in a "Schottky" plot (ln(I) vs V0.5) where the different regimes are highlighted.

IV. CONCLUSIONS

In conclusion, we developed a fabrication procedure that allows to pattern Au and TiN planar nano-vacuum channel emitters with sub-10nm features. Moreover, we studied their emission characteristics, showing that different regimes are present. In particular, we observed that the planar devices tend to turn-on in the Schottky emission regime. Testing at higher biases using the TiN devices reveal a transition from Schottky to Fowler-Nordheim regime at significantly large bias potentials, and then a second transition at even higher bias potentials to a saturation regime that we attribute to Child-Langmuir space-charge effects. Further testing of these devices, including the impact of device temperature, is ongoing. We are also modifying the Au devices fabrication in order to investigate their response at higher potentials for comparison to the TiN devices.

ACKNOWLEDGMENT

This work was funded by the AFOSR MURI (Contract No.: FA9550-18-1-0436).

REFERENCES

[1] J.-W Han, D.-I. Moon, and M. Meyyappan. "Nanoscale vacuum channel transistor." Nano letters 17.4 (2017)

[2] G. Gaertner "Historical development and future trends of vacuum electronics." Journal of Vacuum Science & Technology B, Nanotechnology and Microelectronics: Materials, Processing, Measurement, and Phenomena 30.6 (2012)

[3] S. Nirantar, et al. "Metal–air transistors: semiconductor-free field-emission air-channel nanoelectronics." Nano letters 18.12 (2018)

[4] L. B. De Rose, A. Scherer, and W. M. Jones. "Suspended Nanoscale Field Emitter Devices for High-Temperature Operation." IEEE Transactions on Electron Devices 67.11 (2020)

[5] A. Nardi, M. Turchetti, W. Britton, Y. Chen, Y. Yang, L. Dal Negro, K. K. Berggren, and Phillip D. Keathley. "Nanoscale refractory doped titanium nitride field emitters." Nanotechnology (2021)

Gap in pagination due to unavailable paper.

Page 88

Terahertz Acceleration Technology Towards Compact Light Sources

Arya Fallahi[1,2*]

[1]ETH Zurich, Zurich, Switzerland
[2]IT'IS Foundation, Zurich, Switzerland
*Contact: afallahi@itis.ethz.ch

Abstract— **Despite the already-realized high power radiation sources enabling ultrahigh electric fields, increasing the acceleration gradients above the state-of-the-art values is hampered by the damage threshold of materials. Recent studies on damage mechanisms in accelerators have revealed the strong dependence of operation threshold on the time duration over which fields are influencing the device. Therefore, fast accelerating principles based on short excitations need to be developed for further increasing the accelerating gradient. This contribution will discuss conceptual developments and proof-of-principle studies for THz acceleration using short pulses. The concepts developed here pave the way towards the realization of cheap and compact particle accelerators with control of particles over ultrashort time scales. The introduced concepts comprise three groups focusing on: (1) fast electron sources, (2) THz injectors, and (3) THz linacs. The final goal of the above studies is a fully THz-driven compact light source facility, whose start-to-end simulation will conclude the presentation.**

I. INTRODUCTION

Novel compact accelerating structures are highly favourable compared to conventional devices for studies where high-energy bunches in small but applicable charges are required. Increasing the operation frequency and shrinking the accelerating devices is a suitable path for improving the accelerators' performance. For this purpose, energy transfer to electrons should be realized in shorter distances, which in turn means introducing higher accelerating gradients. Moreover, higher accelerating gradients enable beams with higher quality due to lower emittance growth. However, increasing the operation frequency from RF to optical regimes introduces serious challenges in synchronization, stability and acceleration of considerable charge amount. Consequently, THz acceleration will likely serve as the optimal operation regime for compact accelerators.

The main obstacle in increasing the acceleration gradients above the state-of-the-art values is the damage threshold of materials. This prevents using the newly realized high power radiation sources for efficient particle acceleration. Recent studies on damage mechanisms in accelerators have revealed the strong dependence of operation threshold on the time duration over which fields are influencing the device. Therefore, fast accelerating principles based on short excitations need to be developed for further increasing the accelerating gradient.

The main goal in this contribution is a review on conceptual developments and proof-of-principle studies for THz acceleration using short pulses. The concepts developed here pave the way towards the realisation of cheap and compact particle accelerators with control of particles over ultrashort time scales.

II. RESULTS

The concepts in this presentation comprise three groups focusing on: (1) fast electron sources, (2) THz injectors, and (3) THz linacs. First, the feasibility of ultrafast, high-yield electron emitters based on nanostructured cathodes is demonstrated. Benefitting from field enhancement effects, namely tip-enhancement and plasmonic enhancement, laser-induced field emission is realized over large, dense and highly uniform field emitter arrays. Fig. 1 shows the results of the theoretical studies on these field emitter arrays. The simulations confirm their suitability for pico-Coulomb charge production over femtosecond time-scales. Comparisons with experimental results additionally support these findings [1]–[3].

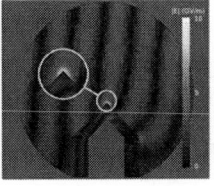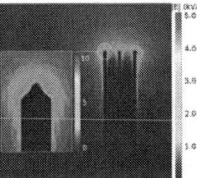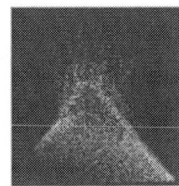

Fig. 1: *(left) The computed field profile using DGTD algorithm. (middle) The computed static field profile using FEM Poisson solver. (right) snap-shot of the emitted charge cloud from the tip.*

In the framework of THz injectors, two ground-breaking concepts including ultrafast single-cycle THz guns (Fig. 2) and segmented THz electron accelerator and manipulator (STEAM) devices are developed and tested [4], [5]. The possibility of using transient fields to realize ultrahigh acceleration gradients close to 0.5 GeV/m is confirmed [6]. Based on this finding, a single-layer device is designed that uses two 50 µJ twin beams at 300 GHz to accelerate electrons to 58 KeV. This device has been recently tested in an independent effort in which the group produced a 2.4 keV electron beam [7]. Specifically, a STEAM

978-1-6654-2590-2/21 $31.00 © 2021 IEEE

device capable of performing multiple high-field operations on the 6D-phase-space of ultrashort electron bunches is demonstrated. With this single device, powered by few-micro-Joule, single-cycle, 0.3 THz pulses, THz-acceleration of > 30 keV, streaking with < 10 fs resolution, focusing with > 2 kT/m strength, compression to ~100 fs as well as real-time switching between these modes of operation are obtained [4].

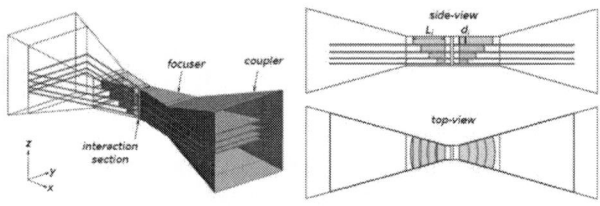

Fig. 2: Schematic illustration of a general con_guration for single-cycle ultrafast electron guns.

Travelling wave THz linacs based on dielectric-loaded metallic waveguides operating under few-cycle excitations are proposed as promising accelerator devices. To show the feasibility of this scheme, optically generated THz pulses were used to accelerate electrons in a simple and practical THz accelerator. An energy gain of 7 keV was achieved over a 3-mm interaction length with good modelled emittance. Performance of these structures improves with an increase in electron energy and gradient making them attractive for compact accelerator applications [8]. The possibility of electron acceleration to tens of MeV with millijoule level THz pulses is theoretically shown [9]. Detailed theoretical analysis of electron bunch acceleration in THz dielectric-leaded metallic waveguides demonstrates the acceleration of a 1.6 pC electron bunch from a kinetic energy of 1 MeV to about 10 MeV over an interaction distance of about 20 mm, using a 20 mJ pulse centered at 0.6 THz in a dielectric-loaded metallic waveguide.

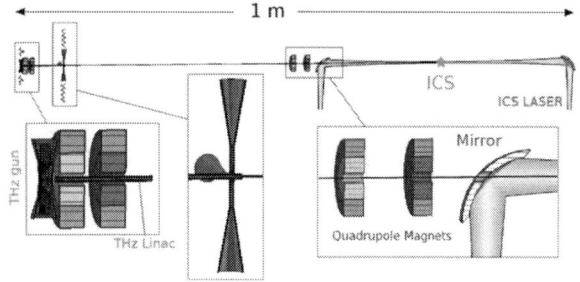

Fig. 3: Layout of the THz-driven compact x-ray source.

The final goal of the above studies is a fully THz-driven compact light source facility, whose start-to-end simulation is fulfilled (Fig. 3). The required THz pulses to excite the light source are categorized under single-cycle and multi-cycle pulses that are generated using laser-driven THz generation concepts. The single-cycle THz pulses feed an ultrafast electron gun, whose output is delivered to a THz linac fed by multi-cycle

pulses. It is shown that 18 MeV beam energy can be produced using two single-cycle THz beams with 400 μJ and one 0.5 ns 300 GHz beam with 20 mJ energy. This beam is then transported to an Inverse Compton Scattering (ICS) section, where the 18MeV electron beam scatters off a 100mJ 1 μm laser beam and generates an X-ray beam with 4 keV central photon energy and 6×104 photons per shot [10].

In a recent study, the possibility to boost the photon number using the concept of confined electron laser is shown [11]. In this work, we showed that by confining the electron beam to the cavity nodes, the detrimental influence of space-charge force in low-energies is compensated, and additionally, the ultrahigh charge density enables high FEL-gain at confinement spots. As a result, the coherence of the output radiation is improved and at the same time the source efficiency is increased by three orders of magnitude.

III. CONCLUSIONS

The studies here underpinned a new research area of laser-driven THz acceleration, which was started by our group around eight years ago and remains still in its infancy stage. However, great progress with transformative impact has been made over the past years demonstrating the promises of this technique for numerous applications. With new inventions for novel applications, new challenges emerge which call for further research and development. This will be an ongoing effort for decades aiming at easy and low-cost access to high energy particles for universities and research centres around the world.

REFERENCES

[1] R. G. Hobbs et al., "High-density Au nanorod optical field-emitter arrays," *Nanotechnology*, vol. 25, no. 46, p. 465304, 2014.

[2] M. E. Swanwick et al., "Nanostructured ultrafast silicon-tip optical field-emitter arrays," *Nano Lett.*, vol. 14, no. 9, 2014.

[3] A. Fallahi and F. Kärtner, "Field-based DGTD/PIC technique for general and stable simulation of interaction between light and electron bunches," *J. Phys. B At. Mol. Opt. Phys.*, vol. 47, no. 23, p. 234015, Dec. 2014.

[4] D. Zhang et al., "Segmented terahertz electron accelerator and manipulator (STEAM)," *Nat. Photonics*, vol. 12, no. 6, pp. 336–342, 2018.

[5] A. Fallahi, M. Fakhari, A. Yahaghi, M. Arrieta, and F. X. Kärtner, "Short electron bunch generation using single-cycle ultrafast electron guns," *Phys. Rev. Accel. Beams*, vol. 19, no. 8, 2016.

[6] W. R. Huang et al., "Terahertz-driven,all-optical electron gun," *Optica*, vol. 3, no. 11, 2016.

[7] N. H. Matlis et al., "THz photogun transversely pumped by twin single-cycle pulses," in *2020 45th International Conference on Infrared, Millimeter, and Terahertz Waves (IRMMW-THz)*, 2020, pp. 1–2.

[8] E. A. Nanni et al., "Terahertz-driven linear electron acceleration," *Nat. Commun.*, vol. 6, 2015.

[9] L. J. Wong, A. Fallahi, and F. X. Kärtner, "Compact electron acceleration and bunch compression in THz waveguides," *Opt. Express*, vol. 21, no. 8, 2013.

[10] A. Fallahi, "Terahertz Acceleration Technology Towards Compact Light Sources," *arXiv Prepr. arXiv1906.06186*, 2019.

[11] A. Fallahi, N. Kuster, and L. Novotny, "Transverse Confinement of Electron Beams in a 2D Optical Lattice for Compact Coherent X-Ray Sources," *arXiv Prepr. arXiv2104.11586*, 2021.

Gap in pagination due to withheld papers.

Pages 91-94

A novel current dependent field emission performance test

Florian Herdl, Michael Bachmann,
Dominik Wohlfartsstätter, Felix Düsberg,
Markus Dudeck, Magdalena Eder,
Manuel Meyer and Andreas Pahlke
KETEK GmbH
81737 Munich, Germany
Email: Florian.Herdl@ketek.net

Simon Edler, Andreas Schels
and Walter Hansch
Institute of Physics
Faculty of Electrical Engineering
and Information Technology
Universität der Bundeswehr München
85577 Neubiberg, Germany

Rupert Schreiner
Faculty of Applied Natural
Sciences and Cultural Studies
OTH Regensburg
93053 Regensburg, Germany

Abstract—**A current dependent performance test for comparison of different field emitter arrays is introduced. Statistical analysis is enabled due to a short measurement time and as a main feature the electric field shift, comparable to the degradation of the emitter is examined. Significance of the test method is shown by a comparison of field emitter arrays with different doping levels.**

I. INTRODUCTION

Long life times of field emitter arrays (FEAs) are mandatory for field emission applications like X-ray sources [1] or electron sources [2]. In the literature life time measurements are performed at an arbitrary voltage or current for a random duration. [3] However, the life time is heavily dependent on the chosen current, as it depends on ion bombardment [4], adsorbates at the tip surfaces [5] and thermal stress [4] at high current densities. To investigate these mechanisms for FEAs, a novel current dependent performance test (CDPT) is presented.

II. MEASUREMENT PROCEDURE

The FEAs are characterized in a vacuum setup with a pressure regulation set to 10^{-5} mbar. This rather high pressure for field emission is used to roughly meet the conditions in a hermetically sealed housing. [2] Multiple constant current measurements (CCMs) with increasing constant current values are performed together with IV-sweeps before and after each step. The current is regulated during CCMs as well as during the IV-sweeps by a regulation circuit presented by Prommesberger et al. [6] This circuit is improved by a voltage divider, enabling the measurement of the applied voltage at the FEA. To ensure that the time and thus the applied load is constant during the IV sweep, 21 currents are predefined according to a logarithmic distribution starting from 1 nA up to the highest current of the last CCM. A typical evolution of the IV-characteristics, prior and after the 1h CCM, is shown in Fig. 1 in Millikan-Lauritsen (ML) coordinates. [7] To analyze the characteristics, a linear regression is realized in Murphy Good coordinates. [8] A plot with the obtained parameters is shown in Fig. 1 as dashed lines. Except for the first three currents, the two characteristics recorded in between two CCMs are largely

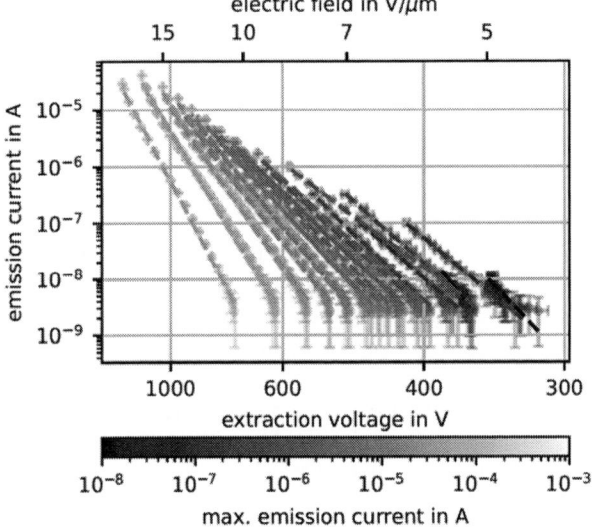

Fig. 1. Exemplary Millikan Lauritsen plot of the characteristics measured prior and after the CCM at different set currents. The dashed opaque lines are obtained by fitting a straight line in Murphy-Good coordinates to the measured characteristics.

coincident. Hence, the degradation mainly occurs during the CCMs.

The evolution of the extraction voltage during 1h of CCM at various currents is shown in Fig. 2. An approximately linear slope of the extraction voltage is observed. To analyze the degradation of the FEA a linear regression (solid lines) is applied. A positive slope implies a degradation of the FEA over time. The degradation shows a dependency on the current. In the shown CCM at 40 μA the maximum voltage of 1.4 kV is reached, hence the CCM is aborted and for the CDPT of this FEA one final IV-sweep is performed.

Due to the rather short measurement time and the steady measurement procedure the CDPT enables a statistical analysis of different FEAs.

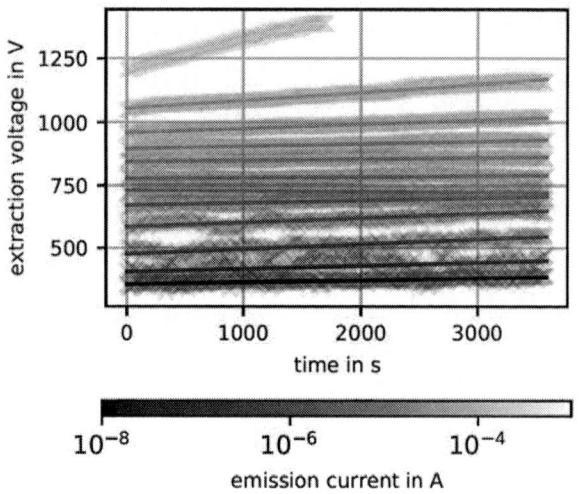

Fig. 2. Exemplary measured evolution of the extraction voltage during the CCM for different set currents. The opaque lines overlaying the measurement points are the fitted lines for the electric field shift computation.

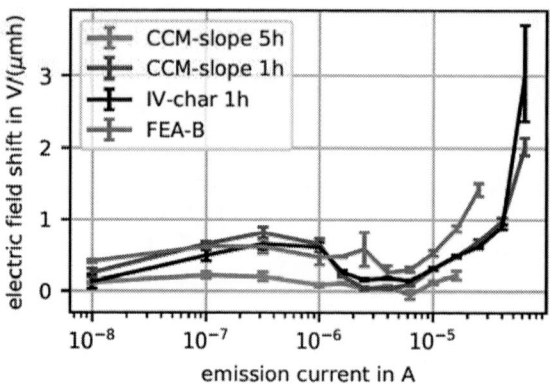

Fig. 3. Mean and error of the obtained electric field shifts dependent on the emission current. From the slopes of the fitted lines at the CCM with 1 h (blue), 5 h (red) and from the characteristics measured during the 1 h measurement (black) all from identical fabricated FEAs. Furthermore the electric field shifts of FEAs with a higher doping level and smaller tips is shown (green).

III. RESULTS AND DISCUSSION

In Fig. 3 the electric field shift during the CCM is presented. The electric field is calculated by the extraction voltage shift (CCM-slope from Fig. 2) divided by the tip to extractor distance. It is used instead of the extraction voltage shift for better comparability due to variations in the tip to extraction distance. For statistical reasons three FEAs of the same wafer are examined under the same conditions and the mean and error estimate of these three samples is shown. Three FEAs are used, because measurements have shown that an increase from three to six measured FEAs does not change the course of the electric field shifts within the error. To compare the results to the data of the IV-sweeps the mean voltage shift of the characteristics is computed and also plotted in the figure (IV-char). Both methods give very similar values for the electric field shift.

To investigate the influence of the randomly chosen operation time during the CCM a measurement with 5 h CCMs is performed and plotted in Fig. 3. A similar shape but lower shift values are obtained for the CCM-slopes as well as for the not shown IV-char. This indicates, that the evolution of the electric field and hence of the voltage is nonlinear with a saturation for longer operation times. Therefore, the obtained shifts are overestimated by this method. More detailed measurements indicate that the initial maxima might be caused by a conditioning of the tips and burn out of the sharpest emitters. However, it can be observed, that the final increase at higher emission currents shows a better agreement. This implies that the conditioning happens faster for higher emission currents. Long term measurements executed at different emission currents lead to the same conclusion. For measurements with even shorter CCM periods the data shows a stronger fluctuation. Therefore 1 h of CCM allows a

high enough throughput to enable statistical analysis of the FEAs, with an overestimated degradation rate. However, this systematic procedure still allows a comparision of different FEAs.

Furthermore, a measurement of a slightly different FEA, with a higher doping concentration and smaller tips, is carried out (FEA B). By measuring IV-sweeps only no significant difference between those two samples can be observed. The degradation rate of these samples is also shown in Fig. 3. Initially a similar degradation is observed. However, at the higher currents a stronger degradation is observed. This results in lower maximum currents and lower lifetimes of this FEA are expected.

IV. CONCLUSION

The presented measurement method gives a valuable tool to systematically compare various FEAs and process variations of FEAs. Therefore, it allows systematic optimization for higher current capability or higher lifetimes.

ACKNOWLEDGMENT

The research work was funded by the Bavarian Research Foundation under project-number AZ-139619.

REFERENCES

[1] K. Kawakita, K. Hata, H. Sato, and Y. Saito, *J. Vac. Sci. Technol. B Microelectron. Nanom. Struct.*, vol. 24, no. 2, p. 950, 2006.
[2] M. Bachmann et al., *J. Vac. Sci. Technol. B*, vol. 38, no. 2, p. 023203, mar 2020.
[3] C. Langer et al., *J. Vac. Sci. Technol. B, Nanotechnol. Microelectron. Mater. Process. Meas. Phenom.*, vol. 34, no. 2, p. 02G107, mar 2016.
[4] W. I. Karain et al., *J. Vac. Sci. Technol. A Vacuum, Surfaces, Film.*, vol. 12, no. 4, pp. 2581–2585, 1994.
[5] S. Edler et al., *J. Appl. Phys.*, vol. 122, no. 12, p. 124503, sep 2017.
[6] C. Prommesberger et al., *IEEE Trans. Electron Devices*, vol. 64, no. 12, pp. 5128–5133, 2017.
[7] R. A. Millikan and C. C. Lauritsen, *Proc. Natl. Acad. Sci.*, vol. 14, no. 1, pp. 45–49, 1928.
[8] R. G. Forbes, *R. Soc. Open Sci.*, vol. 6, no. 12, dec 2019.

Designing Micro-gap Thermionic Energy Harvesters

Ehsanur Rahman[1,2,*] and Alireza Nojeh[1,2]

[1]Department of Electrical and Computer Engineering, University of British Columbia, Vancouver, BC, V6T 1Z4, Canada
[2] Quantum Matter Institute, University of British Columbia, Vancouver, BC, V6T 1Z4, Canada
*Contact: ehsanece@ece.ubc.ca

Abstract— **This work presents a computational study illustrating the complex interactions among various energy transfer mechanisms involved in the thermionic conversion process. It is demonstrated that an optimal gap width exists which leads to maximum device performance. In particular, results relevant to a solar thermionic energy converter are discussed, where the electrode temperatures themselves depend on the gap width.**

I. INTRODUCTION

Static heat-to-electricity converters offer distinct advantages over conventional turbine generators due to their silent operation, lack of moving parts, high specific power, and flexible form factor. These diverse features are particularly attractive for applications that require portability, relatively low maintenance, and high reliability. A prominent example of such static heat engines is thermionic energy converters (TECs). However, a key challenge to implementing a high-performance thermionic energy conversion device is the mitigation of the space charge effect, which results from the mutual Coulombic repulsion of the electrons in the device's interelectrode space. Among various possible configurations of such a TEC device, the micro-gap architecture represents a compact and elegant solution to mitigating the space charge effect. With the advent of microfabrication and nanotechnology, experimental research on micro-gap devices is progressing towards the implementation of chip-scale electrical power generation technology with a power density of several watts per cm^2. However, the optimal design of micro-gap TECs involves a delicate balance among various energy exchange mechanisms, many of which are loss channels and hence need to be minimized.

In this work, we present a comprehensive study of what energy exchange channels are involved in micro-gap TECs and how the strengths of these channels vary with the gap width. To accomplish this, we have developed a self-consistent theoretical model considering the near-field radiative coupling of the electrodes and the space charge effect, as well as other temperature-dependent loss processes, under an overall energy balance formalism. Using this model, we have quantified the gap width dependence of various energy exchange channels, which also reveals the optimal gap width considering the interplay among these energy transfer mechanisms.

II. RESULTS AND DISCUSSION

First, we briefly describe the operation of a TEC with a simple device schematic as shown in fig.1. A conventional thermionic device, in the simplest form, consists of two reservoirs of electrons separated by a vacuum gap between them. One of these reservoirs, i.e. the cathode, is heated using external energy

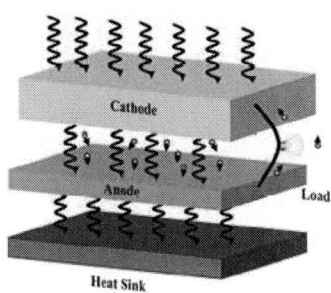

Fig. 1 A simple schematic illustration of the operation of a TEC.

(which might come from various optical or thermal sources) to a higher temperature so that some of its electrons can gain sufficient energy to overcome the vacuum barrier and escape the reservoir. These energetic electrons are collected at the second reservoir (the anode), which is held at a lower temperature, where they thermalize and generate a useful voltage across an electrical load with their remaining energy. Apart from this desirable electron transfer channel, there is also a net photon transfer between the electrodes due to thermal radiation. Moreover, there is parasitic thermal conduction loss via the electrical leads and thermal radiation loss to the surroundings.

For efficient operation of the TEC, one needs to maximize the net electron transfer between the electrodes while minimizing parasitic loss channels. However, such an optimization crucially depends on the vacuum gap width. This is because the thermionic emission barrier for the energized electrons does not solely depend on the material work function of the emitting reservoir. Rather, an additional energy barrier is formed in the vacuum space due to the Coulombic repulsion of the electrons which are transiting the gap region. This space charge barrier opposes further emission of electrons from the cathode and hence can significantly reduce the electrical output from the device if untreated. A simple and efficient solution to remedy this space charge loss is to make the gap between the electrodes smaller. However, maximizing the electron transfer between the electrodes by reducing the gap width does not necessarily lead to optimal conversion performance. This is because the thermal radiation between the electrodes also crucially depends on the gap width and a very small gap can lead to a significant enhancement of the radiative coupling (also known as the near-field effect). Moreover, the temperature of the electrodes strongly depends on the strength of these various energy exchange channels. Thus, the choice of this gap size is not arbitrary. Rather, it plays a fundamental role in optimizing the performance of such a micro-gap TEC and needs to be chosen considering the complex interaction among various energy transfer mechanisms.

To illustrate this effect, in this work, we have utilized a robust computational model that self-consistently determines the electrode temperatures by simulating the physics of various energy exchange processes. Specifically, in our computational model, we have implemented the space charge physics by solving the coupled Poisson-Vlasov equation [1] that governs the electron kinetics in the vacuum gap. As well, the gap dependence of thermal radiative coupling has been implemented using fluctuational electrodynamics [2] that captures the possibilities of photon tunneling at a very small gap. For the illustration of the complex interplay among various energy transfer channels, we have simulated a solar TEC where the solar energy is captured by a selective absorber. The electrodes are made of tungsten with a low work function coating to reduce the thermionic emission barrier. The key results are shown in figs. 2 and 3.

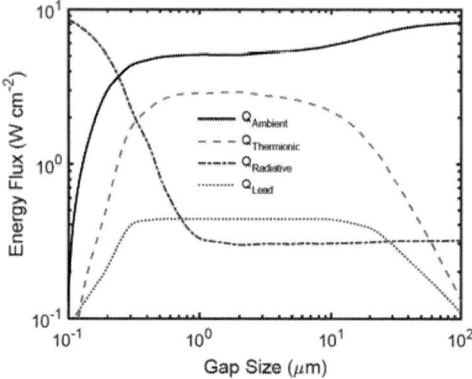

Fig. 2 Various energy transfer channels as a function of the gap size. The data are shown for the AM 1.5 solar spectrum concentrated by a factor of 100.

In fig. 2 we show the various energy exchange channels as a function of the vacuum gap width. Crucially, we see that a very small gap does not necessarily maximize the energy transfer via electrons. (This is counterintuitive given that a smaller gap helps reduce the space charge effect). This is because a very small gap also significantly increases the radiative coupling strength between the electrodes, due to the tunneling of evanescent photons, and thus prevents the rise of the cathode temperature. Therefore, although the space charge effect is reduced at such small gaps, the thermionic electron emission decreases due to lower cathode temperature. As we gradually widen the gap, the radiative coupling strength decreases, which helps raise the cathode temperature. However, this also leads to a stronger space charge effect as the thermionic emission increases at elevated temperature and the electrons spend more time in the gap. Moreover, thermal radiation loss to the ambient monotonically increases with the gap size (due to an increase in cathode temperature) and the thermal conduction via leads shows a local maximum (due to the optimization of the lead

resistance at each gap width). This result shows that the complex interplay among temperature, space charge and radiative coupling leads to an optimal gap size for thermionic conversion. This point is further illustrated in fig. 3 where the output power density is shown as a function of the operating voltage and gap size, where a local maximum is observed for a particular combination of these two important operating parameters.

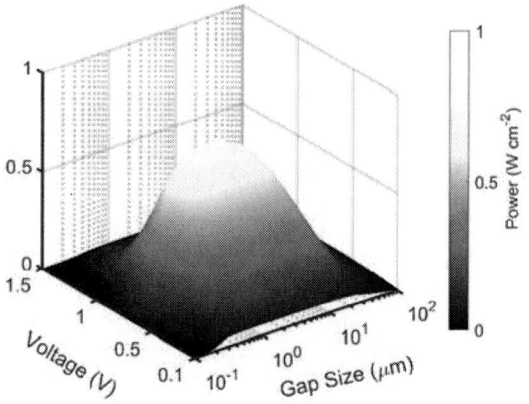

Fig. 3 Output power density as a function of the gap size and operating voltage. The data are shown for the AM 1.5 solar spectrum concentrated by a factor of 100.

III. SUMMARY

We showed how the performance of a TEC is tied to the vacuum gap size dictated by the complex interplay among various energy transfer mechanisms between the cathode and anode. Crucial to the study is the adaptive calculation of the electrode temperatures as the gap is varied. This is particularly important under constant input power, such as in a solar device. Our results show that reducing the gap size to minimize the space charge effect does not necessarily lead to an optimal operation. Rather, an optimal gap size is present due to the net result of this interplay among various energy exchange mechanisms. Therefore, we believe this study may help design the interelectrode gap width in a practical TEC for harvesting thermal and optical energy.

ACKNOWLEDGMENT

We thank the Natural Sciences and Engineering Research Council of Canada and the Canada First Research Excellence Fund, Quantum Materials and Future Technologies Program for supporting this work.

REFERENCES

[1] G. N. Hatsopoulos and E. P. Gyftopoulos, *Thermionic Energy Conversion. Vol. 1: Processes and Devices.* MIT Press, 1973.

[2] M. Francoeur and M. Pinar Mengüç, "Role of fluctuational electrodynamics in near-field radiative heat transfer," *J. Quant. Spectrosc. Radiat. Transf.*, vol. 109, no. 2, pp. 280–293, 2008

Proposal for a Negative Capacitance Vacuum Field Effect Transistors with sub-60mV/dec Subthreshold Swing

N. Hernandez[1], M. Cahay[1], J. Ludwick[2], and T. Back[2]

[1] Spintronics and Vacuum Nanoelectronics Laboratory, University of Cincinnati, Cincinnati, OH 45221, USA
[2] Air Force Research Laboratory, Materials and Manufacturing Directorate, Wright-Patterson Air Force Base, OH 45433, USA

Abstract — **Over the last few years, there has been increased interest in Vacuum Field Effect Transistors (VFETs) with nanoscale features [1-4]. Compared to traditional MOSFETs and FINFETs, VFETs offer advantages in terms of electron velocity, fast switching speed, extreme operating temperatures, and radiation robustness. We calculate the field emission (FE) characteristics of a back-gated nanomembrane VFET making use of the concept of negative capacitance. It is found that the proposed negative capacitance VFET (NC-VFET) has a minimum subthreshold swing (SS) of 34mV/decade, well below the thermionic limit of 60mV/dec at room temperature.**

Keywords—Electron Field emission; Negative Capacitance; Electronstatic Potential; Subthreshold swing; Two Dimensional Monolayer Transition Metal Dichalcogenide Semiconductor Channel .

I. INTRODUCTION/BACKGROUND

Field effect transistors must have a large ratio between the ON and OFF drain currents I_d for adequate performance and power consumption (high I_{ON} / low I_{OFF}). The Boltzmann distribution of electron energies in the source leads to a fundamental lower limit (60 mV/ decade at room temperature) of the SS. Hereafter, we investigate the possibility to reduce the SS of a VFET below this limit by introducing negative capacitance in the device with the use of a ferroelectric layer in the gate stack [5].

We have developed a semi-analytical model of the current-voltage characteristics of the NC-VFET depicted in Fig. 1. The latter makes use of a back-gated 2D transition metal dichalcogenide semiconductor channel. Our model is an extension of the work of Jiang et al. [6] to include the presence of a vacuum gap in the channel with opening in the range of 35-50nm which can be defined by e-beam lithography. The gate stack of the device includes heavily doped Si as a gate electrode, few tens of nm thick HZO (Hafnium Zinc Oxide) as the ferroelectric capacitor, and a few nm thick Al_2O_3 as capping layer and capacitance matching layer.

In the NC-VFET, a portion of the nanomembrane between the source and drain is etched away leading to FE from the edge of the two-dimensional semiconductor across the vacuum gap (region of length D in Fig. 1). The FE characteristics of the NC-VEFT are then obtained by forcing the drift diffusion current through the nanomembrane to match the current emitted across the gap. As in the case of the NC-FET

investigated by Si et al. [7], we anticipate that the presence of negative capacitance in the device will allow a tuning of the subthreshold current over a smaller range of gate voltages, leading to a SS below 60mV/decade.

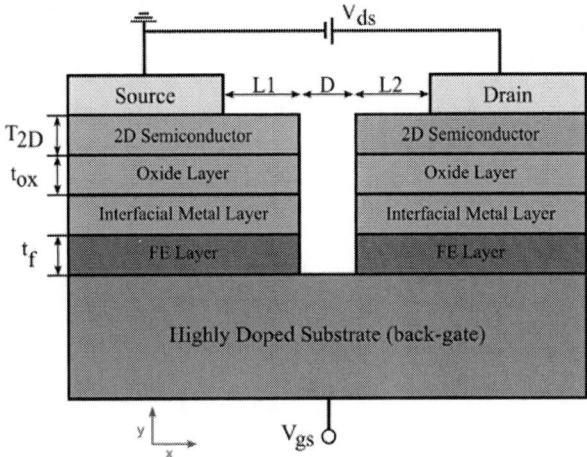

Fig.1: Cross-sectional diagram of the proposed back-gated nanomembrane negative capacitance vacuum field effect transistor (NC-VFET) where a portion of the nanomembrane has been etched away leading to FE from the two-dimensional semiconductor into the vacuum space between the source and drain contacts.

The NC-VFETs were simulated over a wide range of device dimensions and biasing conditions. As shown next, by varying the dimensions of the devices, the gap spacing and location between the source and drain contacts, and the biasing conditions, we show the SS of the proposed NC-VFETs can be substantially reduced below 60mV/decade.

II. RESULTS AND DISCUSSION

The drain current vs. voltage relationship must be determined with an iterative process in which we match the current calculated using the Pao-Sah integral current formulation of the drift-diffusion current through the channel [8] to the emission current emitted at the edge of the 2D semiconductor channel.

$$I_{DD} = \mu \frac{W}{L} \phi \left\{ \frac{\epsilon_{ox}}{T_{ox}} \left[\frac{k_B T}{q} + (V_{mos} - V_{fb0}) - \frac{\phi}{2} \right] + N_d q \right\}_{\phi\,(x=0)}^{\phi\,(x=L_n)}$$

Where k_B is the Boltzmann's constant, T is the temperature, q is the elementary charge, ϵ_{ox} and T_{ox} are the permittivity and thickness of the oxide, W and L are the width and length of the channel, respectively, μ is the effective mobility, V_{fb0} is the flatband voltage, V_{mos} is the electrostatic potential of the interfacial metal gate, N_d is the Areal doping of the channel, and ϕ is the electrostatic potential along the channel.

Assuming that this long-channel device can be described by the gradual channel approximation, we can get an equation describing the electrostatic potential, ϕ, anywhere along the channel. The channel quasi Fermi potential V_{ch} is defined with the following boundary conditions: the source is grounded, V_{L1} and V_{L2} are the electrostatic potentials at the source and drain ends of the vacuum gap, respectively, and V_{ds} is the potential at the drain contact. The channel charge in the NC-VFET is calculated with the Ward-Dutton charge partitioning method [9]. These charges are used in conjunction with the 1D steady-state Landau-Khalatnikov equation [10] to derive the voltage across the ferroelectric.

The FE characteristics of the NC-VFET were calculated by using a numerically efficient Linear Interpolation Lookup Table [11] for a wide range of device dimensions and applied biases. Figure 2 shows the SS dependence on the FE current for a NC-FVFET with the material and device parameters listed in the Table below.

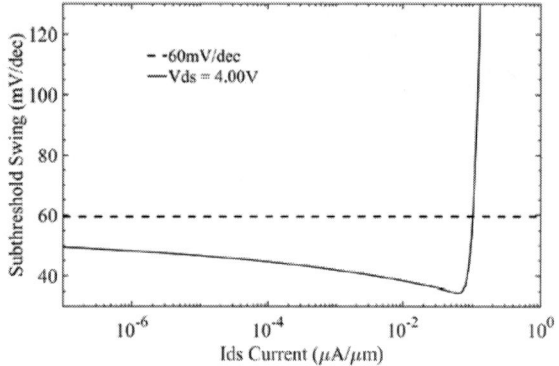

Fig.2: Numerical Solution for the SS of an asymmetrical 2D NC-VFET with a 35nm vacuum gap and a drain voltage Vds of 4.0V. A minimum SS of 34mV/decade is found. The dashed line represents the SS thermionic limit of 60mV/decade at room temperature.

Table 1: List of the material and device parameter of a specific NC-VFET, as schematically shown in Fig. 1

Parameter	Values	Units
Length	1	[μm]
Parasitic Capacitance	3.54e-9	[$CV^{-1}m^{-1}$]
Asymmetrical Cut Offset	70	$L_1/(L_1+L_2)$ [%]
Vacuum Gap Distance	35	[nm]
Field Enhancement Factor	35	
Semiconductor	**β-Ga$_2$O$_3$**	
Thickness	10	[nm]
Relative Permittivity	9.5	[$\varepsilon_r / \varepsilon_0$]

Flatband Voltage	1.1	[V]
Electron Mobility	100	[$cm^2V^{-1}s^{-1}$]
Areal Doping	2.8e16	[m^{-2}]
Effective Electron Mass	0.34	[m_e^*/m_0]
Work Function	4.8	[eV]
Oxide	**Al$_2$O$_3$**	
Thickness	2	[nm]
Relative Permittivity	11	[$\varepsilon_r / \varepsilon_0$]
Ferroelectric	**H$_{0.5}$Z$_{0.5}$O$_2$**	
Thickness	20	[nm]
Alpha	-1.911e8	[VmC^{-1}]
Beta	5.898e9	[Vm^3C^{-2}]
Gamma	0	[Vm^5C^{-3}]

A minimum of 34mV/decade has been found for the range of parameters and biasing conditions investigated so far. The addition of the nanoscale vacuum gap in the NC-FVET shown in Fig.1 requires an increase in drain to source voltage for drain to source current values comparable to those obtained in regular Field Effect Transistors.

REFERENCES

[1] A. Demming, Vacuum Technology Comeback Immunizes Nanoelectronics From Radiation, Physics World, IOP Publishing, 31 Aug 2013.

[2] I.J. Park, S.-G. Jeon, and C. Shin, A New Slit-Type Vacuum-Channel Transistor, IEEE Transactions on Electron Devices 61, 4186-4191 (2014).

[3] J. Han and M. Meyyappan, Introducing the Vacuum Transistor: A Device Made of Nothing, IEEE Spectrum, 23 Jun 2014.

[4] J.-W. Han, D.-I Moon, and M. Meyyappan, Nanoscale vacuum channel transistor, Nano Letters 17, 2146-2151 (2017).

[5] V. V. Zhirnov and R. K. Cavin, "Negative capacitance to the rescue?," Nat. Nanotechnol. 3, 77–78 (2008).

[6] C. Jiang, M. Si, R. Liang, J. Xu, P.D. Ye, and M.A. Alam, A Closed Form Analytical Model Of Back-Gated 2d Semiconductor Negative Capacitance Field Effect Transistors, Journal of the Electron Device Society 6, 189-194 (2018)

[7] M. Si, L. Yang, H. Zhu, and P.D. Ye, "β-Ga2O3 Nanomembrane Negative Capacitance Field-Effect Transistors with Steep Subthreshold Slope for Wide Band Gap Logic Applications ", ACS Omega 2, 7136-7140 (2017).

[8] Pao, H. C., and C. T. Sah. 1966. "Effects of Diffusion Current on Characteristics of Metal-Oxide (Insulator)-Semiconductor Transistors." Solid-State Electronics 9 (10): 927–37.

[9] Ward, D. E., and R. W. Dutton. 1978. "A Charge-Oriented Model for MOS Transistor Capacitances." IEEE Journal of Solid-State Circuits 13 (5): 703–8.

[10] L. D. Landau and I. M. Khalatnikov, "On the anomalous absorption of sound near a second order phase transition point," Dokl. Akad. Nauk SSSR, vol. 96, pp. 469–472, 1954.

[11] J. Ludwick, M. Cahay, H. Hall, J. O'Mara, and T.C. Back, "A New Universal Method for Rapid Semi-Numerical Determination of Local Emission Current Density from Nanoscale Emitters", submitted for presentation at IVNC2021, Lyon, France.

Direct in situ Electron Microscope synthesis of CNTs with applied electric Field and Field Emission

P. Vincent[1], F. Panciera[2], I. Florea[3], M. Ezzedine[3], M.-R. Zamfir[3], S. Perisanu[1], C. Cojocaru[3], N. Blanchard[1], D. Pribat[3], S.T. Purcell[1], P. Legagneux[4]

[1] Univ. Lyon, Univ. Claude Bernard Lyon 1, CNRS, Institut Lumière Matière, F-69622, VILLEURBANNE, France.
[2] University of Paris-Saclay, CNRS, Centre for Nanoscience and Nanotechnology, 91120 Palaiseau, France
[3] LPICM, CNRS, Ecole Polytechnique, IP Paris, 91128, Palaiseau Cedex, France
[4] Thales Research and Technology, 91767 Palaiseau, France
*Contact: pascal.vincent@univ-lyon1.fr,

Abstract— **This work reports experiments on the oriented growth by electric field of free-standing carbon nanotubes(CNTs) by chemical vapor deposition (CVD) for field emission (FE) applications. The growths were observed in a scanning electron microscope (SEM) and in an environmental transmission electron microscope (ETEM) with the ETEM growths observed in real time. The effects of various applied voltage during growth will be presented. For high voltages, videos show that the maximum length of CNTs are limited either by the mechanical breakdown or FE induced thermal evaporation. Finally, the I(V) curves and current densities obtained will be presented with a focus on the different destruction mechanisms.**

I. INTRODUCTION

The principal hurdle for the successful deployment of carbon nanotubes (CNTs) into many of the numerous industrial applications for which this quintessential nanoscience species has been proposed has always been, and remains, the mastering of selected CNT growth of types specific to each application. This is in no small part because of the difficulty in observing time-resolved growth at the necessary nanometer and atomic scales and conjointly of disposing of useful tools for controlling such growth. This article describes our recently achieved breakthrough in the dynamical observation of individual CNT growth on micro-machined chip heaters in Environmental Transmission and Scanning Electron Microscopes (ETEM and SEM). A key innovative aspect is that applied electric fields directs the systematic fabrication of extremely straight and preferentially single-wall nanotubes. This breakthrough allows an unprecedented access to the growth processes of highest quality CNTs, in situ post growth modification and eventually to the metrology of a wide spectrum of their physical properties.

II. EXPERIMENTS

In situ –real time growths (C_2H_2, $10^{-4} – 10^{-3}$ mbar, 700° C) were carried out across a polarizable capacitor gap of specifically-designed micro-chip sample holders. The growth rate varies linearly with the C_2H_2 partial pressure and the growth temperature is very critical. The CNTs are extremely straight over the 2 μm gap used to apply the electric field which we take as an indication of an absence of defects. Preliminary high resolution TEM images of the CNTs near their bases show them to be predominantly single wall carbon nanotubes (SWCNTs).

Fig. 1 a) Observation of SWNTs grown in SEM at low voltage between electrodes (4 Volt). (the SWNT appear as bright lines perpendicular to the electrodes). The growth has been directed by the electric field as TEM videos will show b) Post synthesis TEM observation of the upper part in a) (delimited by the black square). c) Growth on another part of the sample. Carbon nanotubes are visible as straight lines with contrast inversion in the image.

978-1-6654-2590-2/21 $31.00 © 2021 IEEE

Nanotube are still oriented by the electric field and length up to 10 μm are obtained.

Depending the voltage applied during the growth we can distinguish two regimes. At low voltage (typically 4 Volt between the 2-3 μm gap) FE cannot occur and the electrostatic forces are relatively weak. Interestingly we observed that growth of SWNTs is directed by the electric field even at these low voltage values and the nanotube grow until they reach and stick to the opposite electrode (see figure 1).

At higher voltages (between 10 and 50 Volt) the electrostatic forces are stronger and FE can occur for the SWCNTs in the gap. In this configuration the nanotube length is limited to around (1- 1.5 μm) due to mechanical breakdown or thermally activated FE evaporation and we did not observe nanotubes crossing the gap.

After the growth we carried out I(V) measurements of the SWCNTs in order to measure the emitted current and we observe d a maximum field emission current of about 20 μA.

III. CONCLUSIONS

In this talk we will present results and videos on the growth and FE characterization of these cathodes. These experiments show the potentialities of such micro-machined heater in ETEM to better understand the growth and FE properties of CNT cathodes

ACKNOWLEDGMENT

The authors acknowledge financial support from the French state managed by the National Research Agency under the Investments for the Future program under the references ANR-10-EQPX-50 pole NanoMax and the 3DRX-online project. The authors would also like to acknowledge the Centre Interdisciplinaire de Microscopie électronique de l'X (CIMEX) and the Plateforme Nanofils et Nanotubes Lyonnaise of the University Lyon1.

Effect of Substrate Conductivity on Si Self-Assembled Field Emission Arrays

Shabnam Ghotbi*+, Saeed Mohammadi*

*School of Electrical and Computer Engineering, Purdue University, West Lafayette, IN, 47907, USA

+Corresponding author: sghotbi@purdue.edu

Abstract— **Ungated high-density Si field emitter arrays are fabricated using a self-assembly technique based on the Langmuir-Blodgett deposition method. A uniform monolayer of silica nanoparticles is deposited to replace the lithography steps prior to etching vertical Si nanowires to form field emission arrays. High-density arrays are fabricated on Si substrates with two different substrate conductivities. It is observed that both types of arrays operate at low turn-on voltages (<40V) with a measured maximum current of 21 mA limited by measurement capability. Such high current is achieved at average electric fields of 2.5 V/μm and 7 V/μm for the substrate conductivities of 0.2 S.cm⁻¹ and 3.3 mS.cm⁻¹, respectively. The fabricated low operating voltage field emission arrays achieve a maximum current density of 13 A/cm².**

I. Introduction

Vacuum field-effect transistors have the potential to achieve superior performance over their semiconductor counterparts if they can operate reliably under high current densities. Lattice scattering does not occur in the vacuum facilitating ballistic transport. Accordingly, vacuum transistors are expected to achieve high speed and higher breakdown voltages. These transistors are great candidates for high-power mm-wave applications, in the design space that semiconductor devices including GaN-based transistors have difficulty providing high powers (100's of Watts) at mm-wave frequencies. Vacuum devices are also expected to be more tolerant of radiations, which is desired for military and space applications. In addition to vacuum transistors, field emission devices have been used in field emission displays, sensors, detectors, and cold-cathode electron sources for X-Rays and microwave high power amplifiers including Travelling Wave Tubes (TWTs), Magnetrons, and Klystrons [1]. In the late 1920s, Fowler and Nordheim demonstrated that instead of applying high temperatures to generate currents through thermionic emission, high external electric fields can be employed for extracting electrons from the surface of conductors [2]. Applying high electric fields not only decreases the potential barrier height but also shrinks the width of potential barrier which results in electron tunnelling and generating field emission current. To obtain high currents at lower voltages, sharp emitter tips with lower work function materials are desired. The general Fowler-Nordheim (FN) equation can be simplified as:

$$I = A.V^2 \exp\left(\frac{-B}{V}\right) \qquad (1)$$

$$A = \frac{1.56 \times 10^{-6}.S.\beta^2}{1.1\varphi} \exp\left(\frac{10.4}{\sqrt{\varphi}}\right) \qquad (2)$$

$$B = 6.44 \times 10^7 \varphi^{\frac{3}{2}}/\beta \qquad (3)$$

where S is the effective emitting area, φ is the material work function, and β is the field enhancement factor. The higher the β is, the lower the voltage is needed to get the same field strength. However, the field enhancement cannot be measured directly, and to deduce this factor, field emission currents in FN coordinates are plotted. From the FN slope and by knowing the work function of emitters, field enhancement factor is derived. In the following section, the fabrication methodology is presented, and section III demonstrates the measurement and simulation results.

II. Fabrication

Fabrication of field emission devices generally involves multiple optical and E-beam lithography steps, which can be time-consuming and expensive. In [3], we proposed the fabrication method of field emitter arrays using the Langmuir-Blodgett method for deposition of a monolayer of silica particles, which later behave as hard masks against chlorine in the deep reactive ion etching (DRIE) process. In this work, field emission arrays were formed on silicon substrates with two different conductivities of 0.2 S.cm⁻¹ and 3.3 mS.cm⁻¹. For both types of devices, the average length of field emitters is 1.5 μm while the diameter is about 200 nm (Fig.1.). To limit the emission area, 10 nm of aluminum oxide is deposited on the silicon substrates using atomic layer deposition (ALD) technique; then, the substrates are patterned to expose a 400 μm by 400 μm silicon window to etchant gases prior to LB deposition. The spacing between field emitters is controlled by the surface modification of particles and the silica drops volume during the LB process. The spacing can be further increased by shrinking silica particles in the plasma etcher using CF₄ chemistry. The silica particles and the thin oxide layer are stripped away prior to the anode attachment and measurements.

Fig 1. Si field emission array

III. MEASUREMENTS AND SIMULATIONS

A 10 μm film is deposited and patterned on the cathode plate to act as a spacer between the anode and cathode. To obtain IV characteristics, a Keithley 2410 source meter is employed. During the measurements, the devices went under 10^{-5} $Torr$ pressure. The typical IV plots of field emission arrays for low conductivity and high conductivity substrates are shown in Fig.2-a. The turn-on voltage for the low conductivity substrate (in green) is around 40 V, while for the high conductivity substrate, this value reduces to less than 5 V (in grey). IV graphs are also plotted in FN coordinates in Fig.2-b and Fig.2-c with linear fitted lines for higher and lower conductivity nanowires, respectively.

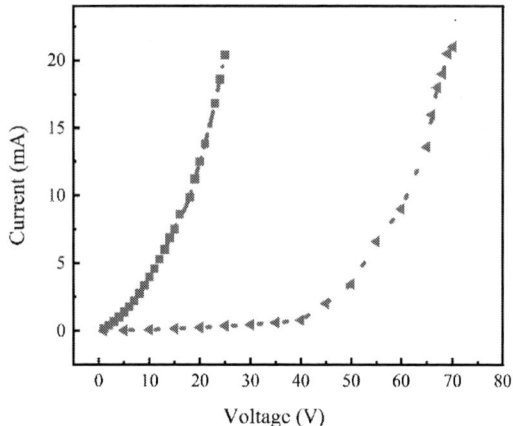

Fig.2-a: IV characteristics for high conductivity (left) and low conductivity (right) substrates

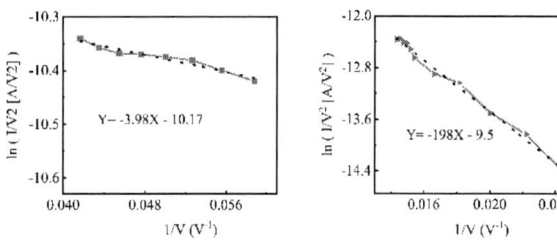

Fig.2-b: FN plot for anode current of the high conductivity substrate   Fig.2-c: FN plot for anode current of the low conductivity substrate

The FN plots have negative slopes of 198 V and 3.98 V for low and high conductivity substrates, respectively. These values correspond to field enhancement factors of 13.6×10^5 cm^{-1} and 12.9×10^7 cm^{-1}, respectively assuming the work functions are about 4 eV. Fig. 3 shows the simulation results for a single nanowire with different substrate conductivities. The average current per nanowire is also calculated from the measurement assuming 1 M nanowires are emitting at the same time. The simulation results for the single nanowire with low conductivity show almost the same turn-on voltage compared to the measurement result. For the high conductivity nanowires, the measurement results show a significantly lower turn-on voltage. This difference is under investigation.

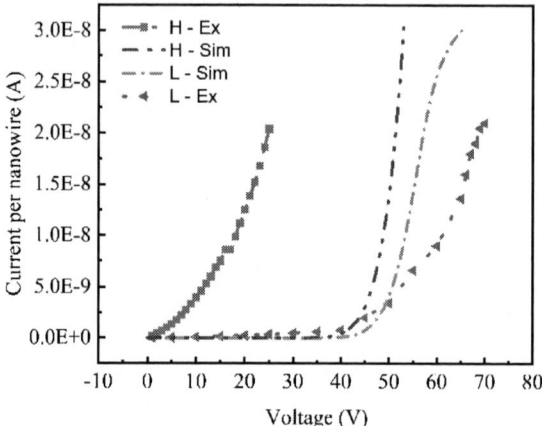

Fig. 3. I-V characteristics of field emission nanowires with substrate conductivity of 0.2 S.cm^{-1} (left trace) and 3.3 mS.cm^{-1} (right trace)

IV. CONCLUSIONS

In this work, self-assembled silicon field emission arrays with two different substrate conductivities were fabricated and the simulation results, as well as the IV measurements, were presented. The simulation result confirmed the measurement results for low conductivity arrays. Using the Ball in Sphere model, the field enhancement factor was expected to be around 2×10^5 cm^{-1}. However, for the field emitters on the high conductivity substrate, the measured slope of the FN plot was significantly lower than the expected value, and the turn-on voltage was smaller than the simulation result. While all the simulation results were performed with the assumption of no electrical interactions between the nanowires, this assumption may not be necessarily valid. More investigations will be conducted to validate the experimental results.

ACKNOWLEDGMENT

The authors would like to acknowledge the funding support from the Air Force Office of Scientific Research under the FERVIN program.

REFERENCES

[1] S. Ghotbi, H. Pajouhi and S. Mohammadi, "A Vacuum Multi-Finger Transistor in CMOS Technology," 2018 76th Device Research Conference (DRC), 2018, pp. 1-2, doi: 10.1109/DRC.2018.8442158.

[2] R. H. Fowler, and L. Nordheim. "Electron emission in intense electric fields." Proceedings of the Royal Society of London. Series A, Containing Papers of a Mathematical and Physical Character 119, no. 781 (1928): 173-181.

[3] S. Ghotbi, N. Opondo, and S. Mohammadi. "A Self-Assembled High Current Si Field Emitter Array", 22nd International Vacuum Electronics Conference, Netherlands, (2021).

Gap in pagination due to withheld papers.

Pages 105-108

Field emitters at atomic scale – insights from order-N density functional theory

C. J. Edgcombe*
Department of Physics, University of Cambridge
*Contact: cje1@cam.ac.uk

I. INTRODUCTION

The description of field emission given by Fowler and Nordheim [1] and developed by later workers (eg [2],[3]), assumes a unidirectional electric field, implying that the emitter surface is smooth and planar. However, emitters used in practice have pointed forms that may be specified down to individual atoms. To make accurate predictions about emitted current and its spatial distribution, it is desirable to have more information about the atomic structure at the emitter surface, the 3D distribution of states and the potential distribution there. Previous work on emission at atomic scale includes that by Han and Ihm [4] and Li [5]. A recent survey by Lepetit [6] describes much further work.

Widely known DFT codes such as Castep and Vasp use plane-wave basis functions which are not convenient for representing pointed structures at atomic scale. Their solution time scales as N^3 for solving structures with N electrons. These properties limit the size of structures that can be analysed swiftly with these codes. Here we describe some results obtained using the code Onetep, which uses basis functions that adapt their shape to model individual orbitals. The code is designed so that much of the solution time scales as N, enabling the modelling of structures containing hundreds of electrons. The results show the distributions of charge density, orbitals and potential for a range of externally applied field.

II. METHOD

The code Onetep gives details of orbitals in the Kohn-Sham ground state. but does not calculate field-emitted current. The code uses adaptive basis functions that model one orbital per electron in the cell. In collaboration with members of the Condensed Matter group at Cambridge University Department of Physics, the code was modified to solve a non-periodic cell, with boundary conditions provided by a separate macroscopic calculation [7]. Our main study was a section of a closed (5,5) carbon nanotube (CNT) of 150 carbon nuclei, terminated with 10 hydrogens and containing 610 valence electrons. The surroundings were modelled by a pair of parallel conducting plates, with the CNT connected perpendicularly to the cathode of the pair. Applying an external field induces charge into the (connected) CNT. This is modelled in the (isolated) DFT cell, a cube of side 5 nm, by inserting charge found by integrating the normal field over the cell boundary surface. This produced good agreement with the macroscopic potential distribution over a wide range of background electric field.

We have also made some initial trials of the same method with a structure consisting of the apex of a 4-faced tungsten (W) pyramid, with 6 pyramidal layers and a smaller base layer. The cell contained 116 W atoms terminated with 36 H atoms, and 1660 orbitals were calculated for the outer 14 electrons of each W atom.

III. RESULTS

Surfaces of constant charge density for the CNT in zero external field (Fig. 1) show the concentration of charge around the C cores, surrounded by a sheath of lower-density electronic charge extending to the equipotential surface for the Fermi level. When accelerating field is applied, the solution provides the additional induced charge by increasing the number of orbitals up to the Fermi level, and the Local Density of States (LDoS) distribution changes accordingly (Fig. 3). Calculation for a longer section of CNT shows a more metallic LDoS, resulting from the more realistic number of unbound electrons in the cell.

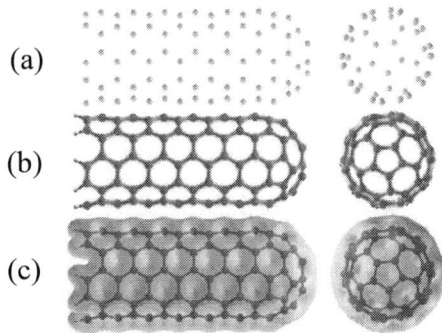

Fig 1. Surfaces of electron density with zero external field, at (a) 1%, (b) 0.25% and (c) 0.02% of maximum charge density;

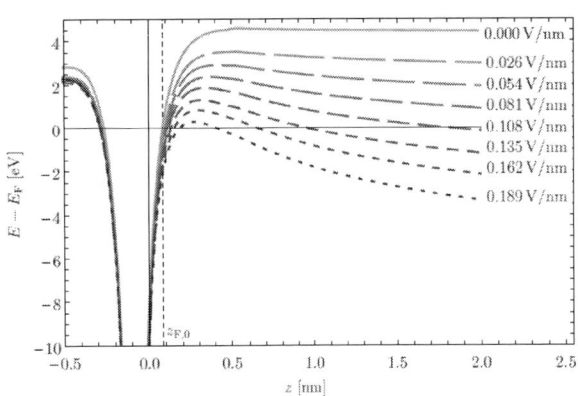

Fig. 2 Potential energy at 100K along the axis of the CNT, for 8 values of applied field. The vertical dashed line shows the axial location of the Fermi level in zero field.

In the CNT, the potential distribution varies differently with applied field on the two sides of the equipotential surface defined by the Fermi level (Fig. 2). Inside the cap ($z < 0$), the potential varies little with the external field, consistent with

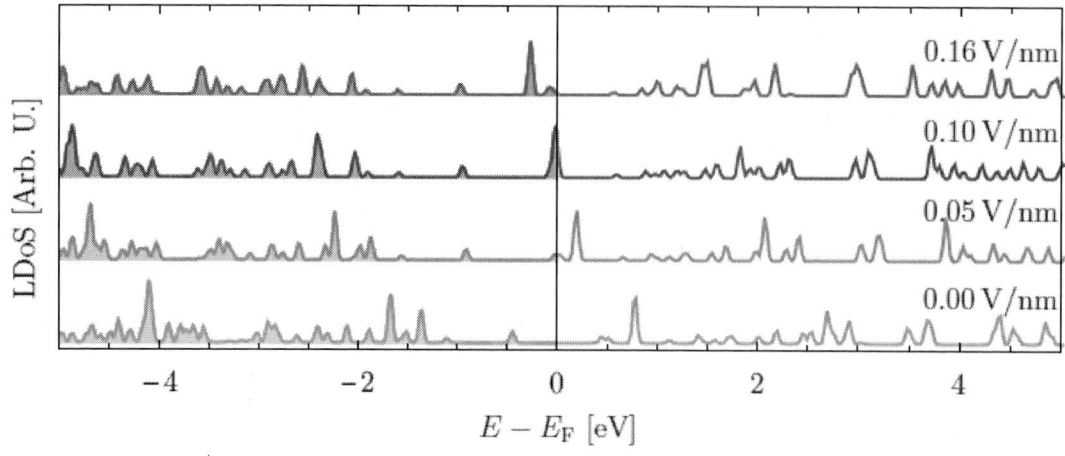

Fig. 3. Variation with external field of the LDoS for the CNT.

charge being free to move along the outer sheath of the CNT. Outside the cap, the barrier is reduced by accelerating field as expected, but the physical location of the equipotential at Fermi level remains close to its position for zero field. This equipotential surface thus forms a natural definition for 'the surface' of the CNT. With this definition, the diameter of the CNT is about 0.37 nm greater than the diameter of the tube of carbon ions forming the skeleton of the CNT.

The work function emerges readily from the calculation as the difference between the Fermi level and vacuum level in zero field. A realistic value is found only when a suitable exchange functional is included [8].

The pyramidal W system did not converge as well as the CNT, so the results are probably less accurate. However, in zero applied field, a conducting electronic sheath like that for the CNT is seen, with charge density in the W sheath over 100 times that in the CNT. In an external field, charge concentrates along the ridges of the pyramid as expected (Fig. 4(a)), but in the occupied orbital nearest to the Fermi level (Fig. 4(b)), the maxima are more localised. Since the system has 4-faced pyramidal symmetry, this orbital has zero amplitude on the axis.

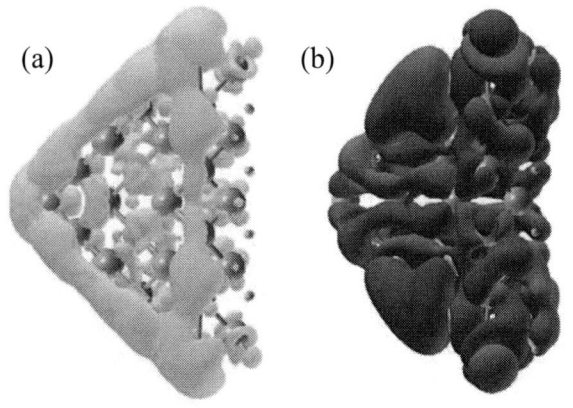

Fig. 4 (a) Change in electron **density** on application of external field of 0.162 V/nm; (b) plot of surface of constant **amplitude** ± 0.003 (e⁻/A²)^(1/2) for occupied orbital nearest the Fermi level, in zero applied field.

IV. CONCLUSIONS

Use of the Onetep code with adaptive basis functions has shown that electronic charge in a CNT extends outside the skeleton of carbon cores as far as the equipotential at the Fermi level, which forms the surface of the CNT. The occupation of Kohn-Sham orbitals changes with the external field, implying that the spatial distribution of emitted current density may also vary with external field.

ACKNOWLEDGMENTS

The results shown were obtained by S. M. Masur in the course of her work at Cambridge towards the PhD degree. All figures shown here are from her thesis and are used with permission.

This research received funding from the People Programme (Marie Curie Actions) of the European Union's Seventh Framework Programme FP7/2007–2013/ under REA grant agreement no. 606988. We are also grateful to Thermo Fisher Scientific for further financial support.

REFERENCES

[1] R H Fowler & L Nordheim, *"Electron emission in intense electron fields"*, Proc. Roy. Soc. A 119 (1928), 173-81, doi: 10.1098/rspa.1928.0091

[2] E L Murphy, R H Good, *"Thermionic emission, field emission and the transition region"*, Phys. Rev. 102 (1956), 1464–1473; doi: 10.1103/PhysRev.102.1464

[3] R G Forbes *"The Murphy–Good plot: a better method of analysing field emission data"*, R. Soc. Open Sci. 6: (2019) 190912; http://dx.doi.org/10.1098/rsos.190912

[4] S. Han and J. Ihm, *"First-principles study of field emission of carbon nanotubes"*, Phys. Rev. B 66 (2002), 241402, DOI: 10.1103/PhysRevB.66.241402

[5] Z. Li, *"Density functional theory for field emission from carbon nano-structures"*, Ultramicroscopy 159 (2015) 162–172, http://dx.doi.org/10.1016/j.ultramic.2015.02.012

[6] B. Lepetit; *"Electronic field emission models beyond the Fowler-Nordheim one"*; Journal of Applied Physics 122, 215105 (2017); doi.org/10.1063/1.5009064

[7] C.J. Edgcombe, S.M. Masur, E.B. Linscott, J.A.J. Whaley-Baldwin, and C.H.W. Barnes, *"Analysis of a capped carbon nanotube by linear-scaling density-functional theory"*, Ultramicroscopy 198 (2019) 26–32; https://doi.org/10.1016/j.ultramic.2018.11.007

[8] S.M. Masur and C.J. Edgcombe, *"Modelling a capped carbon nanotube by linear-scaling density-functional theory"*, J Electron Spectroscopy 241 (2020), 146896.

Thermal-Field Electron Emission from Three-Dimensional Cd$_3$As$_2$

Wei Jie Chan[#], Yee Sin Ang[*], and L. K. Ang[&]

Science, Math and Technology, Singapore University of Design and Technology,
8 Somapah Road, Singapore 487372
[#]Contact: weijie2_chan@sutd.edu.sg
[*]Contact: yeesin_ang@sutd.edu.sg
[&]Contact: ricky_ang@sutd.edu.sg

Abstract — **The emergence of quantum materials with non-parabolic dispersions like graphene have signal the beginning of a paradigm shift in the choice of materials for field emission. Like graphene, the unconventional Dirac conic band structure in cadmium arsenide Cd$_3$As$_2$, a 3D topological Dirac semimetal (TDS) hints a possibility of it being an excellent field emitter. This differing band structure allows the appearance of a non-trivial dual peak feature in its total energy distribution (TED), which can be utilized to garner a larger current density. The commonly used Murphy and Good (MG) model for thermal-field emission could not accurately model Cd$_3$As$_2$ due to its F^3 scaling law in the cold field emission limit. These findings show that Cd$_3$As$_2$, or other similar 3D TDS, has the potential to achieve a low turn-on field, which is a highly sought-after property in field emission physics.**

I. INTRODUCTION

Dirac materials like graphene have been emerging in the study of electron emission physics due to its excellent field emission properties [1], [2]. However, there are many challenges in efficiently fabricating a flat (2D) graphene-based field emitter [2]. This motivates us to source for bulk alternatives amongst Dirac materials that can also exhibit excellent field emission properties.

Cadmium arsenide (Cd$_3$As$_2$), a 3D topological Dirac semimetal (TDS), shares similarities with graphene in its band structure and electronic properties [3], which makes it attractive. Cd$_3$As$_2$ was also predicted to display a non-trivial feature in its total energy distribution (TED), which can be utilized to extract a larger current density [4] during thermal-field emission due to its high fermi level [3]. Additionally, a F^3 scaling law can also be realized for cold field emission. These are due to its Dirac conic band structure, which alters the energy spectrum in the TED when varying the applied field, temperature and fermi level. These reasons reinforce that Cd$_3$As$_2$ can be a promising material for field emission.

In this work, the non-trivial dual peak feature in the TED of Cd$_3$As$_2$ are compared against traditional metals like Tungsten, W and Platinum, Pt. Next, the interplay of this feature and the

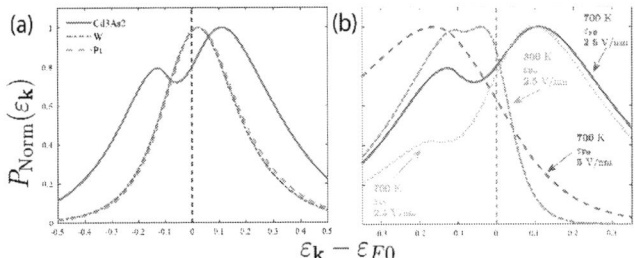

Fig. 1 The normalized (with respect to its peak) TED of metals, Tungsten (W) and Platinum (Pt) in blue dash dotted and green dotted line respectively, and cadmium arsenide (Cd3As2) in red solid line, under $T = 700$ K, $\varepsilon_F = \varepsilon_{F0}$ and $F = 2.5 V/nm$ in (a). The normalized TED is shown as temperature, applied field and the fermi level of Cd3As2 are varied in (b). The intrinsic fermi level is $\varepsilon_{F0} = 0.1$ eV for Cd3As2 and $\varepsilon_{F1} = 0.15$ eV. The black dotted lines here represent the line of ε_{F0} scaled to 0 for these three materials.

emission current density is examined under the variation of temperature, applied field and fermi level. This allow us to extract the key parameters needed to efficiently draw out the largest possible current density. Lastly, we will also be examining the scaling law of Cd$_3$As$_2$ for cold field emission, to determine if the MG equation can be directly applied for Cd$_3$As$_2$. These findings can help pave the wave for not just Cd$_3$As$_2$ as a potential field emitter but also other newly discovered 3D TDS in the future.

II. TOTAL ENERGY DISTRIBUTION AND EMISSION CURRENT DENSITY EQUATIONS

Due to the unconventional band structure of 3D TDS, the TED and current density through a Schottky-Nordheim (SN) barrier [5] have an unconventional form. By taking a single degenerate Dirac cone for emission in a 3D TDS [4], the corresponding equations are

$$P^{TDS}(\varepsilon_k) = \frac{c_{SM}D_F F^2}{\Phi t^2} \exp\left(-\frac{\varepsilon_F}{d_F}\right) f(\varepsilon_k)\lambda\left(\frac{\varepsilon_k}{d_F}\right), \quad (1)$$

$$J_\perp^{TDS} = \frac{c_{SM}D_F F^2}{\Phi t^2} \exp\left(-\frac{\varepsilon_F}{d_F}\right) \int_{-\infty}^{\infty} f(\varepsilon_k)\lambda\left(\frac{\varepsilon_k}{d_F}\right) d\varepsilon_k. \quad (2)$$

978-1-6654-2590-2/21 $31.00 © 2021 IEEE

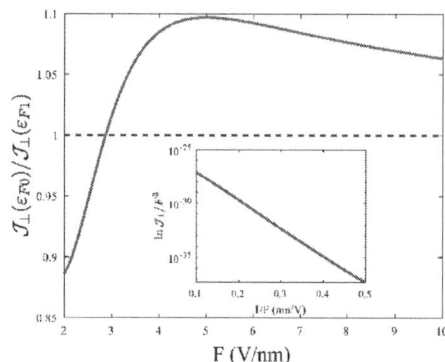

Fig. 2 The ratio of the emission current density with 0.1 eV (ε_{F0}) against 0.15 eV (ε_{F1}) using (4) is shown using a red solid line. The black dotted line represents the point when both these emission current densities are equal. The inset is the F^3 scaling of Cd_3As_2.

Where $c_{SM} = e^3 g/(32 m_e \pi^2 \hbar v_{Fx} v_{Fy})$ is a constant with g being the spin and node degeneracies, D_F is the tunneling exponent constant, Φ is the work function, d_F is the decay width constant, k_B is the Boltzmann constant, t is an image charge correction term, $f(\varepsilon_k)$ is the Fermi-Dirac distribution with m_e being the electron mass, and $\lambda(\varepsilon_k/d_F)$ is a dimensionless tunneling term. Unlike metals, a F^3 scaling law can be achieved in the cold field emission limit ($T = 0$),

$$\ln\left(\frac{J_{\perp,T=0}^{TDS}}{F^3}\right) = -\frac{b_{FN} v \sqrt{\Phi^3}}{F} + \ln\left(\frac{2 c_{SM} \sqrt{\Phi}}{3 b_{FN} t^3}\right)$$

$$+ \ln\left(2\Gamma_- + \frac{\varepsilon_F}{d_F}\Gamma_+\right). \tag{3}$$

Where $b_{FN} = 4\sqrt{2 m_e}/(3 e \hbar)$ is a constant and $\Gamma_\pm = 2 \exp(-\varepsilon_F/d_F) \pm 1$, and is small for cold field emission.

III. DUAL PEAK FEATURE IN THE TED

The material parameters used are work function $\Phi = 5.25$ eV, and $\varepsilon_{F0} = 11.55$ eV for Tungsten W; $\Phi = 5.65$ eV and $\varepsilon_{F0} = 8.78$ eV for Platinum Pt from [6]-[8]; $\Phi = 4.5$ eV and $\varepsilon_{F0} = 0.1$ eV for Cd_3As_2 from [3], [9], [10]. Fig. 1a shows the distinct difference between the TEDs for W and Pt against Cd_3As_2 using (1), under 700 K, 2.5 V/nm and its intrinsic fermi level, ε_{F0}. The appearance of the dual peak feature can be used as a functional control for the energy profile of the TED as shown in [4], absent in traditional materials. This is done by the variation of the applied field; temperature and the fermi level of Cd_3As_2 as shown in Fig. 1b, which is not robust against these parameters. Notably, the increase of F suppresses the right peak of the TED as seen by the red solid line with 2.5 V/nm, to the blue dotted line with 5 V/nm. Similarly, an increase of ε_F suppresses the left peak to become the yellow dotted line with $\varepsilon_{F1} = 0.15$ eV. The increase of T however only serves as an amplification of this feature as shown in the green dash dotted line. Under sufficiently elevated T, it is possible to manipulate the dual peak feature by varying ε_F and F in the TED, to achieve a larger current density as seen below.

IV. EMISSION CURRENT DENSITY AND SCALING LAW

The similar dependences of T and F in [4] and (2) implies that 3D TDS and traditional materials will have no noticeable difference apart from a lower current density in 3D TDS due to its limited electronic carrier density in the thermal-field emission region. The contrasting dependence in ε_F however, stresses the important role of ε_F, while varying F for the emission current density as seen in Fig. 2. With an increased in ε_F, a larger current density can be generated (below 1) at lower fields in Cd_3As_2. This is supported by the disappearance of the dual peak feature as shown in the difference between the red solid and yellow dotted lines in Fig. 1b. With increased ε_F, the suppression of the left peak in the TED by increasing F will reduce the number of carriers in both the conduction (right peak) and valence band (left peak) for emission. This implies that an increased in the fermi level can achieve the same current density for lower applied fields in Cd_3As_2.

The inset in Fig. 2 shows the F^3 scaling of Cd_3As_2 for cold field emission. The linearity of the non-FN scaling law implies that Cd_3As_2 cannot be simply modelled by the MG model.

V. CONCLUSIONS

We have shown that Cd_3As_2 exhibits an unconventional dual peak feature in its TED and a F^3 scaling law, absent from traditional materials. This feature can be manipulated to achieve a larger current density, by the variation of temperature, applied field and the fermi level. Particularly, by increasing the fermi level, a lower applied field is needed to achieve the same current density in Cd_3As_2. This implies that Cd_3As_2 can potentially be a strong candidate for a field emitter with a lower turn-on field, which is a highly sought-after field emission physics.

ACKNOWLEDGMENT

This work is funded by MOE Tier 2 (2018-T2-1-007) and W. J. C. acknowledge MOE PhD RSS.

REFERENCES

[1] Y. S. Ang, Y. Chen, C. Tan, and L. K. Ang, "Generalized High-Energy Thermionic Electron Injection at Graphene Interface," *Phys. Rev. Appl.*, vol. 12, no. 1, p. 014057, Jul. 2019.

[2] L. Chen, H. Yu, J. Zhong, L. Song, J. Wu, and W. Su, "Graphene field emitters: A review of fabrication, characterization and properties," *Mater. Sci. Eng. B*, vol. 220, pp. 44–58, Jun. 2017.

[3] I. Crassee, R. Sankar, W.-L. Lee, A. Akrap, and M. Orlita, "3D Dirac semimetal Cd_3As_2: A review of material properties," *Phys. Rev. Mater.*, vol. 2, no. 12, p. 120302, Oct. 2018.

[4] W. J. Chan, Y. S. Ang, L. K. Ang, "Thermal-Field Electron Emission from Three-Dimensional Topological Semimetals," unpublished.

[5] E. L. Murphy and R. H. Good, "Thermionic Emission, Field Emission, and the Transition Region," *Phys. Rev.*, vol. 102, no. 6, pp. 1464–1473, 1956.

[6] J. W. Gadzuk and E. W. Plummer, "Field Emission Energy Distribution (FEED)," *Rev. Mod. Phys.*, vol. 45, no. 3, pp. 487–548, Jul. 1973.

[7] N. Nicolaou and A. Modinos, "Band-structure effects in field-emission energy distributions in tungsten," *Phys. Rev. B*, vol. 11, no. 10, pp. 3687–3696, May 1975.

[8] A. K. Bordoloi and S. Auluck, "Electronic structure of platinum," *J. Phys. F Met. Phys.*, vol. 13, no. 10, pp. 2101–2105, Oct. 1983.

[9] Z. K. Liu *et al.*, "A stable three-dimensional topological Dirac semimetal Cd_3As_2," *Nat. Mater.*, vol. 13, no. 7, pp. 677–681, Jul. 2014.

[10] Z. Huang *et al.*, "High responsivity and fast UV–vis–short-wavelength IR photodetector based on Cd_3As_2/MoS_2 heterojunction," *Nanotechnology*, vol. 31, no. 6, p. 064001, Jan. 2020.

978-1-6654-2590-2/21 $31.00 © 2021 IEEE

Field emission from two dimensional materials: a quantum mechanical model and its application to graphene

Bruno Lepetit

Laboratoire Collisions Agrégats Réactivité, UMR5589
Université Toulouse III Paul Sabatier, CNRS
118 route de Narbonne, 31062 Toulouse Cedex 09, France
bruno.lepetit@irsamc.ups-tlse.fr

Abstract—**Electron field emission from two dimensional materials and in particular from graphene is modelled using the Bardeen transfer Hamiltonian formalism. In the case of graphene, a full electronic band model of the material (the tight-binding model) is necessary to obtain reasonable results because emission is not restricted to the vicinity of the Dirac points. The emitted current density is small and follows a modified Fowler-Nordheim law with respect to the applied field, with a prefactor exponent for the field $n \approx 1.5$ intermediate between the values for the 2D ($n = 0$) and 3D ($n = 2$) free electron gases. Our study suggests that emission from graphene results almost exclusively from defects.**

I. Introduction

The original Fowler–Nordheim model (FN) still forms nowadays the dominant paradigm to understand electronic field emission [1]. It provides, in particular, an analytical relation beween emitted current density J and applied electric field F:

$$J(F) = \frac{q^3 F^n}{16\pi^2 \hbar W} e^{-\frac{4}{3}\left(\frac{2m}{\hbar^2}\right)^{\frac{1}{2}}\frac{W^{\frac{3}{2}}}{qF}} \quad \text{(eq. 1)}$$

with $n=2$. $\hbar$ is the reduced Planck constant, q the absolute value of the charge of the electron, m its mass, W the work function of the emitting material. Although used widely and successfully over the years, the FN model relies on a crude description of the emitting electrode with a three dimensional (3D) homogeneous electron gas which is obviously questionable if the electrode is a two dimensional (2D) material such a graphene [2-4]. We present here a new model of field emission which is applicable to both 2D and 3D materials [2].

II. Model

The present general model is based on the Bardeen transfer Hamiltonian formalism [5-6] coupled to a detailed description of the electronic structure of the material. We consider a portion of area S (in the (x,y) plane) of an emitting material of thickness L. When subjected to the external electric field F, it emits an averaged current density $J(F)$ which is obtained by integration over the electon momentum k in the first Brillouin zone of the D-dimensional reciprocal space:

$$J(F) = \frac{L^{D-2}}{2^{D-1}\pi^D} \iint d^D\mathbf{k}\, I_k(F) \quad \text{(eq. 2)}$$

$D = 2$ or 3 refers to the dimensionality of the material. The contribution of the electrons with momentum k is:

$$I_k(F) = q\frac{2\pi}{\hbar} \left| \iint_{x,y\in S} dx dy M_k(r, z_0, F) \right|^2 \quad \text{(eq. 3)}$$

where $\mathbf{r} = (x, y)$ and z_0 is the height of the electrostatic potential barrier confining the electrons in the material. M_k is given by:

$$M_k(\mathbf{r}, z_0, F) = \frac{\hbar^2}{2m}\left(\Psi_k(\mathbf{r}, z_0, F)^* \frac{d\Phi_k(\mathbf{r}, z_0)}{dz} \right.$$
$$\left. - \Phi_k(\mathbf{r}, z_0)\frac{d\Psi_k(\mathbf{r}, z_0, F)^*}{dz} \right) \quad \text{(eq. 4)}$$

$\Phi_k(\mathbf{r}, z)$ is a material valence electron orbital and $\Psi_k(\mathbf{r}, z, F)$ the continuum state that describes the electron outside the material. In our implementation for graphene [2], $\Phi_k(\mathbf{r}, z)$ is obtained from a tight-binding model [7] and $\Psi_k(\mathbf{r}, z, F)$ is the product of a plane wave in $\mathbf{r}$ and of a properly normalized Airy function in z.

III. Results

In order to validate the proposed formalism, we first apply it to the material of the FN model which is a 3D free-electron gas. We obtain:

$$J(F) = \frac{q^3 F^2}{16\pi^2 \hbar (WW_B)^{\frac{1}{2}}} e^{-\frac{4}{3}\left(\frac{2m}{\hbar^2}\right)^{\frac{1}{2}}\frac{W^{\frac{3}{2}}}{qF}} \quad \text{(eq. 5)}$$

where W_B is the material valence band width. This result is similar to the FN current density (eq. 1) if we substitute W to $(WW_B)^{\frac{1}{2}}$. In particular, the same field exponent $n=2$ is obtained for the prefactor. As both quantities W and $(WW_B)^{\frac{1}{2}}$ have the same order of magnitude, the present model provides results close to the FN one.

Contrary to the FN model, the present formalism extends naturally and immediately to 2D materials. In the case of a 2D free electron gas : $\Phi_k(\mathbf{r}, z) = \frac{1}{S^{\frac{1}{2}}}e^{ikr}\varphi(z)$ where $\varphi(z)$ is a bound state in z describing the motion of the confined electrons perpendicular to the material plane. Insertion of this expression in eqs. 2-4 provides:

978-1-6654-2590-2/21 $31.00 © 2021 IEEE

$$J(F) = \frac{q(2m)^{\frac{1}{2}}}{\pi\hbar^2} W_B (W + W_B)^{\frac{1}{2}} |\varphi(z_i)|^2 e^{-\frac{4}{3}\left(\frac{2m}{\hbar^2}\right)^{\frac{1}{2}}\frac{(W+W_B)^{\frac{3}{2}}}{qF}}$$

(eq. 6)

In going from the 3D to the 2D free electron gas, the F^2 prefactor is lost as the field exponent changes from $n=2$ to $n=0$. $\log J(F)$ is linearly dependent with respect to $1/F$, with a slope controlled by $W+W_B$ in the 2D case, the energy necessary to extract an electron from the band bottom, instead of by W, the work function necessary to extract an electron from the band top, in the 3D case.

In the case of graphene, there is no compact analytical form for the emitted current density but a numerical solution to eqs. 2-4 can be found and it can be fitted to an analytical form akin to eq. 1, keeping the field exponent n in the prefactor and work function W as free fitting parameters. Figure 1 provides current densities obtained for the tight binding model for graphene [2, 7] as well as fit results. The best fit is obtained for $n=1.53$ and $W=12.78$ eV. A power law with $n\approx1.5$, intermediate between the 2D and 3D free-electron gas cases, is thus found to provide the best model for graphene. A similar $n=3/2$ exponent has already been obtained experimentally [8] as well as in a theoretical study of emission from a graphene nanowall, where the graphene flake is mounted perpendicularly to its substrate [9].

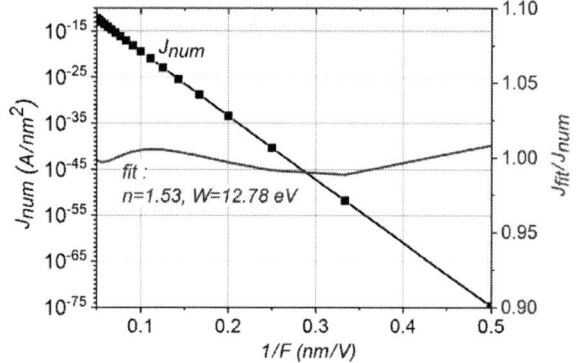

Fig. 1 Full black line connecting symbols (left scale, $A/nm2$): emitted current density J_{num} as the function of the applied electric field F (V/nm), obtained by numerical integration [eqs. 2-4]. The parameters of the model are the same as in [2]. Blue line (right scale): ratios between fitted and numerical results J_{fit}/J_{num}. The fit parameters are: $n=1.53$ and $W=12.78$ eV.

We know which regions of reciprocal space are the main contributors to emission from the dependence of $I_k(F)$ (eq. 3) with respect to $\boldsymbol{k}$. Interestingly, emission from the vicinity of the K point - the so called Dirac point at the Fermi level where the valence band touches the conduction one – is negligible. Instead, emission is maximum in an annular region around the Γ point ($\boldsymbol{k}=0$). In fact, for both 2D and 3D materials, emission is expected to be maximum for $\boldsymbol{k}$ vectors such that the energy

available to overcome the tunnelling barrier – i.e. the energy associated to the the z motion (perpendicular to the potential barrier) - is maximum. This energy is $\Delta e_k = \varepsilon_k - \frac{\hbar^2 k_{xy}^2}{2m}$ where ε_k is the material electron energy and $\boldsymbol{k}_{xy}$ its momentum component parallel to the surface. Whereas Δe_k maximum is located for 3D materials on the Fermi surface where 2 conditions (ε_k maximum – by definition of the Fermi surface – and $\frac{\hbar^2 k_{xy}^2}{2m}$ minimum – i.e. $\boldsymbol{k}_{xy}= 0$) can be simultaneously satisfied, this is not this case in general for 2D materials where the maximum of ε_k does not correspond necessarily to $\boldsymbol{k}$ vectors for which $\frac{\hbar^2 k_{xy}^2}{2m}$ is minimum (i.e. $\boldsymbol{k}_{xy}= 0$). In this case, Δe_k maximum is obtained for $\boldsymbol{k}$ values which do not belong to the Fermi surface. As a consequence, a full band model is necessary to study emission from 2D materials like graphene and linear approximations valid only in the vicinity of the Dirac points are not sufficient.

IV. CONCLUSIONS

We have presented an extension of the Fowler-Nordheim model to 2D materials based on the Bardeen formalism. The model has been applied to free electron gas and to graphene. We have shown that the usual Fowler-Nordheim expression has to be modified to be valid for such materials. Notice finally that the emitted current levels are low because the electron kinetic energy corresponds to a motion parallel to the emitting surface, which is not efficient in promoting emission from the surface. This suggests that the significant emission levels [8, 10-11] measured on graphene result almost exclusively from deviations from the perfect graphene 2D structure considered here.

REFERENCES

[1] R. Diehl, M. Choueib, S. Choubak, R. Martel, S. Perisanu, A. Ayari, P. Vincent, S. T. Purcell, and P. Poncharal, Phys. Rev. B **102**, 035416 (2020).

[2] B. Lepetit, J. Appl. Phys. **129**, 144302 (2021).

[3] Y. S. Ang, S.-J. Liang, and L. K. Ang, MRS Bull. **42**, 505 (2017).

[4] Y. S. Ang and L. K. Ang, in 31st International Vacuum Nanoelectronics Conference (IVNC), Kyoto (IEEE, 2018), p. 1.

[5] J. Bardeen, Phys. Rev. Lett. **6**, 57 (1961).

[6] R. Ramprasad, L. R. C. Fonseca, and P. von Allmen, Phys. Rev. B 62, 5216 (2000).

[7] A. H. Castro-Neto, F. Guinea, N. M. R. Peres, K. S. Novoselov, and A. K. Geim, Rev. Mod. Phys. **81**, 109 (2009).

[8] Z. Xiao, J. She, S. Deng, Z. Tang, Z. Li, J. Lu, and N. Xu, ACS Nano **4**, 6332 (2010).

[9] X.-Z. Qin, W.-L. Wang, N.-S. Xu, Z.-B. Li, and R. G. Forbes, Proc. R. Soc. A **467**, 1029 (2011).

[10] L. Chen, H. Yua, J. Zhong, L. Song, J. Wuc, and W. Su, Mater. Sci. Eng. B **220**, 44 (2017).

[11] F. Giubileo, A. D. Bartolomeo, L. Iemmo, and G. Luongo, Appl. Sci. 8, 526 (2018).

Gap in pagination due to withheld paper.

Pages 115-116

General scaling laws of space charge effects in field emission

A. Kyritsakis[*†‡], M. Veske[†‡], V. Zadin[†] and F. Djurabekova[‡§]

[†]Institute of Technology,University of Tartu, Nooruse 1, 50411 Tartu, Estonia
[‡]Deparment of Physics and Helsinki Institute of Physics, University of Helsinki, PO Box 43, 00014 Helsinki, Finland
[§]National Research Nuclear University MEPhI, Kashirskoye sh. 31, 115409 Moscow, Russia
[*]Contact: andreas.kyritsakis@ut.ee; akyritsos1@gmail.com

Abstract—**The suppression of field electron emission by space charge (SC) is described for planar diodes by the one dimensional (1D) theory. Here we generalize in 3D by deriving the scaling behavior of field suppression in the weak SC regime. We propose a corrected equivalent planar diode model, which describes the SC effects for any geometry in terms of the 1D theory, utilizing a correction factor that adjusts the scaling characteristics of the field suppression. We validate our theory by comparing it to both numerical calculations and existing experimental data, either of which can be used to obtain the correction factor.**

I. Introduction

Space Charge (SC) effects [1], i.e. the suppression of the extracting field due to the space charge formed by the emitted particles, significantly limits the function of electron emitters. This effect plays a very important role in all forms of electron and ion sources and is of paramount importance for the understanding of the ignition of vacuum arcs [2]. The 1D planar-diode model has been widely used as reference to estimate the SC effects on field emission [3], [4] due to its simplicity and intuitiveness. The connection of the planar diode model to an experimental emitter geometry is typically done via Barbour's equivalency [5], i.e. assigning the real values of voltage and cathode field to a planar diode, also known as the Equivalent Planar Diode (EPD) model. However, recent Particle-In-Cell (PIC) calculations [2] have shown that this equivalence is not valid. Here tackle this problem by developing a general 3D theory for SC-suppressed emission.

II. Theory

The standard formulation of the SC problem considers a continuous charge density and zero initial particle velocity, leading to the the Poisson equation taking the form $\nabla^2 \Phi = kJ(\mathbf{r})\Phi^{-1/2}$, with $\Phi = 0$ at the emitter and $\Phi = V$ at the collector electrodes. Here $k = \epsilon_0^{-1}\sqrt{m/2q}$ is a constant that depends on the mass to charge ratio m/q of the emitted particles and the vacuum electric permittivity ϵ_0; V is the applied voltage. The current density $J(\mathbf{r})$ obeys continuity $\nabla \cdot \mathbf{J} = 0$, with boundary condition $\mathbf{J}(\mathbf{r}_s) = J_s(\mathbf{r}_s)\hat{n}(\mathbf{r}_s)$, where $J_s(\mathbf{r}_s)$ is the emitted current density and $\hat{n}(\mathbf{r}_s)$ is the normal unit vector at the emitting surface point $\mathbf{r}_s$. In order to obtain $\mathbf{J}$ and Φ, the above equations have to be solved self-consistently, along with the surface emission laws that give $J_s(\mathbf{r}_s)$ as a function of the local electric field $F(\mathbf{r}_s)$.

This problem can be solved analytically only for the planar diode case, where J is constant. The solution of the 1D Poisson equation for gap distance d can be approximated for small ζ as $\theta = 1 - \frac{4}{3}\zeta + O(\zeta^2)$, in terms of the reduced dimensionless variables [3] $\theta \equiv F/F_L$ and $\zeta \equiv\equiv kJV^{1/2}/F_L^2$, where $F_L = V/d$ is the Laplace field, θ is the field reduction factor, and ζ the space charge strength. The above asymptotic approximation is valid for $\zeta < 0.25$ and $\theta > 0.6$, i.e. the weak SC regime, which covers most cases in field emission.

We now express the Poisson equation in terms of the reduced variables $\phi \equiv \Phi/V$, $J(\mathbf{r}) = J(\mathbf{r}_s)\xi(\mathbf{r})$, with $\xi(\mathbf{r})$ being a unitless variable expressing the variation of the current density from $\mathbf{r}_s$. Approximating that $\xi(\mathbf{r})$ depends only on the emitter geometry and not on $J(\mathbf{r}_s)$ and using the reduced space coordinate $\tilde{\mathbf{r}} = \mathbf{r}/\chi$, where $\chi \equiv V/F_L(\mathbf{r}_s)$ is the conversion length, it yields $\tilde{\nabla}^2\phi(\tilde{\mathbf{r}}) = \zeta\xi(\tilde{\mathbf{r}})(\phi(\tilde{\mathbf{r}}))^{-1/2}$, where $\tilde{\nabla}$ denotes derivatives with respect to $\tilde{\mathbf{r}}$ and $\zeta \equiv kJ(\mathbf{r}_s)V^{1/2}/F_L(\mathbf{r}_s)^2$ is the generalized version of the SC strength.

It can be shown [6] that the field reduction factor $\theta = F/F_L$ can be expanded as $\theta = 1 - \frac{4}{3}\omega\zeta + O(\zeta^2)$, where

$$\omega \equiv \frac{3}{4}\int_\Omega \tilde{\nabla}_\mathbf{r}G(\tilde{\mathbf{r}},\tilde{\mathbf{r}}')\Big|_{\tilde{\mathbf{r}}_s}\frac{\xi(\tilde{\mathbf{r}}')}{\sqrt{\phi_0(\tilde{\mathbf{r}}')}}d^3\tilde{\mathbf{r}}', \tag{1}$$

$G(\tilde{\mathbf{r}},\tilde{\mathbf{r}}')$ is the geometry-dependent Green's function for Dirichlet problems and Ω is the vacuum gap space. This demonstrates the inadequacy of the standard planar diode equivalence of Refs [3], [5]. The scaling of $\theta(\zeta)$ for $\zeta \ll 1$ is the same as the planar diode only if $\omega = 1$, which does not hold in general. However, ω can be incorporated as a correction to the planar model. Thus, for a given geometry and a surface point $\mathbf{r}_s$, we define the corrected equivalent planar diode model (CEPD), for which the SC strength is $\zeta_c = \omega\zeta$.

III. Validation

Although ω is formally given by (1), it is more convenient to calculate $\theta(\zeta)$ numerically using PIC for a given geometry, and fit ω to the results, thus yielding a more general and computationally cheap approximate solution of the SC problem. Here we used the finite element (FEM)-based tool FEMOCS, which has been recently enhanced with PIC capabilities [2] (see [6] for details). Note also that ω is obtained analytically for the cylindrical and spherical diodes [6].

978-1-6654-2590-2/21 $31.00 © 2021 IEEE

Our theory and methods are applicable to any electrode geometry and emitted particle type. However, here we focus on the specific case of [5] to validate our theory against experimental data. Barbour *et. al.* [5] took $I-V$ measurements from a W emitter coated with Ba in various coverages to reduce its work function W and reach SC-relevant currents. The shape of the tip was emulated by the sphere-on-a-cone (SOC) model [7] shown in Fig. 1. We used our computational method to solve the SC problem on this geometry for various values of V and then fitted ω to the results. Although resulting values vary slightly on the emitting surface, this variation is small within the emission area (less than 10%), allowing to use a single effective value ω_{eff} that describes the whole emitter. The value of ω_{eff} is fitted by minimizing the deviation between the numerical $(I-V)$ curves and the ones of the CEPD model.

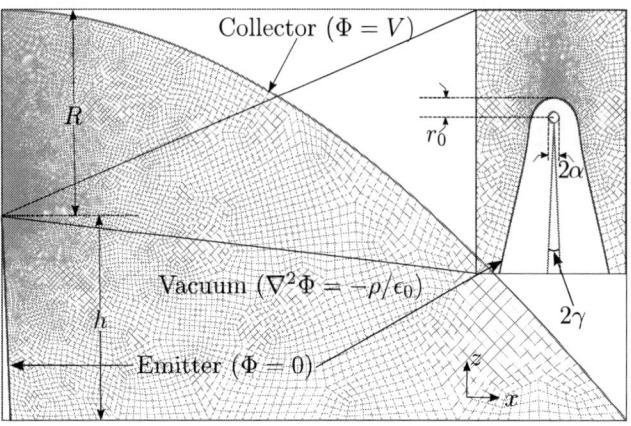

Fig. 1. Schematic of the simulated system along with the quadrangular tesselation used in FEM. The emitter (blue), and collector (green) are shown along with the internal SOC generator (magenta). The used parameters are: $r_0 = 315$ nm, $\gamma = 0.78^o$, $R = 6.5$ cm, $h = R$ and $\alpha = 0.235 r_0$.

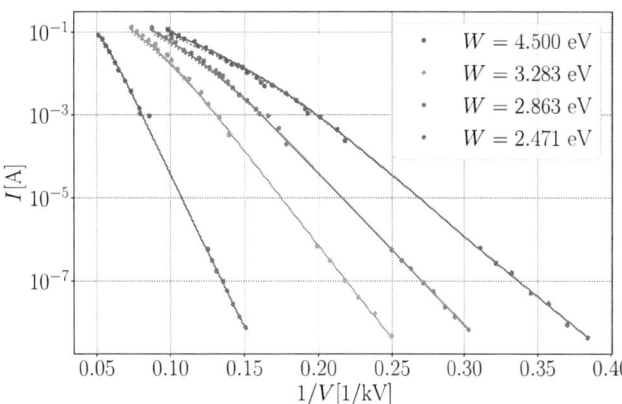

Fig. 2. Experimental $I-V$ curves from [5] (dots) and the corresponding CEPD calculations with $\omega_{\text{eff}} = 0.8$ (solid lines) and EPD ones (equivalent to $\omega_{\text{eff}} = 1$). A uniform work function W was assumed for all curves, with the indicated values being determined by assuming $W = 4.5$ eV for clean W and then fitting the values for the coated cathodes to match the measurements in the low field regime. For the red curve, I was multiplied by a fitted correction factor 0.78 to account for the reduction of the effective emission area due to the increased non-uniformity of the work function.

In Fig. 2 we compare the experimental data of [5] against the CEPD model, showing that the latter predicts accurately the SC-induced curvature of the experimental plots at high fields. Surprisingly, the EPD model gives also good agreement with the measurements. However, this is due to the fact that ω_{eff} is very close to 1, resulting in minor differences between the EPD and CEPD models, as seen by the hardly distinguishable dashed curves. However, this is rather a coincidence, specific to this particular geometry. We performed the same calculations for different geometries, varying the cone aperture γ, while all other geometrical parameters are kept as in Fig. 1. As γ increased from 0.78^o to 12.5^o, ω_{eff} decreased from 0.8 to 0.31, revealing a significant deviation of EPD from PIC, while CEPD is kept in excellent agreement.

These results demonstrate the utility of the CEPD model. If the value of ω_{eff} and the field conversion length χ are available for a certain electrode geometry, the complex problem of calculating the SC suppressed emission from a 3D emitter is reduced from running full PIC simulations, to evaluating an algebraic formula self-consistently with the emission characteristics. Although to obtain ω_{eff} theoretically PIC simulations are required, this needs to be done only once for a given geometry. Then the CEPD model can be used to calculate the emission current for different emission characteristics, such as work function and temperature. ω_{eff} values can be also tabulated for typical emitter geometries, becoming readily available for use. Note that ω of eq. (1) is scale invariant, i.e. remains unchanged under a geometrical scaling $\mathbf{r} \to a\mathbf{r}$ for any a, which further facilitates tabulation. Finally, ω can be also fitted directly to experimental data, in a similar manner as the enhancement factor is typically obtained.

IV. CONCLUSIONS

We have developed a three-dimensional theoretical model, describing the scaling laws of space charge limited charge emission at high electric fields. Our model generalizes the one-dimensional planar model to be applicable for any geometry, using a geometry-specific correction factor.

ACKNOWLEDGMENTS

We acknowledge CERN's CLIC K-contract No. 47207461 and the European Union's ERA Chair MATTER (No 856705).

REFERENCES

[1] C. D. Child, "Discharge from hot cao," *Phys. Rev.*, vol. 32, p. 492, 1911.
[2] M. Veske, A. Kyritsakis, F. Djurabekova, K. N. Sjobak, A. Aabloo, and V. Zadin, "Dynamic coupling between particle-in-cell and atomistic simulations," *Phys. Rev. E*, vol. 101, p. 053307, 2020.
[3] R. G. Forbes, "Exact analysis of surface field reduction due to field-emitted vacuum space charge, in parallel-plane geometry, using simple dimensionless equations," *J. Appl. Phys.*, vol. 104, no. 8, p. 084303, 2008.
[4] P. Chen, T. Cheng, J. Tsai, and Y. Shao, "Space charge effects in field emission nanodevices," *Nanotechnology*, vol. 20, no. 40, p. 405202, 2009.
[5] J. Barbour, W. Dolan, J. Trolan, E. Martin, and W. Dyke, "Space-charge effects in field emission," *Phys. Rev.*, vol. 92, no. 1, pp. 45–51, 1953.
[6] A. Kyritsakis, M. Veske, and F. Djurabekova, "General scaling laws of space charge effects in field emission," *New J. Phys.*, 2021. [Online]. Available: doi.org/10.1088/1367-2630/abffa8
[7] W. P. Dyke, J. K. Trolan, W. W. Dolan, and G. Barnes, "The field emitter: Fabrication, electron microscopy, and electric field calculations," *Journal of Applied Physics*, vol. 24, no. 5, pp. 570–576, 1953.

Absence of space-charge-limited current from field emission due to non-FN law

Cherq Chua, Chun Yun Kee, Yee Sin Ang [a)], Lay Kee Ang [b)]

Science, Mathematics and Technology (SMT), Singapore University of Technology and Design, Singapore

[a)] Electronic Mail: yeesin ang@sutd.edu.sg
[b)] Electronic Mail: ricky ang@sutd.edu.sg

Abstract— **With the emergence of non-FN based field emitters, the establish transition from source-limited field emission (FE) at low voltage to space-charge-limited current (SCLC) at high voltage is no longer valid. In this paper, we predict a critical pre-factor field scaling exponent $k_c = 3/2$, where for emitters with $k < k_c$, unconventional transition behaviours could occur, which includes: (i) exclusively FE process with absent of SCLC; and (ii) three-stage FE-to-SCLC-to-FE transition, depending on gap spacing D. While for emitters with $k > k_c$, classical two-stage FE-to-SCLC transition is expected. Several existing non-FN based FE models are used as examples in this paper to illustrate the unconventional transition behaviours.**

I. INTRODUCTION

The electron emission from bulk material to free space under low external voltage is known to be governed by the classical Fowler-Nordheim (FN) law [1]. As the emitted current density J increases to sufficiently large amount, the electron transport is limited by space charge effects. The space-charge-limited current (SCLC) is typically described by one-dimensional (1D) Child-Langmuir (CL) law [2, 3]. This two stage FN-to-CL transition is based on the assumptions that the source limited field emission (FE) follows the classical FN law. The recent emergence of materials not obeying the scaling of the classical FN models (ex. Graphene [4, 5], 2DEG [4]), however, raises questions to the validity of this classical two-stage transition framework.

Hence, in this work, we aim to study the transition between FE and SCLC for arbitrary values of pre-factor field scaling exponent k over a wide range of applied voltage V and gap spacing D. Intriguingly, we found that there is a critical value $k_c = 3/2$ that for emitters with $k \leq k_c$, SCLC at high field is no longer warranted. As a consequence, two unconventional transition behaviours arise: (i) exclusively FE process with the absence of SCLC; and (ii) three-stage FE-to-SCLC-to-FE transition.

II. METHOD

We consider a generalized FE equation which has an arbitrary scaling exponent k for the pre-factor surface electric field E_s. The generalized FE equation is transformed into normalized form to be applicable over wide range of materials and parameters. The normalized equation is then linked to Llewellyn form of Poisson equation to account for space charge effects [6]. From here we can numerically obtain the $J - V$

characteristic. Note that the voltage scaling of $3/2$ in CL law suggests that $k_c = 3/2$ is a critical factor in classifying the transition behaviours of the current emission. Based on this critical value, we separate the scaling exponent k into three different regimes: i) supercritical regime ($k > 3/2$); ii) critical regime ($k = 3/2$); and iii) subcritical regime ($k < 3/2$). For each regime, we study the asymptotic behaviours of the current emission at low field and high field. We found that other than the scaling exponent k, the gap spacing D is another determining factor for the transition behaviours. There is a critical gap spacing D_c in critical and subcritical regimes, where SCLC could not exist for $D < D_c$.

III. RESULT

In the supercritical-regime [Fig. 1(a)], the current emission transits from FE to SCLC in two-stage manner at all D. This is similar to the conventional understanding of the FN-to-CL transition framework when classical FN law ($k = 2$) is considered. The two-stage FE-to-SCLC transition is, however, not warranted in critical and subcritical regime. In critical regime [Fig. 1(b)], we have the two-stage FE-to-SCLC transition only at $D > D_c$. For $D < D_c$, the emission is exclusively FE process with the absence of SCLC. Lastly in the sub-critical regime [Fig. 1(c)], we have the exclusively FE process at $D < D_c$, and an unconventional three-stage FE-to-SCLC-to-FE transition at $D > D_c$. The absence of SCLC at high field in subcritical case is due to the inability of the FE to generate enough current density to reach SCLC regime. This property might be favourable in certain operating condition where low and stable current output is required even at high voltage.

Using several FE models with different scaling regimes as examples, the current density-voltage characteristics at $D = 100$ nm and 1 mm are presented in Fig. 2. Fig. 2(a) shows the $J-V$ characteristic of classic FN model ($k = 2$) in normalized form [1]. It can be seen that the expected two-stage FE-to-SCLC transition is obtained. For critical case, we consider nanowall emitter [7], which has $k = 1.5$ due to the quantum confinement effect leading to a quantized energy component along the confinement direction [Fig. 2(b)]. Similar to Fig. 2(a), a two-stage transition from FE to SCLC is found, suggesting that both gap spacings chosen are smaller than the critical D_c. In Fig. 2(c), we show the Schottky-Nordheim (SN) barrier model [8], which is reported to have a scaling of $k = 1.23$ for

978-1-6654-2590-2/21 $31.00 © 2021 IEEE

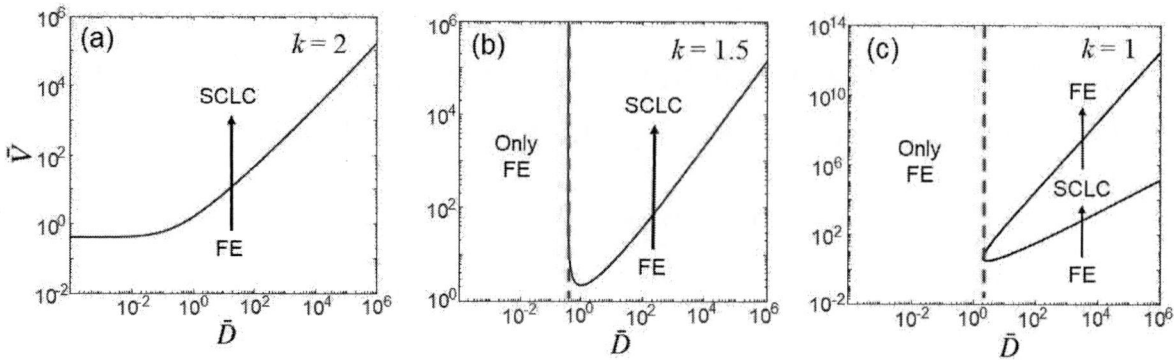

Fig. 1 Susceptibility diagram of field emission (FE) and space-charge-limited current (SCLC) for normalized voltage $\bar{V}$ and normalized gap distance $\bar{D}$ with (a) supercritical case ($k = 2$); (b) critical case ($k = 1.5$); and (iii) subcritical case ($k = 1$). The red dashed line in (b) and (c) represent the critical $\bar{D_c}$

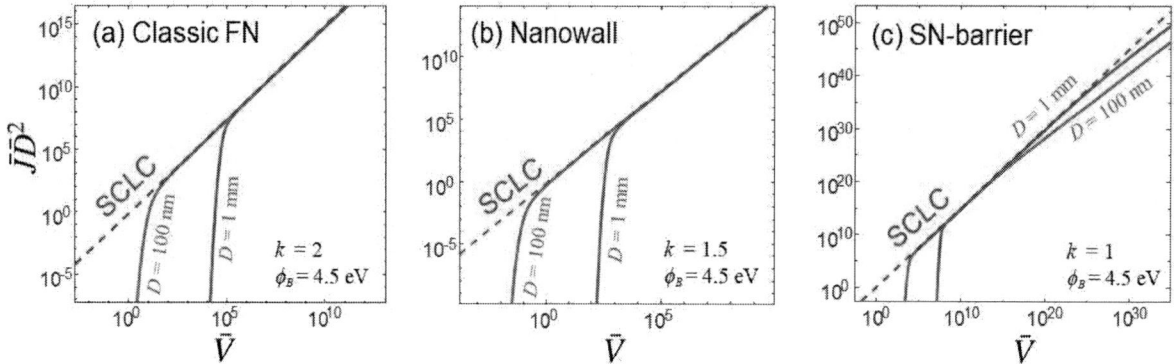

Fig. 1 Normalized $\bar{J} - \bar{V}$ curves of (a) classic FN law; (b) nanowall; and (c) Schottky-Nordheim (SN) barrier with $\phi_B = 4.5$ eV for $D = 100$ nm and 1 mm. The red dashed lines represent the space-charge-limited current (SCLC).

material with work function $\phi_B = 4.5$ eV. In contrast to Fig. 2(a), unconventional three-stage FE-to-SCLC-to-FE transition process is obtained.

IV. CONCLUSIONS

We show that for emitters beyond the classical Fowler-Nordheim (FN) law, the conventional transition from field emission (FE) to space-charge-limited current (SCLC) is no longer warranted. Our model shows that for emitter model with arbitrary field scaling k, there is a critical scaling exponent $k_c = 3/2$, which can classify emitters into three regimes: (i) super-critical ($k > k_c$); (ii) critical ($k = k_c$); and subcritical ($k < k_c$). Each scaling regime has different transition behavior, depending on the gap spacing. The unified picture of the emission physics from FE to SCLC has not been reported to our best knowledge. Thus, this paper offers a theoretical foundation for modelling different unconventional field emission as high current electron sources, which are important for many applications in high-power microwave sources, compact vacuum electronics, beam physics and plasma.

ACKNOWLEDGMENT

This work is supported by Singapore Ministry of Education Tier 2 grant (2018-T2-1-007), A*STAR AME IRG (RGAST2007) and SUTD SRG (SRT3CI21163).

REFERENCES

[1] R. H. Fowler and L. Nordheim, "Electron emission in intense electric fields," *Proc. R. Soc. Lond.*, vol. A119, pp. 173-181, 1928.

[2] C. D. Child, "Discharge from hot CaO," *Phys. Rev. (Series I)*, vol. 32, pp. 492-511, May 1911.

[3] I. Langmuir, "The effect of space charge and residual gases on thermionic currents in high vacuum," *Phys. Rev.*, vol. 2, pp. 450-486, 1913.

[4] B. Lepetit, "A quantum mechanical model of electron field emission from two dimensional materials. Application to graphene," *J. Appl. Phys.*, vol. 129, p. 144302, 2021.

[5] Y. S. Ang, C. H. Lee and L. K. Ang, "Universal scaling and signatures of nodal structures in electron tunneling from two-dimensional semimetals," 2020, arXiv:2003.14004.

[6] Y. Y. Lau, Y. Liu and R. K. Parker, "Electron emission: From the Fowler–Nordheim relation to the Child–Langmuir law," *Phys. Plasmas*, vol. 1, pp. 2082-2085, 1994.

[7] X.-Z. Qin, W.-L. Wang, N.-S. Xu, Z.-B. Li and R. G. Forbes, "Analytical treatment of cold field electron emission from a nanowall emitter, including quantum confinement effects," *Proc. R. Soc. Lond.*, vol. A467, pp. 1029-1051, 2011.

[8] R. G. Forbes, "Use of Millikan–Lauritsen plots, rather than Fowler–Nordheim plots, to analyze field emission current-voltage data," *J. Appl. Phys.*, vol. 105, p. 114313, 2009.

[9] C. Chua, C. Y. Kee, Y. S. Ang and L. K. Ang, "Absence of space-charge-limited current in unconventional field emission," 2021, arXiv:2105.10462.

Behavior of notional cap-area efficiency (g_n) for hemisphere-on-plane and related field emitters

S.V. Filippov*, A.G. Kolosko, E.O. Popov
Div. of Plasma Physics, Atomic Physics and Astrophysics
Ioffe Institute
St Petersburg, Russia
*s.filippov@mail.ioffe.ru

Richard G. Forbes
University of Surrey, Advanced Technology Institute
and Dept. of Electrical and Electronic Engineering
Guildford, Surrey GU2 7XH, United Kingdom

Abstract—We obtained and compared numerical and analytical results for the hemisphere-on-plane notional cap-area efficiency. For various emitter shapes, the behavior of the notional emission area has been analyzed. The shape contributon to the pre-exponential voltage exponent was obtained by plotting the notional cap-area efficiency against the apex value of dimensionless local surface field.

Keywords—emission area, emitter shapes, hemisphere-on-plane, hemisphere-on-cylindrical-post, kernel current density, notional emission area, cap-area efficiency, pre-exponential voltage exponent, k-power, k_A shape contribution power.

I. Introduction

The so-called empirical field electron emission (FE) equation writes an expression for the measured FE current I_m in terms of the measured voltage V_m, in the form:

$$I_m = CV_m{}^\kappa \exp[-B/V_m], \qquad (1)$$

where B, C and κ (also written k) are parameters discussed below. This equation was introduced by Abbott and Henderson in 1939 [1], was used by Forbes in 2008 [2] and has recently been used by Forbes, Popov et al. [3].

If the measured current-voltage $I_m(V_m)$ characteristics of a FE system depend only on well-defined emission characteristics, and do not depend on any other system features (such as series resistance in the current path, or space-charge) then the FE system is said to be *ideal*. For ideal systems involving post or needle-shaped emitters, it is thought that B can be treated as constant, and that the parameters k and C vary weakly with voltage.

Reference [3] has shown, by simulations, that the best mean value of k found from a plot of $\ln\{I_m/V_m{}^k\}$ vs $1/V_m$ (a so-called *power-k plot*) is a sensitive function of both (a) the details of emission theory and (b) the shape of the field emitter. Hence, if $I_m(V_m)$ measurements of high quality were available, and if the effects of emitter shape could be disentangled, then a measured value of k should provide a method of comparing different FE theories with experiment. Here we report some of our first steps towards disentangling emitter-shape effects.

Our initial thinking derives from Murphy-Good FE theory, and uses the so-called *kernel current density for the Schottky-Nordheim barrier* ($J_k{}^{SN}$), given as usual by:

$$J_k{}^{SN} = a\phi^{-1}F^2 \exp[-v_F b\phi^{3/2}/F], \qquad (2)$$

where the symbols have their usual meanings [4]. By using the *planar transmission approximation* to integrate expression (2) over the surface of a post-like emitter (either analytically where possible, or numerically), we can write an expression for the related emission current $I_e{}^{kSN}$ in the form:

$$I_e{}^{kSN} = A_n{}^{kSN} J_{ka}{}^{SN}, \qquad (3)$$

where $J_{ka}{}^{SN}$ is the apex value of $J_k{}^{SN}$, and the *notional emission area* $A_n{}^{kSN}$ is defined by (3). In principle, we now wish to investigate how $A_n{}^{kSN}$ depends on measured voltage.

In practice, it is better to investigate the behaviour of the *notional cap-area efficiency* g_n defined by:

$$g_n = A_n{}^{kSN}/(2\pi r_a{}^2), \qquad (4)$$

where r_a is the emitter apex radius. It is also better to investigate how g_n varies with the apex value f_a of the dimensionless local scaled-field f as usually defined. (For an ideal system, f_a is proportional to V_m.)

For the special case of a hemisphere on a plane (HSP), Jensen [5, see formula (30.19)] has obtained an analytical approximation, which he writes in the form $g(F) \approx 1/[b+4-v]$ In our notation this becomes:

$$g_n(f_a) \approx 1/[\eta/f_a+4-\eta/6]; \quad \text{or} \quad 1/g_n \approx (4-\eta/6) + \eta/f_a, \quad (5)$$

where η is the Murphy-Good exponent scaling parameter [4].

Here, using finite element methods, we numerically checked eq. (5) for the hemisphere-on-plane emitter. Analogous numerical analyses were then carried out for related emitter shapes.

II. Simulation Details

To solve the Laplace equation numerically, a commercial software package (COMSOL v5.3) was used. We studied three widely-known emitter shapes, namely: hemisphere-on-cylindrical post (HCP), hemispherically rounded cone (hSoC) and ellipsoidal tip (Elli). Recently we have shown that the apex field enhancement factors for ellipsoidal, paraboloidal and hyperboloidal (with half-angle 5°) tips are very similar, so we ignored paraboloidal and hyperboloidal here [6]. Details of the simulation procedure can be found in [3, 6, 7]

The emitters have the following geometric parameters: height h= 4 µm, apex radius of curvature r_{apex}=50 nm. The hSoC emitter has an additional parameter θ, which defines the half-angle of the vertex (here θ = 5°). The local work fuction of the emitters was set to 4.50 eV.

III. RESULTS AND DISCUSSION

For the hemisphere-on-a-plane (HSP) emitter, Fig.1. compares values of the notional cap-area efficiency g_n cacluated numerically and obtained analytically, using (5).

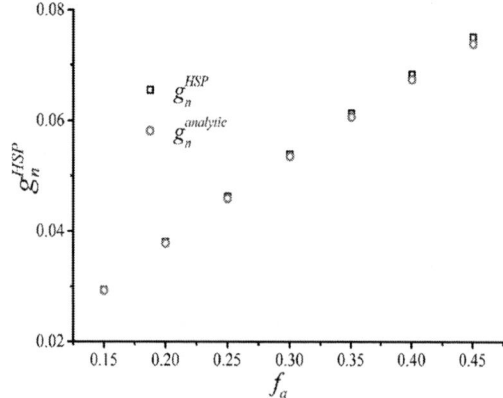

Fig.1. Numerically and analytically calculated values of g_n for HSP emitter.

Fig. 2 shows the notional cap-area efficiency g_n as a function of $f_a \in [0.15;0.45]$ for all above-mentioned emitter shapes.

It should be noted that the highest g_n value was obtained for the HCP model. This result was not obvious, as the HSP and Elli shapes seem to be closer to the planar emitter. It can be seen, from the calculations for the HCP model, that the emitter not only has the largest FEF at the top, but also has higher field values along the surface. This leads to greater integral current values, and, therefore, a larger notional emission area.

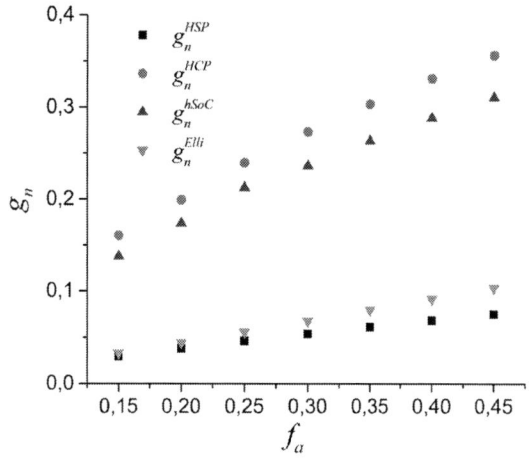

Fig. 2. To show dependence of the notional cap-area efficiency g_n as a function of dimensionless apex field f_a, in the range of $f_{apex} \in [0.15;0.45]$.

In [3], a hypothesis about functional dependence was made

$$CV_m^{\ k} = C_1 V_m^{\ k1}\ C_2 V_m^{\ k2} \qquad (6)$$

where $k_1 \equiv k_p$ – planar power (e.g. $k_p \approx 1.227$ for 4.5 eV), $k_2 \equiv k_A$ – exponent contributionn due to emitter shape .

It can be shown that k_A can be obtained from the slope of the graph, which in general form can be represented as $\ln(I_m/J_{ka})$ vs $\ln(V_m)$. In its usual form, this graph is presented in Fig. 3.

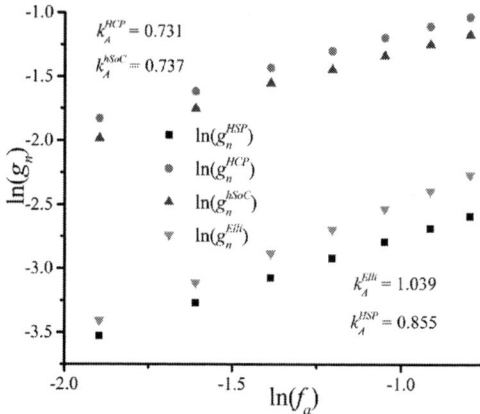

Fig.3. Slope k_A derived using the (dimensionless) notional cap-area efficiency as a function of the dimensionless apex field f_a, for $f_a \in [0.15;0.45]$.

IV. CONCLUSIONS

We have verified by a numerical treatment that analytical formula (5) is a very good approximation for HSP emitter. For both numerical and analytical results, it is found that a plot of $\ln(g_n)$ vs $\ln(f_a)$ has slight curvature downwards with slope (and hence the apparent contribution of emitter shape to k_A lying in the range 0.90 to 0.76 for $0.15 \leq f_a \leq 0.45$. For other forms of emitters, the k_A also slightly depend on the f_a range.

Comparisons with another, independent method of current-voltage characteristics processing - the least-residual method (LR) [3] were made. The LR method yielded the following results $k_{A,LR}^{HCP} = 0.518$, $k_{A,LR}^{hSoC} = 0.540$, $k_{A,LR}^{HSP} = 0.729$ and $k_{A,LR}^{Elli} = 1.079$. Despite some discrepancy between the results shown in Fig. 3, the best agreement of the results for the two methods was obtained using a wider range of fields f_a .

REFERENCES

[1] F R. Abbott and J. E. Henderson, Phys. Rev. 56, 113 (1939).

[2] R. G. Forbes, Appl. Phys. Lett. 92, 193105 (2008).

[3] R. G. Forbes, E. O. Popov, A. G. Kolosko, and S. V. Filippov, Roy. Soc. Open Sci. 8, 201986 (2021).

[4] R.G. Forbes, "Renewing the mainstream theory of field and thermal electron emission", Chapter 9 (pp. 387–447) in: G. Gaertner, W. Knapp and R.G. Forbes (eds) Modern Developments in Vacuum Electron Sources (Switzerland: Springer Nature, 2020).

[5] K. L. Jensen, Introduction to the Physics of Electron Emission (Chichester, UK: Willey, 2018).

[6] S. V. Filippov, E. O. Popov, A. G. Kolosko and F. F. Dall'Agnol, "Modeling basic tip forms and its field emission," 2020 33rd International Vacuum Nanoelectronics Conference (IVNC), 2020, pp. 1-2.

[7] T. A. de Assis and F. F. Dall'Agnol, "Minimal domain size necessary to simulate the field enhancement factor numerically with specified precision," J. Vac. Sci. Technol. B 37, 022902 (2019).

Does a banal tungsten field emitter obey the field emission theory?

Anthony Ayari[*+], Pascal Vincent[*], Sorin Perisanu[*], Philippe Poncharal[*], Stephen T. Purcell[*]

[*]Univ Lyon, Univ Claude Bernard Lyon 1, CNRS, Institut Lumière Matière, F-69622, VILLEURBANNE, France.

[+]Corresponding author: anthony.ayari@univ-lyon1.fr

Abstract—**Prompt by new theoretical propositions challenging the Fowler-Nordheim theory, we carried out field emission studies on the archetypal tungsten tip in a UHV environment. We tried different characterization methods in order to extract physical parameters more reliably. However, the different methods proposed so far rely on sweeping the voltage and fitting a straight line. Such methods suffer from the lack of long term stability of cold field emitter due to adsorption of electronegative residual gazes, ion bombardment and surface migration. We show that even with a moderate long term drift of the current, obtaining reproducible results is challenging. We propose another method based on a dual lock-in measurement of the current that could improve the extraction of physical parameters.**

Keywords—basic field emission, tunneling

I. INTRODUCTION

The Fowler-Nordheim (FN) [1] plot is a graphical representation widely adopted among experimentalist to analyze the properties of field emitters. Its main advantage is to give most of the time a reassuring and nice linear curve that allows to obtain two experimental coefficients (the slope and intercept). If the curve is not linear, it is a clear indication that some additional physical mechanisms need to be taken into account such as a voltage drop due to an electrical resistance or space charge effects. However, a linear curve is not a proof that the FN model describes correctly the behavior of the field emitter. It only confirms that the emission current has some sort of exponential dependence as function of the applied voltage. Practically, it is rather difficult to extract quantitative parameters from the analysis of the FN plot. One of the reason is that the FN model needs 3 independent physical quantities the work function ϕ, the emission area S and the field enhancement factor β relating the applied voltage to the electric field, but the fit gives only 2 parameters. Several attempts to verify quantitatively the FN model with or without the image charge potential on a tungsten emitter have been made and at best the field is estimated with an accuracy of 3 to 5 % and the current with a factor of 5 to 20 [2]. Such an experimental uncertainty doesn't allow to really test the field emission theory and to check the form of the tunneling barrier or to verify that ϕ, S and β are voltage independent. Recently, Forbes [3] has proposed a new analytical form of the field emission current and a method to test it. This work has motivated some theoretical studies [4-5] but, so far, it remained experimentally not very conclusive [6]. In this paper, we describe a new method to test the field emission theory and present our first attempt on a tungsten field emitter.

II. HOW TO CHARACTERIZE FIELD EMISSION?

A. The Standard approach

In a field emission experiment the emitted current is measured as a function of the applied voltage. According to the standard theory the current is given by:

$$I = aS(\beta V)^2/\phi \ exp(-bv(y)\phi^{3/2}/\beta V) \qquad (1)$$

where a and b are universal constants and $v(y)$ is the barrier shape correction factor that depends on the applied electric field. Then, from a linear fit of the logarithm of I/V^2 as a function of $1/V$, it is possible to deduce two physical parameters, if some hypotheses are made about the third physical parameter and about the value of $v(y)$. For instance, finite element numerical calculations can give the value of β if the geometry of the emitter is perfectly known (which is never the case in practice). It is also possible to obtain an independent measure from the electron energy distribution to avoid such doubtful hypothesis.

B. Measuring the Derivative of the Current

Recently an interesting analytical approximation of $v(y)$ was obtained in the case of a triangular tunneling barrier with a classical charge image correction. It appeared that a plot of the logarithm of I/V^κ as a function of $1/V$ (where κ~1.2 for a tungsten field emitter) should give a better fit of the experimental data. It was then proposed either to plot experimental data with different κ values to check which one gives the best fit or to obtain the voltage derivative of the current [3] because the new analytical approximation gives the following expression:

$$(V^2/I) \ dI/dV \ = b\phi^{3/2}/\beta+\kappa V \qquad (2)$$

So a plot of this ratio as a function of V should be linear with a slope giving directly the value of κ. Until now a reliable measurement of κ is still missing.

C. Measuring the Second Derivative of the Current

The derivative of the current can be obtained either by numerical derivation of the experimental I-V characteristic or by direct measurement with a lock-in amplifier when an additional AC voltage is applied. However, performing one of these measurements for different applied voltages might be difficult because of the lack of long term stability of field emitter. It is rather common to observe jumps and hysteresis by sweeping up and down the voltage. We propose to obtain the value of κ in a measurement requiring only a fixed DC applied voltage. This measurement consists of recording at the same time the DC current and AC current at the driving frequency of the look-in (which gives the first derivative of the current) as well as twice the driving frequency (which is

978-1-6654-2590-2/21 $31.00 © 2021 IEEE

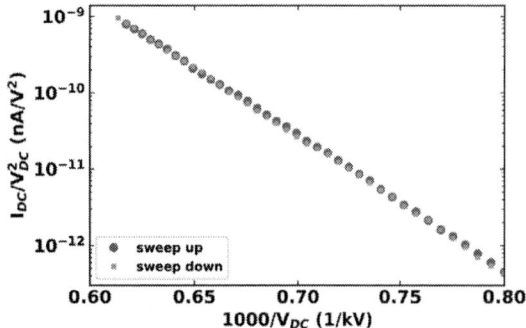

Fig. 1. Fowler-Nordheim plot of a tungsten field emitter.

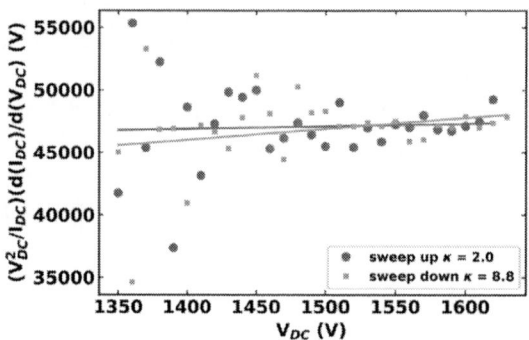

Fig. 3. Left hand side of (2) as a function of the applied voltage.

proportional to the second derivative of the current). Then a simple calculation gives:

$$\kappa = (V^2/I) \left[(2/V) \, dI/dV + d^2I/dV^2 - (1/I)(dI/dV)^2 \right] \quad (3)$$

III. CAN FIELD EMISSION BE EASILY CHARACTERIZED ?

A. Experimental set-up

We have performed field emission experiments on an electrochemical etched <111> tungsten tip cathode in ultra-high vacuum (base pressure ~ 3.10^{-10} Torr). Before each experiment, the tip was heated for 30 second with a resisting loop at a temperature of 1400°C. An AC voltage of 1 VRMS and a variable DC voltage was applied on the tip with a bias tee. A quadrupole at zero bias was in front of the tip. The current is collected with a coupled microchannel plate/phosphor screen system connected to a current amplifier and a lock-in amplifier. The role of the quadrupole is to shield the AC signal in order to avoid crosstalk with the phosphor screen by capacitive coupling.

B. Results

An I-V characteristic was performed for a voltage between 1250 V and 1630 V and current between 0.7 fA and 3 pA. The experiment was performed at such low current in order to reduce the current instability that generally increase with the square of the DC current. As shown in Fig. 1, the current is rather stable with moderate changes (~2%) in current between sweeping up and down the voltage. The AC current is noisier at low voltages than at high voltages. This is clearly observed for instance in Fig. 2 which represents the ratio of the lock-in

current on the DC current. At a low voltage the difference in current between the sweep up and the sweep down is about 10%. Nevertheless, it can be seen that this ratio tends to depend linearly with the applied voltage. The discrepancy between the two slopes is about 15 %. Finally, when the data are plotted according to (2) from [3], the linearity is much less convincing. The V^2 term amplify the uncertainty in the measurement. A fit of the data gives a value of κ that can fluctuate from -3 to 12. In Fig. 3, for instance the value of the slope is equal to 2 for the up sweep of the voltage and nearly 9 for the down sweep. The measurement of the second derivative of the current is too noisy to be usable for the moment.

IV. CONCLUSIONS

In conclusion, a new method to test experimentally field emission theory has been proposed by measuring the second derivative of the emission current. The experiment performed on a tungsten field emitter requires a reduction of the instrumental background noise of the detector as well as a better stability of the emitter, before giving a a result with sufficient accuracy. We expect a major improvement in the near future.

ACKNOWLEDGMENT

The authors would like to thank the Plateforme Nanofils et Nanotubes Lyonnaise of the University Lyon1.

REFERENCES

[1] R. Fowler, and L. Nordheim. "Electron emission in intense electric fields." Proceedings of the Royal Society of London. Series A, Containing Papers of a Mathematical and Physical Character 119.781, pp. 173-181, 1928.

[2] C. D Ehrlich, and E. W. Plummer. "Measurement of the absolute tunneling current density in field emission from tungsten (110)." Physical Review B 18.8, pp. 3767, 1978.

[3] R. G. Forbes, "Call for experimental test of a revised mathematical form for empirical field emission current-voltage characteristics." Applied Physics Letters 92.19, pp. 193105, 2008.

[4] M. Zubair, Y. S. Ang, and L. K. Ang. "Fractional Fowler–Nordheim law for field emission from rough surface with nonparabolic energy dispersion." IEEE Transactions on Electron Devices 65.6, pp. 2089-2095, 2018.

[5] B. Lepetit, "A quantum mechanical model of electron field emission from two dimensional materials. Application to graphene." Journal of Applied Physics 129.14, pp. 144302, 2021.

[6] R. G. Forbes, E. O. Popov, A. G. Kolosko, and S. V. Filippov, "The pre-exponential voltage-exponent as a sensitive test parameter for field emission theories." arXiv preprint arXiv:2012.02095, 2020.

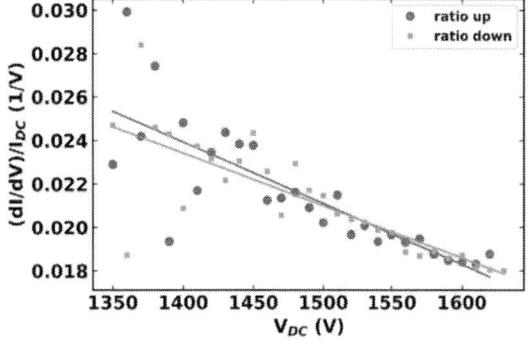

Fig. 2. Ratio of the derivative of the lock-in current on the DC current at fixed AC voltage and variable DC voltage.

A Generalized Formula for Barrier Strength (Gamow Factor), applicable to various field ion and electron emission contexts

Richard G. Forbes

University of Surrey, Advanced Technology Institute & Dept. of Electrical and Electronic Engineering, Guildford, Surrey GU2 7XH, UK
Permanent e-mail alias: r.forbes@ trinity.cantab.net

Abstract—Starting from the "scaled" form of the expression for barrier strength (Gamow factor) used in Murphy-Good field electron emission (FE) theory, this Poster develops (without detailed proof here) a generalized barrier-strength formula that can be applied to tunneling/transmission problems in several different field electron and field ion emission contexts. Tables show how this formula can be customized to the different barriers concerned, and how it can adapted to apply to older equation systems. This generalization means that modern developments in the theory of the special mathematical function v(x) can now be applied in physical contexts other than FE from free-electron metals.

Keywords—*Barrier strength, Gamow factor, transmission theory, tunneling theory, field ion and electron emission.*

I. Introduction

To discuss the theory of field electron emission (FE) current-field characteristics, the author now uses an *Extended Murphy-Good (EMG)* formulation [1]. This puts all theoretical uncertainty into a *formal* area-parameter that can be measured experimentally (and interpreted as and when theory allows).

In EMG-type FE theory, formulas for predicted emission current I_e can be written in any of the following *formats* (amongst others). In an *abstract format*, one can write

$$I_e = A_f^{SN} Z_F^{el} D_F^{SN} = A_f^{SN} Z_F^{el} \exp[-G_F^{SN}]. \quad (1)$$

Here, Z_F^{el} is the effective incident electron current density "at the Fermi level" ("$_F$"), as given by elementary theory based on an exactly triangular barrier. A_f^{SN} is the formal emission area for the Schottky-Nordheim (SN) barrier. D_F^{SN} is the tunneling probability through a SN barrier "at the Fermi level" (i.e., with total-energy component normal to the surface—the *normal-energy E_n*—equal to the Fermi level). G_F^{SN} is the *Gamow factor* (or *barrier strength*) (Jensen [2] denotes this by ϑ).

In the usual *direct format*, one replaces Z_F^{el} and G_F^{SN} by appropriate expressions and writes

$$I_e = A_f^{SN} (a\phi^{-1}F_C^2) \exp[-v_F b\phi^{3/2}/F_C], \quad (2)$$

where a and b are the Fowler-Nordheim constants as usually defined (e.g., [3]), ϕ is the relevant local work function, and F_C is (the magnitude of) a characteristic local surface electrostatic field (usually taken in modelling as the apex field). v_F is the appropriate particular value (for a barrier defined by ϕ and F_C) of the *principal field emission special mathematical function* v(x), where x is the *Gauss variable*.

To convert this to so-called *scaled format*, one defines a new variable (the *scaled field f*) by

$$f \equiv c^2\phi^{-2}F, \quad (3)$$

and scaling parameters $\eta(\phi)$ and $\theta(\phi)$ by

$$\eta(\phi) \equiv bc^2\phi^{-1/2}; \quad \theta(\phi) \equiv ac^{-4}\phi. \quad (4)$$

Using these expressions to substitute for a, b and F_C in (2) yields a formula for the barrier strength (Gamow factor)

$$G_F^{SN} = v(f_C) \cdot \eta/f_C, \quad (5)$$

and thus: $\quad I_e = A_f^{SN} (\theta f^2) \exp[-v(f_C) \cdot \eta/f_C]. \quad (6)$

Here: f_C corresponds to F_C, and the correction factor v_F in (2) has been written explicitly as a function of f_C, but the ϕ-dependences of η and θ have not been shown explicitly.

A merit of (5) and (6) is that, for given work-function ϕ, there is only one independent variable (f_C) in the equation. This makes various basic algebraic manipulations easier [4].

A second merit of form (5) is that all parameters appearing in it are dimensionless. This means that they can be used whatever the choice of independent variable (field or voltage), and whatever convention is used to represent electrostatic field, and in any equation system (see below).

A third merit is that (5), and the related modern theory of v(x) [4], can be generalized to apply to a range of field dependent transmission contexts. This Poster illustrates this.

II. Generalized Formula for Gamow Factor

A. Background: Electron Motive Energy and Gamow Factor

The theory presented in this Section is for a barrier of arbitrary zero-field height H, rather than for height ϕ.

The form of a transmission barrier is given by the related *motive energy M(z)*, defined in terms of the energies in the one-dimensional Schrödinger equation (the potential energy $U(z)$ and the normal-energy E_n, both measured from the same total-energy reference level) by

$$M(z) = U(z) - E_n. \quad (7)$$

It is convenient to measure z outwards from the emitter's electrical surface. The *barrier strength (Gamow factor) G* is a mathematical modeling parameter defined by

$$G \equiv 2(2m/\hbar^2)^{1/2} \int M^{1/2}(z)\, dz, \quad (8)$$

where m is the mass of the tunneling particle (not necessarily

978-1-6654-2590-2/21 $31.00 © 2021 IEEE

an electron), $\hbar$ is Planck's reduced constant, and the integral is taken "across the barrier", where $M(z) \geq 0$. The transmission probability D is then given by the chosen tunneling formalism. For the Kemble formalism [5], the formula is

$$D \approx 1/(1+\exp G), \qquad (9)$$

which reduces to the common simple (first-order) JWKB formalism $D \approx \exp[-G]$ when G is sufficiently large.

B. Analysis of the Basic Laurent-Form Barrier

The barriers of interest here have the mathematical form

$$M(z) = H - Cz - B/z, \qquad (10)$$

where H is the barrier's zero-field height, and B and C are constants. This barrier form seems to have no well-recognized name, so I have called it the *basic Laurent-form barrier*.

The roots of the equation $M(z)=0$ are

$$z = \frac{H \pm \sqrt{H^2 - 4BC}}{2C} = \frac{H \pm H\sqrt{1 - 4BC/H^2}}{2C} \qquad (11)$$

i.e.,
$$z = (H/2C)\left[1 \pm (1-\mu)^{1/2}\right], \qquad (12a)$$

where
$$\mu = 4BC/H^2. \qquad (12b)$$

Thus, the width of the barrier is given by

$$Width = (H/C)\,(1-\mu)^{1/2} \qquad (13)$$

The parameter μ has been called the "barrier parameter"; it now seems better to call it the *barrier width parameter*.

It can be shown that, for the basic Laurent-form barrier ("LB") of zero-field height H, a generalized form of the scaling parameter η defined by (4) is

$$\eta^{LB}(H) \equiv \left[(4/3)(2m)^{1/2}/\hbar\right] \cdot 4BH^{-1/2}, \qquad (14)$$

where m is the mass of the escaping particle. It is convenient to call η^{LB} the *barrier height parameter*.

It can be shown, by a lengthy mathematical argument (currently spread over several papers, using several different notations) that the barrier strength G^{LB} for barrier (10) is

$$G^{LB} = v(\mu) \cdot \eta^{LB}/\mu. \qquad (15)$$

III. Applications

We initially apply the above theory using the modern system of equations, now formally called the *International System of Quantities (ISQ)*. As Table I shows, different processes are described by different values for B and for C, thus defining different tunneling barriers, which may either be exact or a "useful simple approximation". Barriers are pulled down "on the outside" by an electrostatic field, and "on the inside" by the potential energy due to a model point charge, which may model a localized real charge or an image charge

The following notations apply in all cases in the Table: e is the ISQ elementary positive charge, F is the magnitude of the ISQ electrostatic field that defines the tunneling/transmission barrier, and ε_0 is the vacuum electric permittivity.

Proposed labels and names for the barriers of interest, and the contexts in which they occur, are as follows.

SN: *Schottky-Nordheim barrier* [6], in FE from metals.

MS: *Morgulis-Stratton barrier* [7], in FE from semiconductors.

HK: *Haydock-Kingham barrier* [8], in field ionization (FI) of gas atoms (used in field ion microscopy and the gas field-ion source), and in metal-ion post-field ionization (PFI) (as used in field evaporation theory and atom probe microscopy).

ID: *Ion-desorption barrier* [9], in field desorption, field evaporation, and high-temperature/field surface ionization.

TABLE I. FORMULAS FOR ISQ FORMS OF PARAMETERS B AND C

Barrier	C	$B \times (4\pi\varepsilon_0/e^2)$
SN	eF	$1/4$
MS	eF	$\lambda_s/4$
HK	eF	Z_{eff}
ID	neF	n^2

In Table I: λ_s is a correction factor arising because it is not accurate to model a semiconductor as a free-electron metal; Z_{eff} denotes the "effective nuclear charge" for the purposes of defining a tunneling barrier for a FI or PFI process; and n is the charge-number of the ion involved in ion tunneling.

The rules for deriving formulae applicable to the Gaussian and Hartree-unit equation systems are as shown below. ("$_s$" labels a Gaussian quantity, and "$_H$" a Hartree-units quantity.)

Gaussian system: replace eF by $e_s F_s$; $(4\pi\varepsilon_0/e^2)$ by $1/e_s^2$.

Hartree-units system: replace eF by F_H; $(4\pi\varepsilon_0/e^2)$ by 1.

These relations are provided so that the equivalences of older formulas with modern formulas can be checked. All modern work should be formulated using ISQ equations.

Formula (15) has the same tactical advantages and merits as does formula (5), but applies to a much wider range of physical phenomena. Hence, recent developments in the mathematical understanding of $v(x)$ can now be used in these application areas too. It will be particularly important to re-visit field ionization (FI) theory, because—as shown by Oppenheimer [10]—there are interesting analogies between (a) metal-atom FI and (b) FE from surface metal atoms.

References

[1] R. G. Forbes, Roy. Soc. Open Sci. 6, 190912 (2019).

[2] K. L. Jensen, Introduction to the Physics of Electron Emission Chichester, UK: Wiley, UK, 2018.

[3] R. G. Forbes and J. H. B. Deane, Proc. R. Soc. Lond A 467, 2927 (2011). See electronic supplementary information.

[4] R. G. Forbes, Chapter 9 (pp. 387–447) in: Modern Developments in Vacuum Electron Sources, G. Gaertner, W. Knapp and R. G. Forbes, Eds. Switzerland: Springer Nature, 2020).

[5] E. C. Kemble, Phys. Rev. 48, 549 (1935).

[6] L. W. Nordheim, Proc. R. Soc. Lond. A 121, 626 (1928).

[7] L. N. Dobretsov and M. V. Gomoyunova, Emission Electronics,. Moscow: Nauka, 1966. (In Russian.) Translated into English by Israel Program for Scientific Translations, 1971.

[8] R. Haydock and D. R. Kingham, J. Phys. B.: At. Mol. Phys. 14, 385 (1981).

[9] R. Gomer and L. W. Swanson, J. Chem. Phys. 38, 1613 (1963).

[10] J. R. Oppenheimer, Phys. Rev. 13, 66 (1928).

Gap in pagination due to formatting issues.

Pages 127

Electron energy analysis in Scanning Field Emission Microscopy using a Bessel box energy analyzer.

M. Bodik[1], M. Demydenko[1], C.G.H. Walker[1,2*], T. Bähler[1], T. Michlmayr[1], A.-K. Thamm[1], U. Ramsperger[1], A. Pratt[2], S.P. Tear[2], M.M. El Gomati[3], D. Pescia[1]

[1]*Laboratorium für Festkörperphysik, Auguste-Piccard-Hof 1, ETH Zürich, 8093 Zürich, Switzerland.*
[2]*Department of Physics, University of York, Heslington, York, YO10 5DD, UK.*
[3]*York Probe Sources Ltd, 7 Harwood Rd, York YO26 6QU, UK.*
Contact: chwalker@phys.ethz.ch, phone +41-4463 32506

Abstract— **In this study, we use Scanning Field Emission Microscopy (SFEM) combined with a miniature electron energy analyzer known as a Bessel box to measure electron energy spectra emitted from a sample. Previous studies using SFEM have revealed that the work function (ϕ) of the material under study has a significant role to play in the formation of the signal intensity. Hence, in order to understand the role of ϕ in greater detail, a sample of W(110) (ϕ = 5.25 eV) and a sample of Cs deposited on W(110) ($\phi \approx 1.7$ eV) were investigated. STM images show that the Cs covered surface has a speckled appearance indicating small Cs islands. The electron energy loss spectra obtained (which are the first using the Bessel box in SFEM) show differing structure in the elastic peak region. Monte Carlo (MC) simulations including quantum mechanical "bouncing" have been carried out. The results are consistent with MC simulations of the electrons escaping from the tip-sample junction.**

Keywords: Field Emission, Tungsten, Cesium, STM, EELS, Work Function

I. INTRODUCTION

The invention of Scanning Tunnelling Microscopy (STM) [1] led to a plethora of Scanning Probe Microscopy (SPM) techniques. Scanning Field Emission Microscopy (SFEM) [2] is an SPM technique whereby the microscope is operated in the STM manner, but the tip is retracted a few nm from the surface and a higher potential applied to the tip. This enables the tip to field emit electrons towards the surface. This in turn causes secondary electrons (SEs) to be emitted from the surface which can provide extra characterization of the surface such as the plasmon losses and spin polarization of the elastic peak [3]. Previous work has shown that the work function (ϕ) of the sample plays an important role in the emission of SEs [4]. Despite this, a simulation study found that ϕ should play a relatively minor role [5]. Hence, it is important that this aspect of SFEM should be studied in greater detail using samples with different work functions.

It was also found that SEs emitted from the tip-sample (t-s) junction had a low probability of escape due to the electric field from the tip forcing the emitted SEs back on to the sample [5]. In the case of low energy SEs, only those SEs generated up to 1 mm from the tip had any reasonable chance of escape. Higher energy electrons such as those in the elastic peak would also undergo interactions with the surface and thus lose any "memory" of the conditions at the t-s junction. The simulations show that a "halo" of electron emission will surround the t-s junction. More recently, further simulations by Walker et al. [6] which include electron bouncing due to quantum mechanical effects [7] suggest that in the case of the elastic peak electrons (and energies just below), there is a much higher probability of escape than previously estimated [5]. In order to test the conflicting simulation models, a miniature Electron Energy Analyzer (known as a Bessel box - BB) has been inserted into the SFEM Ultra High Vacuum (UHV) chamber. The BB can also be used to explore the effects of ϕ on the emitted electron energy spectrum.

II. SAMPLE AND TIP PREPARATION

A W(110) surface was cleaned in UHV using a procedure described in [4]. Subsequently, the sample was exposed to a Cs evaporation source to deposit ~0.5 ML Cs. This enabled two surfaces with large differences in ϕ to be used – W(110), ϕ = 5.25 eV and ~0.5 ML Cs on W, $\phi \approx 1.7$ eV [8]. Tips for the SFEM were made in-house using a method described in [4].

III. BESSEL BOX IMPLEMENTATION

The BB was designed and simulated by Suri [9]. It had been previously characterized using a standard electron gun [10]. Insertion into the SFEM was carried out by placing the BB vertically above the SFEM (see Fig. 1.). The BB position was controlled by a three-axis manipulator. Dimensions of the BB are shown in Fig. 1. After insertion of a new SFEM tip, the BB position is adjusted to determine the optimum position by acquiring counts from the elastic peak. Typically, the distance from BB front aperture to sample edge was adjusted to be about 2 mm.

978-1-6654-2590-2/21 $31.00 © 2021 IEEE

IV. METHOD OF DATA ACQUISITION

The field emission tip is initially operated in STM mode so that the precise tip-sample distance is known. STM images of the surface were acquired to ensure good surface cleanliness and to check for islanding of the Cs on the surface.

Process of data acquisition is: 1. Withdraw tip to ~40 nm distance from sample surface, 2. Apply a higher tip potential until field emission occurs, 3. Acquire SFEM data.

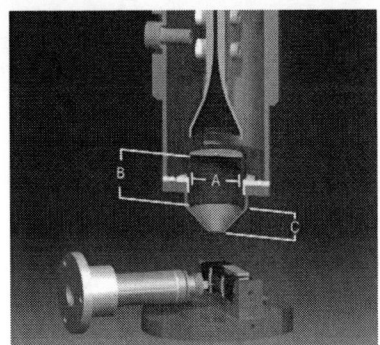

Fig 1. The Bessel box above the SFEM sample block (grey) and tip assembly (gold). Blue: BB cylinder (A = 10 mm, B = 10 mm), red: BB front cone (C = 5 mm) containing input aperture (dia 0.5 mm), cyan: BB back plate containing output aperture (dia. = 0.5 mm), yellow and green: channeltron (type: Dr. Sjuts KBL10RS).

V. MONTE CARLO SIMULATIONS

The Monte Carlo (MC) simulations were carried out using a similar geometric model to that used previously [5]. However, in this study, a COMSOL [11] model was used to simulate the electron trajectories in vacuum and a MC model [12] using Geant4 [13] was used to simulate the electrons in the sample. When an electron strikes the sample surface, the electron may bounce (either externally or internally). In this case, the program reverses its vertical direction but remains in the same program (either COMSOL or Geant4). If the electron passes through the surface, the parameters describing the trajectory (but including refraction effects) are transferred between the two programs. 10^5 electrons were simulated in each run. Further details of the simulation will be made available in [6].

VI. RESULTS

STM images of the clean W(110) and 0.5 ML Cs covered W(110) surfaces are shown in Fig. 2. Fig. 3 (Left) shows the electron energy loss spectra acquired when in the SFEM mode. Figure 3 (middle/right) shows the simulated spectrum from W/Cs surfaces. Note that with 10^5 initial electrons, only very few are able to emerge from the t-s junction. The MC simulations show that elastic peak reduces in size for lower ϕ in reasonable accord with the experimental results shown in Fig. 3. The MC simulations also show that many electrons from the t-s junction can escape and contribute to the detected signal. A small feature at ~2 eV loss in the experimental Cs spectrum is possibly replicated in the MC results. Means of improving the proportion of electrons from the t-s junction will be given in [6].

VII. CONCLUSION

STM images of Cs on W(110) show a speckled appearance in agreement with the disordered structure reported previously [8]. There is good agreement between simulations and experiment from materials with low and high values of ϕ suggesting that the MC model including reflection of the electrons at the surface due to a Quantum Mechanical effect are important. This allows many electrons originating from the tip-sample junction to escape the sample and be detected.

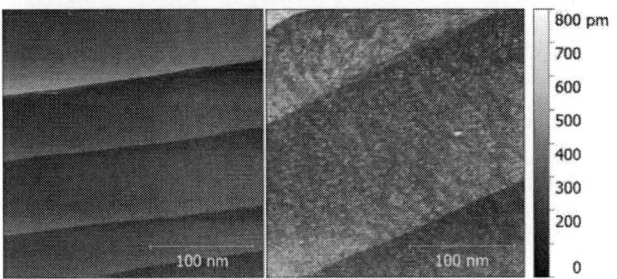

Figure 2. STM images of W(110) (Left) and 0.5 ML Cs on W (110) (Right).

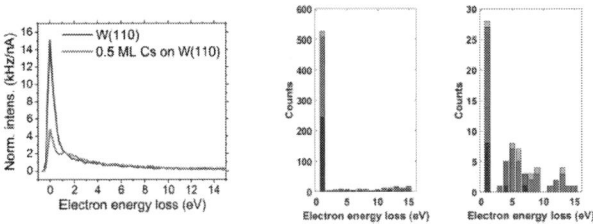

Figure 3. Left: Electron energy loss spectra acquired from clean W(110) substrate (black) and 0.5 ML Cs on W(110) (orange). t-s bias (U_B) = -50 V, t-s distance ≈ 40 nm. Middle: Simulated spectrum from W, Right: Simulated spectrum from Cs. For both simulations: U_B = -50 V and 100 nm t-s distance, Blue bars: electrons that have originated from the t-s junction. Red/orange bars: electrons that have emerged after returning into the material once/twice.

ACKNOWLEDGMENT

Support given by COMSOL Inc. and by the ISG at ETH and the financial support by the FP7 People: Marie-Curie Actions Initial Training Network (ITN) SIMDALEE2 (Grant No. PITN606988) is gratefully acknowledged.

REFERENCES

[1] G.Binnig and H.Rohrer, Surf. Sci., 126, 236-244 (1983).
[2] G.Bertolini et al., J. Elect. Spect. Rel. Phen. Volume 241, 146865, (2020).
[3] A-K Thamm et al., "Scanning Field Emission Microscopy with Spin and Energy Analysis" IVNC2021.
[4] G. Bertolini, Diss. ETHZ Nr. 27240, https://doi.org/10.3929/ethz-b-000476660.
[5] W.S.M. Werner et al., Appl. Phys. Lett., vol. 115, 251604, (2019).
[6] C.G.H. Walker, U. Ramsperger, H. Cabrera, D. Pescia, "Simulation of electron trajectories in Scanning Field Emission Microscopy", unpublished
[7] J. Cazaux, J. Appl. Phys. 111, 064903 (2012).
[8] A.G.Fedorus and A.G.Naumovets, Surf. Sci., 21, 426-439, (1970).
[9] A. Suri, (2020). PhD thesis, Univ. of York, http://etheses.whiterose.ac.uk/26527/
[10] A. Bellissimo et al., IVNC2020, pp. 1-2, doi: 10.1109/IVNC49440.2020.9203493.
[11] COMSOL Multiphysics® v. 5.5. www.comsol.com. COMSOL AB, Stockholm, Sweden.
[12] E. Kieft, E. Bosch, J. Phys. D: Appl. Phys., 41, 215310 (2008).
[13] S. Agostinelli et al., Nucl. Instr. Meth. Phys. Res. A, 506, 250-303, (2003).

978-1-6654-2590-2/21 $31.00 © 2021 IEEE

Fowler-Nordheim Slope Dependence on Pressure in Controlled Poor Vacuum

Girish Rughoobur, Olusoji O. Ilori and Akintunde I. Akinwande

Microsystems Technology Laboratories
Massachusetts Institute of Technology
Cambridge, MA 02139, USA
Email: grughoob@mit.edu

Abstract—We present the emission characteristics from a single Si emitter with integrated nanowire, obtained at different pressures by the influx of gases. This is enabled by a nano-positioning stage and a sharp tungsten anode. We compared emission characteristics in ultra-high vacuum to operation in Ar, O_2 and H_2 up to 10^{-5} Torr. Our results indicate that emission characteristics improved in H_2 but severely deteriorated in O_2. Original characteristics are recovered after measurements in O_2 by performing multiple sweeps, implying the degradation is not permanent and due to adsorbed gas molecules.

Keywords—single emitter, poor vacuum, FN slope

I. INTRODUCTION

Poor reliability is a major limitation of field emission cathodes, despite their instantaneous response and their exponential current-voltage ($I-V$) dependence. One source of poor reliability is adsorption/desorption of gas molecules, which makes ultra-high vacuum (UHV) essential for field emission devices [1]. In addition, Si emitters have stability and reliability issues as they could be contaminated with adsorbed molecules either from the ambient, or desorbed from an anode, or segregated from the bulk. The properties of this surface layer can have a dramatic impact on the emission properties. While surface adsorption can be beneficial for field ionization, [2] this causes significant variation of the work function (or the surface barrier), and hence the emission current in poor vacuum. Moreover, clean Si is reactive and easily contaminated, even at low pressures. Si forms a natural oxide SiO_x with uncontrolled properties; and to prevent the oxidation of Si, protective films coating with controlled properties or H_2 plasma clean are often used [3]. Several reports have attempted to characterize the effects of gases on field emitters, however they were limited by the fact that a single emitter tip could not be isolated [3], [4]. Here, we use the capability of a nano-positioning stage to move a single emitter very close to a sharp anode to enable emission at low voltage. We investigate the effects of different gases (Ar, O_2, and H_2) on the absolute value of the Fowler-Nordheim (FN) plot slope, b_{FN}, from a single Si emitter with integrated nanowire.

Sponsors: AFOSR/DURIP: FA9550-16-1-0244 and AFOSR/MURI: FA9550-18-1-0436.

II. EXPERIMENTAL

A. Emitter Fabrication and Measurement Set-up

Arrays of nanowires were fabricated by the following steps: (1) Wet-oxidation of n-type Si at 1000 °C to form 300 nm thermal oxide; (2) photolithography using a maskless aligner (MLA150, Heidelberg Instruments, Heidelberg, Germany) to form circular discs with diameter of 5 µm and pitch of 100 µm, (3) dry-etching of oxide-cap (4) semi-isotropic etch using SF_6/He, (5) deep reactive ion etching to form pillars with heights of 20 µm; (6) wet oxidation at 950 °C for 3 hrs to sharpen the emitter radius, $r_{TIP} < 20$ nm; and (7) oxide stripping in buffered oxide etch.

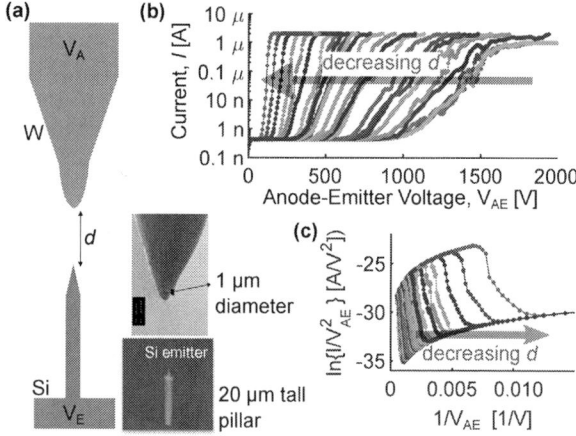

Fig. 1. (a) Schematic of measurement of a single Si emitter using a scanning W anode positioned at a distance d from the emitter tip; including scanning electron microscope images of the emitter and anode; (b) $I-V$ characteristics measured with decreasing d; and (c) FN plot of the $I-V$ curves.

Measurements are conducted in a scanning anode field emission microscope that has a W tip (radius ~500 nm) as the anode, and the Si emitter placed on a nano-positioning stage [Fig. 1(a)]. We first characterized the devices in UHV (10^{-9} Torr). $I-V$, characteristics are recorded using a Keithley 2657A with varying z-axis distance, d, between the anode and the emitter [Fig. 1(b)]. The current is limited to 2 µA to prevent early tip burn-out as we found in prior experiments that at high emission current the tip melted and its radius changed. The corresponding FN plots are shown in Fig. 1(c).

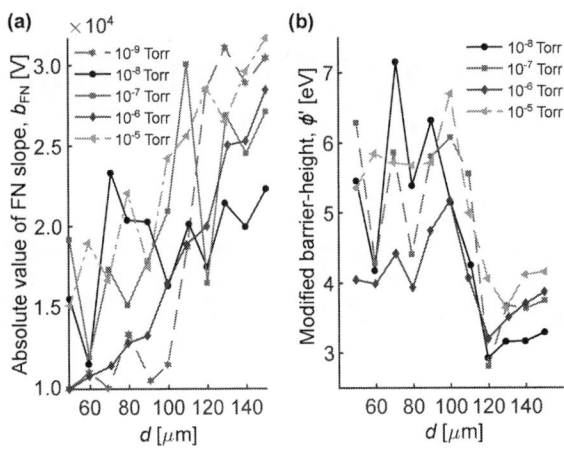

Fig. 2. (a) Absolute value of FN slope extracted in UHV (10^{-9} Torr) with decreasing distance, d, compared with different pressures of Ar and; (b) Modified barrier-height values calculated from the b_{FN} obtained in UHV.

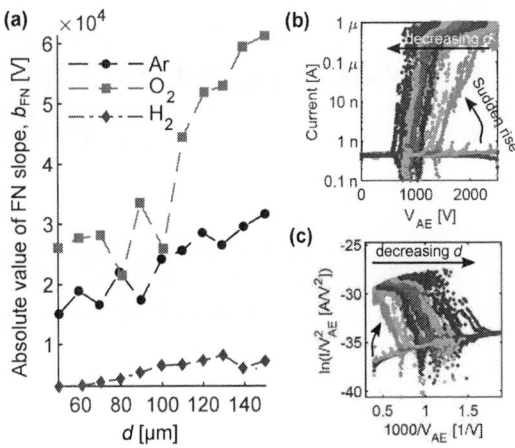

Fig. 3. (a) Comparison of b_{FN} for Ar, H_2 and O_2 at 10^{-5} Torr demonstrating the increase in the case of O_2, and a decrease in the case of H_2, at all the distances scanned; (b) Recovery in UHV after O_2 measurements showing sudden increase in currents after several sweeps at the larger distances; and (c) Corresponding FN plots showing performance recovery.

B. Effects of Pressure with Inert Gas, Ar

In poor-vacuum (10^{-8} Torr -10^{-5} Torr) using an inert gas, in this case Ar, we measure the $I-V$ characteristics at different d and extract b_{FN}. As pressure increases [Fig. 2], the values of b_{FN} in Ar oscillate around the b_{FN} obtained for UHV at the distances scanned. Using the measurement in UHV as reference, we extract the geometrical field-factor, β from b_{FN}. From the FN plot of type ($\ln\{I/V^2\}$ vs $1/V$), the value of b_{FN} can be extracted to estimate β in UHV from (1):

$$\beta = s(f)B\phi^{3/2}/b_{FN} \qquad (1)$$

where $s(f)$ is a special mathematical function [$s(f) \approx 0.95$ for metals], B is the second FN constant (6.83 eV$^{-3/2}$·V·nm^{-1}) and ϕ is the local work-function [5]. From these values of β, we then calculate the values of the modified barrier-height, ϕ', in the pressure range measured [Fig. 2(b)]. Significant deviations in the barrier-height are observed at all pressures, possibly due to Ar molecules impinging randomly on the anode or emitter.

C. Comparison with Oxidizing and Reducing Gases

The measurements are repeated in reactive gases: O_2 as oxidizing gas and H_2 as reducing gas. Before each gas, the measurements at UHV are performed to ensure that the performance is the same. The gas is allowed to stabilize for 1 hr in the chamber before each pressure measurement. We show the comparison of the values of b_{FN} extracted at the highest pressure of 10^{-5} Torr tested for all three gases in Fig. 3(a).

With the different gases investigated, we find that O_2 is the most detrimental to the performance as shown in Fig. 3(a). As expected, operation in O_2 resulted in substantial increase in b_{FN}, and hence $\phi' > 2\phi$ for $d \sim 150$ µm; however, in H_2, we measured a decrease in b_{FN}, suggesting a reduction ($\phi' < 0.5\phi$) for $d \sim 150$ µm. After field emission in gases, and overnight pump-down, we re-measured the emission characteristics in UHV. While recovery of the original emission characteristics after measurements in Ar and H_2 is immediate, several sweeps

($\sim$3-5) are necessary to obtain the $I-V$ characteristics at the outset as shown in Fig. 3(b) and (c) after characterization in O_2. Possible explanations are the adsorption of O_2 molecules on the emitter surface, which are electrostatically scrubbed from the surface by the $I-V$ sweeps. Nonetheless, since the measurements lasted over several hours in the presence of O_2, the emitter could also have been oxidized thus inhibiting electron emission. With initial sweeps in UHV, the oxide could be removed, which could cause the sudden recovery of the emission. In the future, we will compare these measurements with Si emitters decorated with Pt, Ir and Au.

III. CONCLUSION

Using an isolated Si nanowire with sharp emitter, we have performed a study on the field emission characteristics in poor vacuum. Our results show that the emission characteristics are dependent on the gas, with O_2 causing the most severe deterioration while H_2 causing improvement. Original characteristics are recovered after the deterioration in O_2. Through these measurements, we aim to understand the mechanism for the reliability issues of field emitters in poor vacuum.

REFERENCES

[1] G. Rughoobur, J. Zhao, L. Jain, A. Zubair, T. Palacios, J. Kong, and A. I. Akinwande, "Enabling atmospheric operation of nanoscale vacuum channel transistors," in *2020 Device Research Conference (DRC)*. IEEE, Jun. 2020.

[2] G. Rughoobur, A. Sahagun, O. O. Ilori, and A. I. Akinwande, "Nanofabricated Low-Voltage Gated Si Field-Ionization Arrays," *IEEE Transactions on Electron Devices*, vol. 67, no. 8, pp. 3378–3384, Aug. 2020.

[3] P. R. Schwoebel and I. Brodie, "Surface-science aspects of vacuum microelectronics," *Journal of Vacuum Science & Technology B: Microelectronics and Nanometer Structures*, vol. 13, no. 4, p. 1391, Jul. 1995.

[4] S. Itoh, T. Niiyama, and M. Yokoyama, "Influences of gases on the field emission," *Journal of Vacuum Science & Technology B: Microelectronics and Nanometer Structures*, vol. 11, no. 3, p. 647, May 1993.

[5] J. H. B. Deane and R. G. Forbes, "The formal derivation of an exact series expansion for the principal Schottky–Nordheim barrier functionv, using the Gauss hypergeometric differential equation," *Journal of Physics A: Mathematical and Theoretical*, vol. 41, no. 39, p. 395301, Aug. 2008.

Collector dependence of field emission in the Scanning Field Emission Microscopy

H.J. Gotsis[1], N.C. Bacalis[2], and J.P. Xanthakis[3,*]

[1]48-52 Konstantilieri Str 16231 Athens Greece

[2] National Hellenic Research Foundation Vasileos Constantinou Str Athens 11635 Greece

[3]Electrical and Computer Engineering Dept, National Technical University of Athens, Athens 15700 Greece

*Contact: jxanthak@central.ntua.gr, phone +30 2107723845

Abstract— **We have performed density functional VASP calculations of a carbon covered tungsten surface under the presence of an electric field F directed away from the surface i.e., pushing the electrons into the surface. We have found that under the presence of this field the extrapolated to the surface straight line which represents the increase of the vacuum energy with distance cuts the energy axis on the surface below the value E_{vac} at F = 0, by an amount ΔW = 0.1 - 0.15eV when F = 1V/nm. This effectively constitutes an F dependent decrease of the work-function W(F), which at higher F can explain the reported collector dependence of current in Scanning Field Emission Microscopy.**

I. INTRODUCTION

Experiments at ETH [1] have shown that in the near-field emission regime, that is when the tip is only a few nm away from the collector, (henceforth also sample), see Fig. 1, the current does not depend only on the tip characteristics but also on those of the sample. This phenomenon is by no means new and has been attributed to the work-function difference between emitter and collector. However, when the later was taken into account in either the Fowler-Nordheim equation (FN) or the FN appropriate for nanotips [2] the required work-function difference ΔW was approximately 1eV, an order of magnitude greater than the experimental 0.1eV deduced from Gundlach oscillations. It became clear that the observed experiments could not be interpreted by the traditional free-electron FN equation or by any of its modifications.

II. METHOD

In an attempt to understand these experiments we have performed density functional VASP calculations of surfaces with an electric field F directed away from the surface, i.e., pushing the electrons into the solid, which represents the situation at the collector side. The actual surfaces studied in [1] was 1) a reconstructed 15x12 carbon epitaxial monolayer on the (110) tungsten surface which was then compared to 2) a pure tungsten surface. In this work we compare the simpler completely carbon covered (100) surface of tungsten to the (100) surface of pure tungsten. Our intention is to uncover any new physics involved in these experiments rather than to obtain high numerical accuracy. Seven layers of tungsten were included and the top three were allowed to relax. In all our

calculations once convergence near the surface was achieved, the correction described in [3] was applied.

III. RESULTS

The calculated band diagram of the clean (100) tungsten surface is shown in Fig. 2 and that of carbon covered tungsten (C-W for short) is shown in Fig. 3. The clean-W work-function W(F) = E_{vac} - E_F was found to be 3.9eV. This is lower than the experimental value of 4.4 - 4.5eV. This difference is usually attributed to the exchange and correlation PBE functional that VASP is employing [4]. The work function of C-W comes out to be 4.0eV, i.e., 0.1eV higher than that of clean-W. The experimental value measured by the ETH group was 0.1eV lower than that of the pure tungsten surface. However this value was for the (15x12) reconstructed carbon covered tungsten surface.

We now examine what happens if an electric field pushing the electrons into the solid is applied to the clean-W (Fig. 4) and to the C-W material (Fig. 5). An examination of Fig. 1 reveals that the barrier at the emitter side is also determined by the slope of the line in the intermediate region between emitter and collector. This potential variation is in fact what we have calculated by VASP and is shown on the right side of Fig. 4 and Fig. 5. We observe that the extrapolation of this line to the surface of C-W cuts the energy axis at the surface at 0.15eV below the work-function = 4.0eV at an electric field F = 1V/nm. This effect appears also in pure tungsten but is not so prominent there. The above essentially constitute an electric field dependent work-function as initially predicted by Jensen [5] by a semi-analytic method but for an emitting and not for a receiving surface.

Furthermore, we observe that near a few nm away from the clean-W collector, the potential has lower slope (is more "horizontal") than the nominal 1V/nm (we checked that this also happens for 0.9 and 0.8V/nm), whereas for C-W it has exactly the nominal slope. Thus, the velocities of the incoming electrons near clean-W are more abundant in the x-, y-, directions compared to near C-W. Then, near clean-W, more electrons parallel to the xy surface will not be measured. Hence near C-W the current along z would increase. However, this effect would not be as significant as the observed ΔW since the latter has an exponential effect on the current.

978-1-6654-2590-2/21 $31.00 © 2021 IEEE

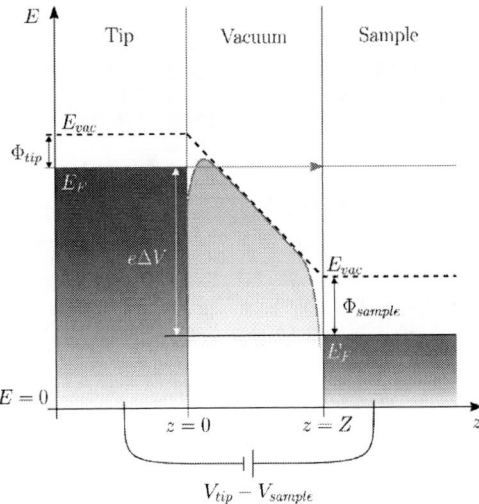

Fig. 1 The potential under electric field between tip and collector, linearly extrapolated toward the surfaces, ideally connects their work-functions. (In practice this is approximately true.)

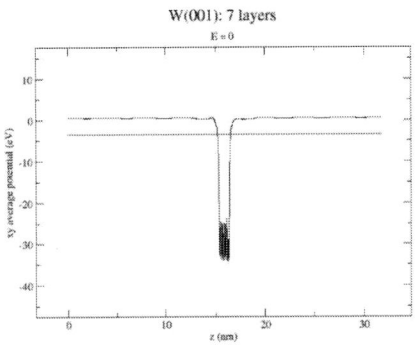

Fig. 2 The potential of clean-W under E=0 along with Fermi level (red line)

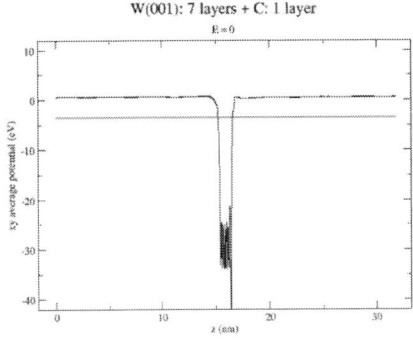

Fig. 3 The potential of C-W under E=0 along with Fermi level (red line)

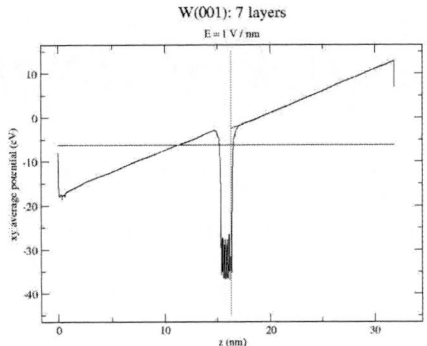

Fig. 4 The potential of clean-W under E=1V/nm along with Fermi level (red line). The green line is the energy axis at the surface.

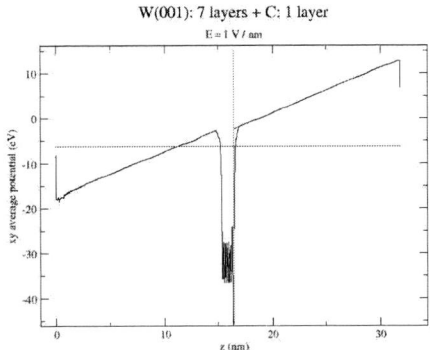

Fig. 5 The potential of C-W under E=1V/nm along with Fermi level (red line). The green line is the energy axis at the surface.

IV. CONCLUSIONS

Our VASP computations offer an explanation for the collector dependent field emission current in Field Emission Microscopy by revealing a field dependent work-function.

REFERENCES

[1] G. Bertolini, "Spectro-microscopy in the field emission regime of Scanning Tunneling Microscopy," PhD thesis ETH, 2020.

[2] A. Kyritsakis and J.P. Xanthakis, "Derivation of a generalized Fowler-Nordheim equation for nanoscopic field-emitters," Proc R Soc A 471 20140891, 2015.

[3] H.J. Gotsis, N.C. Bacalis and J.P. Xanthakis, "Density functional calculations with lattice relaxation of field emitted currents", 33rd International Vacuum Nanoelectronics Conference (IVNC), 2020.

[4] H. Toijala, K. Eimre, A. Kyritsakis, V, Zadin and F. Djurabekova, "Ab initio calculation of field emission from metal surfaces with atomic scale defects," Phys Rev B 100, 165421, 2019.

[5] K.L. Jensen "Semianalytical model of electron source potentialbarriers," Journal of Vacuum Science and Technology B 17, 515, 1999.

A Study of Self-Heating Effects in Looped Carbon Nanotube Fibers

Geet Tripathi[1], Kartik Sharma[1], Marc Cahay[1], Jonathan Ludwick[2], F. F. Dall' Agnol[3], T. A. de Assis[4]

[1]Spintronics and Vacuum Nanoelectronics Laboratory, University of Cincinnati
Cincinnati, OH 45221, USA
[2] Air Force Research Laboratory, Materials and Manufacturing Directorate, WPAFB, OH 45433, USA
[3] Department of Exact Sciences and Education (CEE), Universidade Federal de Santa Catarina
Campus Blumenau, Rua João Pessoa, 2514, Velha, Blumenau 89036-004, SC, Brazil
[4] Institute of Physics, Federal University of Bahia,
Campus Universitário da Federação, Barão de Jeremoabo St, 40170-115, Salvador, BA, Brazil
Corresponding author: cahaymm@ucmail.uc.edu

Abstract—Recently, Dall'Agnol et al. [1] have studied the field electron emission (FE) characteristics of carbon nanotube (CNT)-based looped fibers and shown that these cathodes have very high effective field enhancement factors (FEFs). Based on SEM images taken after FE measurements from looped CNT fibers, they proposed a two-stage model for FE from these cathodes consisting of a looped fiber with CNT fibrils near its apex. The modeling of a combined two-stage structure (looped CNT fiber + fibrils) based on COMSOL simulations lead to apex FEF values in excellent agreement with an orthodoxy theory analysis [2-4] of FE experiments performed on these fibers. In this work, we perform extensive computational simulations of the temperature distribution through the fiber and in some of the fibrils near it apex, which are eventually responsible for their observed large FEFs. In order to determine the condictions for thermal runaway in these looped CNT two-stage emitters, we determined the temperature at the end of a fibril located at the apex of the fiber as a function of the fiber and fibril geometries, their electrical and thermal conductivities, and the distance between the apex of the fiber and the anode.

Keywords—Field electron emission; electrostatic depolarization; field enhancement factor; self-heating effects; thermal runway

I. INTRODUCTION/BACKGROUND

Recently, Zhang et al. [5] showed that there is added thermal management benefit to using a looped fiber geometry for cold cathode applications. When compared to a vertical emitter, a semi-analytical treatment of self-heating effects through the CNT-fibers shows that the temperature at the apex of a looped fiber emitter is approximately half the temperature at the apex of a vertical CNT fiber of the same height when producing the same amount of emitted current [5]. The looped CNT fiber was modeled as a parabolic arc of height h_{fiber} = 4mm, base length b_{fiber} = 2 mm and diameter $2r_{fiber}$ = 200 μm, as shown in Fig.1(Left). To investigate the importance of the fibrils, we studied self-

heating effects in the the two-stage emitter model schematically illustrated in Fig.1(Right). The fibril consists of a "hemisphere on a post" model located at the apex of the main fiber [1].

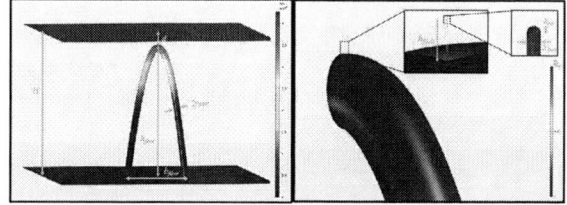

Figure 1: (Left) Looped CNF sandwiched between cathode and anode plates; d is the distance between the apex of the looped CNF and the anode; (Right) Illustration of the simulation domain used in the COMSOL Multiphysics 5.5 simulations to study the FE properties and self-heating effects in a looped CNF with a fibril at the apex of fiber. The fibril is modeled as a cylinder of height h_{fibril} = 20 μm and diameter $2r_{fibril}$ = 200 nm with a hemispherical cap of the same diameter at the apex.

In our simulations, the temperature dependence of the thermal conductivity κ(T) (in W/mK) of both the looped fiber and the fibrils is assumed to be of the form [6]:

$$\kappa = 392.5 - 0.4884\,T. \qquad (1)$$

The electrical resistivity of the fiber and fibril is assumed to be constant and equal to 73.95 μΩ-cm [6].

The FE characteristics of the looped fiber + fibril combinations are determined as a function of the fiber and fibril dimensions (h_{fiber}, h_{fibril}, r_{fiber}, r_{fibril}, and b_{fiber}), anode to fiber apex distance (d), and DC applied bias on the anode (V_{anode}). Conditions for thermal runaway of the fibrils are determined as a function of the cathode parameters and biasing conditions. We introduce a new approach to calculate the FE properties of the two-staged emitter based on a numerically efficient Linear Interpolation Table by Ludwick et al. to calculate the local emission current

978-1-6654-2590-2/21 $31.00 © 2021 IEEE

density over a wide range of applied external electric fields and temperatures [7]. The importance of including the temperature dependence of thermal conductivity of the fibers and fibrils to determine the FE characteristics of the looped fibers and the onset of their thermal runaway is illustrated by comparing our results to those when the thermal conductivity is assumed to be constant. Our simulations provide guidance for determining the optimum design and biasing conditions needed to avoid thermal runaway in looped CNT fibers to take full advantage of their impressive field enhancement factors.

II. RESULTS AND DISCUSSION

We first computed the FE characteristics of the looped CNT fiber+fibril cathode described above while varying the gap between the anode and apex of the fiber [distance d in Fig.1(Left)] from 250 to 450 μm, in steps of 50 μm. The FE characteristics are shown in Fig.2. This figure shows that an emission current I_{em} of 80 μA is achieved at an anode DC bias increasing from 5665 V to 7119 V when the anode to fiber apex is increased from 250 μm to 450 μm.

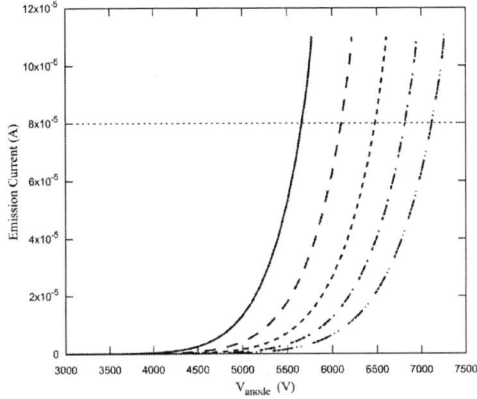

Figure 2: FE characteristics of the looped CNT fiber+fibril cathode with the parameters listed in the main text. From left to right, the FE characteristics were computed for an anode to fiber apex spacing equal to 250 μm, 300 μm, 350 μm, 400 μm, and 450 μm, respectively. An emission current of 80 μA is achieved at a DC bias on the anode equal to (from left to right) : 5665 V, 6106 V, 6482 V, 6819 V, and 7119 V, respectively.

Next, we varied the electrical and thermal conductivites of the fibril (while keeping the thermal conductivity of the fiber as given in Eq.(1) and its electrical resistivity fixed at 73.95 μΩ-cm) and calculated the temperature of its tip while varying its themal and electrical resistivities. Figure 3 shows the temperature at the end of a fibril (with the same dimensions as listed above) as a function of the applied anode voltage for an anode to fiber apex distance set equal to 250 μm. From left to right, the curves correspond to a fibril with a thermal conductivity equal to 1x, 2x, 4x and 8 times the $\kappa(T)$ temperature dependence given in Eq.(1). The resistivity ρ of the fibril was simultaneousy set equal to (from left to right) 1x, 0.5x, 0.25x, and 0.125x the electrical resistivity $\rho = 73.95$ μΩ-cm. For a constant bias of $V_{anode} = 5700$ V, the fibril apex-temperature was found to be equal

to (from left to right) 476.40 K, 337.44 K, 309.12 K, and 302.33 K..

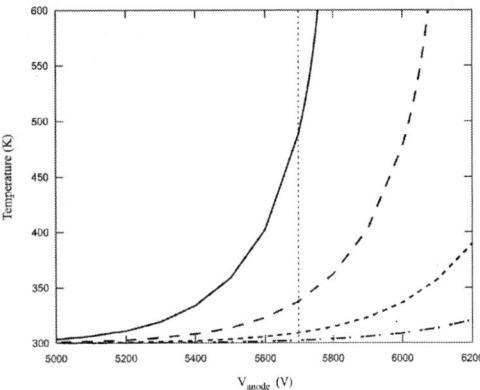

Figure 3: Temperature at the tip of the fibril located at the apex of a looped CNT fiber as a function of the anode bias for an anode to apex of the fiber spacing equal to 250 μm. The dimensions of the looped fiber and fibril are listed in the main text. The thermal conductivity of the looped fiber was modeled using Eq.(1) and its electrical resistivity was set to 73.95 μΩ-cm [6].

III. CONCLUSIONS

We have perfomed COMSOL simulations of self-heating effects in a looped CNT fiber+fibril cathode as a function of the anode to fiber apex distance. The FE characteristics of the cathode and the temperature at the tip of the fibril were determined for different models of the thermal and electrical conductivities of the fibril. These simulations shed some new light on the design of looped CNT fibers as efficient cold cathodes.

REFERENCES

[1] F. F. Dall'Agnol, T. A. de Assis, S. B. Fairchild, J. Ludwick, G. Tripathi and M. Cahay, Applied Physics Letters, 117, 253101 (2020).

[2] R. G. Forbes, Proceedings of the Royal Society of London A: Mathematical, Physical and Engineering Sciences **469**, 20130271 (2013).

[3] R. G. Forbes, Nanotechnology **23**, 095706 (2012).

[4] R. G. Forbes, J. H. B. Deane, A. Fischer, and M. S. Mousa, Jordan Journal of Physics **8**, 125 (2015).

[5] P. Zhang, J. Park, S. B. Fairchild, N. P. Lockwood, Y. Y. Lau, J. Ferguson, T. Back, Appl. Sci. 8, 1175 (2018).

[6] W. Zhu, M. Cahay, J. Ludwick, K.L. Jensen, R.G. Forbes, S.B. Fairchild, T.C. Back, P.T. Murray, J.R. Harris, D.A. Shiffler, "Chapter 22 - Multiscale Modeling of Field Emission Properties of Carbon-Nanotube-Based Fibers", In Micro and Nano Technologies, Nanotube Superfiber Materials (Second Edition), Chapter 22, Pages 541-572 (2019).

[7] J. Ludwick, M. Cahay, H. hall, J.O'Mara, and T.C. Back, "A new universal method for rapid semi-numerical determination of emission current density", submitted for presentation at IVNC2021

Influence of Contact Resistance on the Field Emission Characteristics of a Carbon Nanotube

Geet Tripathi[1], Marc Cahay[1], Jonathan Ludwick[2], and Kevin L. Jensen[3]

[1] Spintronics and Vacuum Nanoelectronics Laboratory, University of Cincinnati
Cincinnati, OH 45221, USA

[2] Air Force Research Laboratory, Materials and Manufacturing Directorate, WPAFB, OH 45433, USA

[3] Code 6362, Naval Research Laboratory, Washington, DC 20375, USA

Corresponding author: cahaymm@ucmail.uc.edu

Abstract— **Recently, we introduced an accurate algorithm to calculate the spatial dependence of the temperature distribution along a carbon nanotube (CNT) during field emission (FE). The algorithm considers the effects of Joule heating in the CNT, radiative losses from the CNT sidewall and tip, as well as the rate of heat exchange per unit area at the CNT tip due to either Henderson-cooling or Nottingham-heating effects. The previous work assumed that the temperature at the base of the CNT was the same as the temperature of the chuck to which it is attached and that the latter acts as a perfect heat sink. In this work, we consider the effects of contact resistance between the base of the CNT and the chuck and investigate its influence on the CNT FE properties. All other parameters being equal, our simulations show that the temperature profile along a CNT and its emission current are actually found to be larger those determined assuming a constant temperature $T_0 = 300$ K at the CNT/chuck contact [1].**

Keywords—Field electron emission; emission current; contact resistance.

I. INTRODUCTION/BACKGROUND

Recently, Tripathi et al. calculated the spatial dependence of the temperature along a single carbon nanotube (CNT) during field emission (FE) via a finite difference method [1]. They showed the importance of including the temperature dependence of the electrical resistivity $\rho(T)$ and thermal conductivity $\kappa(T)$ to determine the spatial dependence of the temperature along the CNT and its influence on its FE properties. The model assumed that the temperature at the CNT/heat-sink chuck contact was equal to the temperature of the heat sink and was set equal to $T_0 = 300$ K. With these improvements, the spatial dependence of the temperature profile along the CNT was found to be higher than the one predicted using constant thermal and electrical conductivity models.

In the past, Huang et al. investigated the FE properties of a CNT assuming a temperature dependent resistance in the contact between the CNT and the chuck to which it is attached [2]. Our work shows that the inclusion of this temperature-influenced contact resistance in our iterative model alters the I-V characteristics previously presented. The spatial dependence of the temperature along the CNT is determined using the one-dimensional model originally developed by Vincent et al. [3]:

$$\pi R^2 \frac{\delta}{\delta x}\left(\kappa(T)\frac{\delta T}{\delta x}\right)\delta x - 2\pi R\sigma\,\delta x\left(T^4 - T_o{}^4\right)\frac{I^2\rho(T)}{\pi R^2}\,\delta x = 0 \quad (1)$$

The latter is solved using the boundary condition proposed by N. Y. Huang et al. [2] to include the effects of contact resistance at the CNT/chuck interface:

$$T(x = 0) = \lambda\pi r^2\kappa(T)\left(\frac{\delta T}{\delta x}\right) + T_0\,, \quad (2)$$

where r is the CNT radius, x is the position along the CNT, κ is the temperature dependent thermal conductivity of the CNT, and λ is a tunable parameter representing the effects of contact resistance between the CNT and the heat sink. In this work, we consider values of λ similar to those considered by Huang et al. [3] and W. Wei et. al. [4].

We calculated the spatial dependence of the temperature profile along a CNT for a fixed CNT radius and aspect ratio, as a function of the applied external electric field, E_{ext}, while varying the parameter λ, which we take to be of the order of 10^7 K/W [3,4]. The algorithm used is a modification of the one outlined in ref. [1] modified to implement the boundary condition given in Eq. (2). To speed up the convergence of the iterative procedure needed to determine the spatial dependence of the temperature along the CNT, an efficient *Linear Interpolation Table* [5] was used to determine the emission current as a function of temperature and applied external electric field.

As shown in the next section, for a given set of CNT dimensions and physical parameters, our simulations show that the temperature profile along a CNT is actually higher than the one determined assuming a constant temperature $T_0 = 300$ K at the CNT/chuck contact [1]. As a result, the emission current is predicted to be larger at the tip of the CNT when the value of the parameter λ is increased. This

978-1-6654-2590-2/21 $31.00 © 2021 IEEE

is a counterintuitive result since the effects of contact resistance would be expected to reduce the emission current.

II. RESULTS AND DISCUSSION

We first calculated the temperature profile along a 3 μm long CNT with a diameter of 2 nm. The CNT thermal and electrical conductivities of this CNT were modeled using the analytical expressions detailed in ref. [1]. In this case, the critical applied external electric field (E_{crit}) for the onset of thermal runaway of the CNT with a base temperature (T_0 = 300 K) was found to be 532.691 V/cm [1]. Keeping all other parameters the same, the implementation of the contact resistance-boundary condition given in Eq. (2) (the value for λ was set equal to 1.774x10^7 K/W) resulted in a higher temperature at the CNT/chuck contact, as well as an overall upward shift of the spatial dependence of the temperature profile along a CNT, as illustrated in Fig.1.

With the inclusion of the contact resistance, the corresponding critical electric field before thermal runaway, E_{crit}, was found to be equal to 558.478 V/cm. For a value of the externally applied electric field of E_{ext} = 558.477 V/cm slightly below E_{crit}, the emission current ($I_{emission}$) value with and without contact resistance was found to be equal to 1.339 μA and 1.259 μA, respectively.

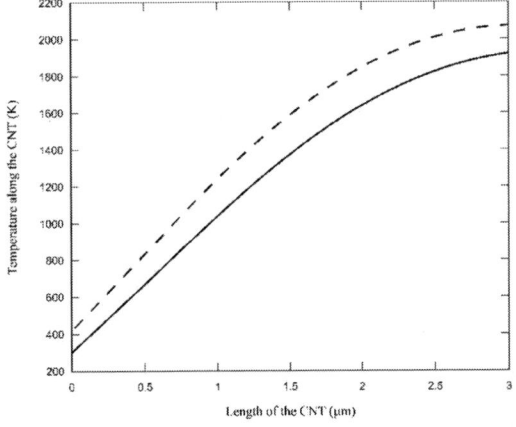

Fig.1: Temperature variation along a 3 μm long CNT with a diameter of 2 nm. The CNT thermal and electrical conductivities of this CNT were modeled using the analytical expressions detailed in ref. [1] The dashed line corresponds to the temperature profile in the presence of a contact resistance at the CNT/chuck interface, whereas the solid line corresponds to the temperature profile obtained with the temperature at the base of the CNT equal to the temperature of the chuck (300K). The curves are plotted for E_{ext} = 558.477 V/cm, a value slightly below the critical electric field E_{crit} value for thermal runaway in the presence of a contact resistance. The emission current ($I_{emission}$) was found to be equal to 1.339 μA and 1.259 μA, when the contact resistance is included and the temperature at the base of the CNT is set equal to 300K, respectively.

Next, we studied the influence of the parameter λ on the spatial dependence of the temperature profile along a CNT and on its FE charactetistics. The CNT parameters were the same as above but the value of λ was set equal to 1x, 3x, and 5x 1.774x10^7 K/W. The curves labelled (i), (ii), and (iii) show the temperature profiles along the CNT

for λ equal to 1x, 3x, and 5x 1.774x10^7 K/W, respectively. All curves were computed for a value of the externally applied electric field, E_{ext} = 557 V/cm. Figure 2(a) shows that the temperature at the CNT/chuck interface and the overall temperature profile in the CNT is larger when the value of λ increases. The CNT FE-characteristics for the 3 different values of λ mentioned above are shown in Fig. 2(b). For E_{ext} = 557 V/cm, the CNT emission current was found to increase with the value of λ: 1.265 μA, 1.341 μA, and 1.377 μA, respectively.

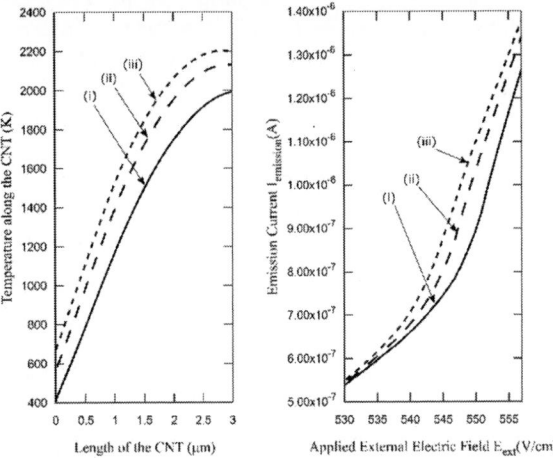

Fig.2: (a) Spatial temperature dependence along the CNT for a value of applied external electric field of E_{ext} = 557 V/cm. (b) FE characteristics of the CNT. In both figures, the curves labelled (i), (ii), and (iii) show the temperature profile along the CNT for λ equal to 1x, 3x, and 5x 1.774x10^7 K/W, respectively.

III. CONCLUSIONS

In this work, we have studied the spatial temperature profile along a CNT and its FE characteristics in the presence of a contact resistance at the CNT/chuck interface. There is an overall upward shift of the temperature profile as the effects of contact resistance increases. Also, the FE current at a given value of the external applied electric field was found to increase when the effects of contact resistance increased.

REFERENCES

[1] G. Tripathi, J. Ludwick, M. Cahay and K. L. Jensen, Journal of Applied Physics 128, 025107 (2020).

[2] N.Y. Huang, J. C. She, Jun Chen, S. Z. Deng, N. S. Xu, H. Bishop, S. E. Huq, L. Wang, D. Y. Zhong, E. G. Wang, and D. M. Chen, Phys. Rev. Lett. 93, 075501 (2004).

[3] P. Vincent, S. T. Purcell, C. Journet, and Vu Thien Binh, Phys. Rev.B 66, 075406 (2002).

[4] W. Wei, Y. Liu, Y. Wei, K. Jiang, L. M. Peng, and S. Fan, Nano Lett. 7, 64 (2007)

[5] J. Ludwick, M. Cahay, H. hall, J.O'Mara, and T.C. Back, "A new universal method for rapid semi-numerical determination of emission current density", submitted for presentation at IVNC2021, Lyon, France.

User-friendly method for testing field electron emission data: Technical report

Mohammad M. Allaham[1,2*], Alexandr Knápek[1], Marwan S. Mousa[3], and Richard G. Forbes[4]

[1]*Institute of Scientific Instruments of CAS, Královopolská 147, 612 64 Brno, Czech Republic*
[2] *Central European Institute of Technology, Brno University of Technology, Purkyňova 123, 612 00 Brno, Czech Republic*
[3]*Physics Department, Mu'tah University, Al-Karak 6170, Jordan*
[4]*Advanced Technology Institute & Department of Electrical and Electronic Eng., University of Surrey, Guildford, Surrey GU27XH, UK.*
Contact: allaham@isibrno.cz, Phone +420 541 514 318

Abstract—In field electron emission (FE) studies, the current/voltage or the macroscopic current-density/field characteristics of single tip or large area field emitters can be expressed in a nearly linear form using one of a small number of standard data-analysis plots. Usually, the chosen plot is a Fowler-Nordheim plot, a Murphy-Good plot or a Millikan-Lauritsen plot. The FE orthodoxy test can be applied to any of the three types of plots, to test the reasonability of the obtained experimental data. A difficulty of using the orthodoxy test is that there is no technical procedure or simple platform that can be used to apply the test to the experimental data. This report describes a simple web-tool that is designed to apply the FE orthodoxy test to any of these data-analysis plots, and then to use the test results to extract the emitter characterization parameters if the data passes the orthodoxy test. The web-tool is used by specifying the nature of the plot, the emitter's local work function, relevant system macroscopic parameters, and the coordinates of two "end-of-range" points on a line fitted to the data-analysis plot. The web-tool simplifies the data processing related to FE studies and experiments by: determining the value of the pre-exponential voltage/field exponent κ for Murphy-Good plots; evaluating the scaled-field parameters in FE theory that correspond to the ends of the working range; determining the status of the tested data before publishing it; determining the status of the emitter or experiments. Hopefully, the web-tool can help to develop basic understanding of the different behaviors of emitters.

Keywords—Fowler-Nordheim plot, Murphy-Good plot, orthodoxy test, characterization parameter, data-analysis web-tool.

I. INTRODUCTION

In field electron emission (FE) theory, Fowler-Nordheim (FN), Murphy-Good (MG) and Millikan-Lauritsen (ML) data-analysis plots are methods for presenting measured current/voltage $I_m(V_m)$ or the macroscopic current-density/field $J_M(F_M)$ characteristics of FE experiments in a nearly linear form. Advantages of using Murphy-Good plots are that they are predicted to be "very nearly straight" [1], and that there are fewer correction factors in the mathematics of the data-analysis process.

A web-tool has been developed that allows researchers to easily test and analyze experimental results by first applying the

so-called orthodoxy test and then (if the test is passed) extracting characterization parameters for the emitter used.

The three types of plots can be presented in the general form $\ln(Y/X^\kappa)$ vs X^{-1}. Here, Y is either the measured total emission current I_m or the macroscopic current density J_M, X is either the measured voltage V_m or the macroscopic electrostatic field intensity F_M, and κ is the pre-exponential voltage exponent in the *empirical FE equation*

$$I_m = CV_m{}^\kappa \exp[-B/V_m], \qquad (1)$$

where B can be treated as a constant and C can often be approximated as a constant.

In the expanded form of the Murphy-Good FE equation, κ is given by $2 - \eta/6$, where η is the scaling parameter for the Schottky-Nordheim (SN) barrier and is given by $\eta(\phi) = bc_S^2\phi^{-1/2}$, where b is the second FN constant, c_S is the Schottky constant and ϕ is the local work function [2]. Thus, $\kappa=2$ for FN plots, 0 for ML plots, and $(2 - \eta/6)$ for MG plots [2].

The field emission orthodoxy test is a quantitative test that can be applied to any of the data-analysis plots and to any geometrical emitter shape for which MG FE theory is an adequate approximation. The test is based on extracting values of a specific and important parameter in FE theory. This parameter is the *characteristic scaled field* $f_C = c_S^2\phi^{-2}F_C = c_S^2\phi^{-2}\zeta_C^{-1}V_m$, where F_C is the local electrostatic field at a characteristic location on the emitter surface (usually taken as its apex), and ζ_C [$\equiv V_m/F_C$] is the related *characteristic voltage conversion length (VCL)*. The extracted f_C-values are then compared to a set of internal analyzed historical data, taken from metal emitters between 1926 and 1972. These data provide the orthodoxy test criteria as listed in [3].

The test is an "engineering triage" test and provides three results. (1) Pass: the data are reasonable. (2) Fail: the data are unreasonable, and the extracted parameters are likely to be spurious. (3) Inconclusive; the data need more study and analysis. If the orthodoxy test is passed, then the characterization parameters of the emitter can validly be extracted. Typically, these are: the formal emission area (for a Schottky-Nordheim barrier) A_f^{SN}, and the related formal area

978-1-6654-2590-2/21 $31.00 © 2021 IEEE

efficiency α_f^{SN}; and the VCL ζ_C, and the related characteristic field enhancement factor (FEF) γ_C [2].

II. METHODOLOGY

The FE analysis web-tool [4] had been developed for the purpose of applying the orthodoxy test. It does this by making precise calculations for the parameter-extraction process from any of the three types of data-analysis plot. It is necessary to precisely calculate the values of the FE theory scaled parameters and (for MG plots) the κ value as defined in [1]. The main tool requires only the plot form, the assumed value of ϕ, and the coordinates of the upper and lower limits for the line fitted to the data. To extract values for γ_C and α_f^{SN}, where relevant-usually only for large area field electron emitters (LAFEs)-further macroscopic system parameters are needed, namely a cathode-anode macroscopic distance d_M (there are several types) and the macroscopic area A_M of the LAFE.

To test the performance of the web-tool, simulated current-voltage data plots were generated, for each of the three types of plot, using the input parameter-values: ϕ=4.65 eV, A_f^{SN}=100 nm^2, ζ_C=180 nm, d_M=100 µm, A_M=100 mm^2, with the tested range set to be $0.17 \leq f_C \leq 0.43$ (which is chosen to pass the orthodoxy test).

III. RESULTS AND DISCUSSION

The simulated data are presented in Fig. 1. For the MG plot, the value of κ has been calculated using the formula given earlier, which yields κ=1.2398. With each of the plots, a line fitted to the plot would have slope S^{fit} and intercept $\ln(R^{fit})$ on the vertical ($1000/V_m$=0) axis. Both these values can in fact be obtained from the coordinates that correspond to the range of the fitted lines. The resulting values of S^{fit} and $\ln(R^{fit})$, as evaluated by the web-tool, are shown in Table I. [Note that the Neper (Np) is the SI recognised unit of natural logarithmic difference, for amplitude-type quantities].

The range of f_C-values extracted from each of the plots, using the relevant formula in [3], coincides with the chosen input range, thereby demonstrating consistency.

Table I also shows values of extracted characterization parameters, except that no reliable theory currently exists for extracting area-like quantities from a ML plot.

All the data-analysis plots have "nearly straightline behavior", but each plot is slightly curved in a different way. This leads to a noticeable variation and percentage error in the extracted values of the characterization parameters but does not significantly affect the results of the orthodoxy test.
With this set of chosen data the FN plot appears to work slightly better than the MG plot, but our general impression is that for extracting area-like quantities the MG plot is expected to be more reliable and easier to use.

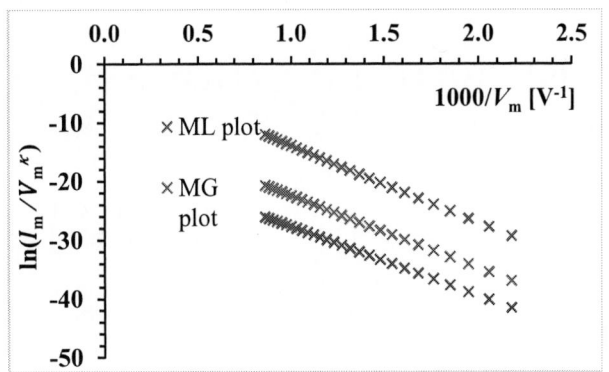

Fig. 1. Simulated data-analysis plots for the three plot types shown, for ϕ=4.65 eV, A_f^{SN}=100 nm^2, ζ_C=180 nm, and $0.17 \leq f_C \leq 0.43$.

IV. CONCLUSIONS

There is a need for a user-friendly tool that can apply the field emission orthodoxy test to experimental data, and (if appropriate) easily extract characterization parameters. Information of this kind is useful in the research and development of electron sources, in our case improved electron microscope sources and hybrid-design sources involving dielectric layers on metal point sources, but more generally sources based on large area field electron emitters. Our web-tool [4], now in its final stages of development, seeks to provide this facility, for all the forms of data-analysis plot commonly used.

Notwithstanding this, our strong recommendation is that best engineering and scientific practice is to *always* plot the *raw measured current/voltage data* [5], to use a Murphy-Good plot rather than a Fowler-Nordheim plot, and to apply the orthodoxy test before attempting to extract characterization parameters.

ACKNOWLEDGMENT

The research described in this paper was financially supported by the Ministry of the Interior of the Czech Republic (project. No. VI20192022147)

REFERENCES

[1] R. G. Forbes, R. Soc. Open Sci. 6, 190912 (2019).
[2] M. M. Allaham, R. G. Forbes, A. Knápek, and M. S. Mousa, J. Electr. Eng. Slovak 71, 37 (2020).
[3] R.G. Forbes, Proc. R. Soc. Lond. A 469, 20130271 (2013).
[4] M. M. Allaham, R. G. FORBES, A. Knápek, and M. S. MOUSA. Field emission analysis software [online]. Mu'tah University [cit. 2021-6-11]. Available from: https://fieldemissionanalysis.weebly.com.
[5] R. G. Forbes, "Using the parameter 'formal area efficiency (α_f^{SN})' to analyze current-voltage measurements on large-area field electron emitters, 34th IVNC, Poster/Paper this conference.

TABLE I

THE EXTRACTION RESULTS FOR EACH OF THE FIG. 1 DATA-ANALYSIS PLOTS

Plot type	S^{fit} [Np V]	$\ln(R^{fit})$ Np	f_C	ζ_C nm	γ_C	A_f^{SN} nm^2	α_f^{SN}
FN	−11720	−16.00	0.17 – 0.43	180.11	555.21	94.21	9.42×10^{-7}
MG	−12258	−10.18	0.17 – 0.43	178.96	558.79	94.72	9.47×10^{-7}
ML	−11813	−0.68	0.17 – 0.43	181.54	550.83	-	-
Simulation Input	-	-	0.17 – 0.43	180.00	555.56	100	10^{-6}

Testing the performance of Murphy-Good plots when applied to current-voltage characteristics of Si field electron emission tips

Mohammad M. Allaham[1,2,*], Philipp Buchner[3], Rupert Schreiner[3] and Alexandr Knápek[2]

[1] Institute of Scientific Instruments of CAS, Královopolská 147, 612 64 Brno, Czech Republic
[2] Central European Institute of Technology, Brno University of Technology, Purkyňova 123, 612 00 Brno, Czech Republic
[3] Faculty of Applied Natural Sciences and Cultural Studies, OTH Regensburg, 93053 Regensburg, Germany
*Contact: allaham@isibrno.cz, Phone +420 541 514 318

Abstract— **Murphy-Good plots are the most recent type of the analysis methods in the field electron emission theory, this type of plots has several useful characteristics such as: having the very-nearly straight line to represent the current-voltage characteristics and the absence of the correction factors in the mathematical procedure of the analysis process. In this study, n-type <111> and <100> Si chips containing four individual emitters are used as base emitters and mounted in a diode configuration field emission set-up where the experiments are operated in an ultra-high vacuum (~10^{-7} Pa). Each chip has four individual controllable emitters with 5 µm distance between the Si tips and the same material grid. Laser micromachining and subsequent wet chemical etching technique is used to structure and polish the tips. Murphy-Good plots are used to study the behavior of the Si individual tips and compare the results with the array current by extracting the field emission characterization parameters of the emitters.**

Keywords—*Murphy-Good plot, characterization parameter, n-type Si chip, electron source.*

I. INTRODUCTION

In field electron emission theory, Murphy-Good analysis plots; is a methodology to present the current-voltage characteristics of field electron emission experiments in a very-nearly straight-line curve. This allows to easily test and analyze the obtained results using the well-known field emission orthodoxy test and to extract the characterization parameters of the corresponding used emitter in the case the tested data passed the orthodoxy test. This type of plots has the form of $\ln(I_m/V_m^\kappa)$ vs V_m^{-1}. V_m is the measured voltage and I_m is the measured current. It is equal to the emission current when canceling out the effect of any noise currents in an orthodox system. $k = 2 - \eta/6$ and η is a scaled parameter defined in [1].

Using the slope and the exponent of the vertical axis intercepts of a Murphy-Good plot, several characterization parameters can be extracted for the used emitter such as: the formal emission area from a Schottky-Nordheim barrier A_f^{SN}, the characteristic voltage conversion length ζ_C and the characteristic field enhancement factor γ_C [1].

Field emission analysis webtool [2] is used in this work to easily apply the orthodoxy test and to extract the corresponding

characterization parameters if the practical data pass the test as described in [1].

The cathodes were manufactured as described in [3] with the exception that other types of silicon were used. (<111>, ϕ=4.67 eV and <100>, ϕ=4.82 eV [4]) the resulted Si tips are shown in Fig. 1-a and the grid is shown in Fig. 1-b.

II. PROCEDURE AND RESULTS

The resulting electron source is then mounted in a diode-configuration field emission setup and operated under a vacuum level of ~10^{-7} Pa. Each tip is connected to 1 MΩ resistance and a grounded picoamperemeter. The grid is connected to a DC and a grounded picoamperemeter as described in Fig. 1-c. The cathode current was measured for each tip individually in addition to the grid current. Fig. 2-a presents the resulted current-voltage characteristics for each tip in comparison with the total cathode and grid currents of the <100> electron source where the vertical red line describes the maximum voltage

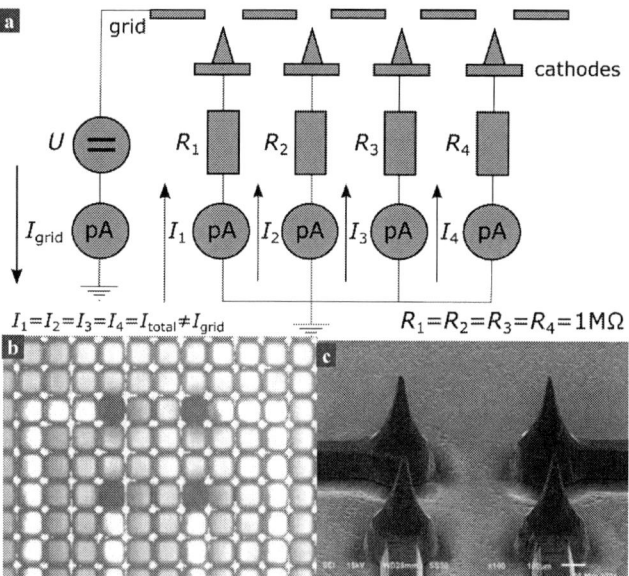

Fig. 1 (a) the cathode tips, (b) the grid. Both are prepared from the same material and (c) Schematic diagram for the diode configuration setup.

value before the noisy behavior started to appear. Fig. 2-b Shows the obtained results from the <111> electron source. Fig. 2-c shows the Murphy-Good plot for the <100> electron source results and Fig. 2-d shows the Murphy-Good plot for the <111> electron source results. Table I presents the extracted values for the characterization parameters of the used emitters as obtained from the webtool for the tips that passed or have an inconclusive result of the orthodoxy test only.

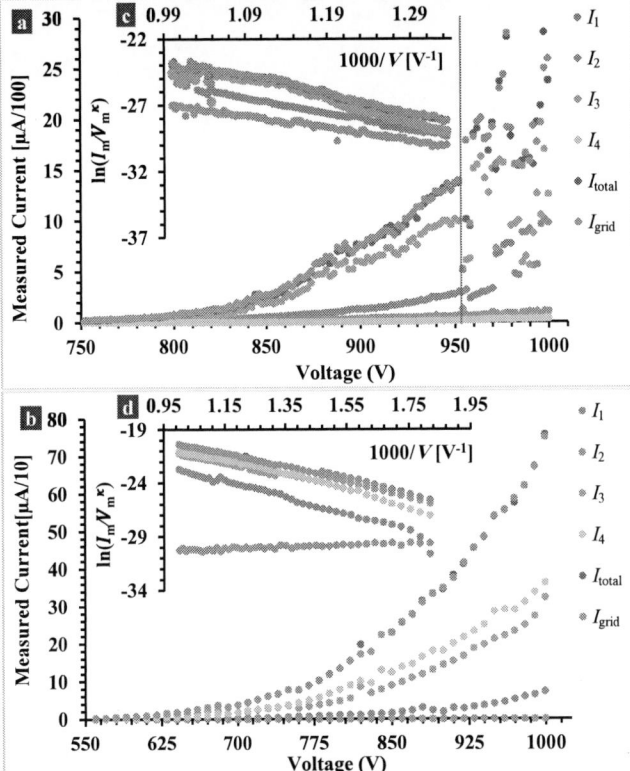

Fig. 2 The obtained results from the n-type Si electron sources (a) <100> current-voltage characteristics, (b) <111> current-voltage characteristics, (c) <100> Murphy-Good plot and (d) <111> Murphy-Good plot.

TABLE I
EXTRACTION RESULTS

Source	Current source	ζ_C [nm]	γ_C	A_f^{SN} [nm²]
<100>	I_1	160.19	31.07	2.66
	I_2	122.32	40.88	0.02
	I_3	208.24	24.01	490.24
	I_4	155.82	32.09	0.10
	I_{total}	189.3	26.41	139.58
	I_{grid}	182.03	27.47	77.23
<111>	I_1	100.18	49.91	0.11
	I_2	N/A	N/A	N/A
	I_3	81.14	61.62	0.15
	I_4	90.66	55.15	0.44
	I_{total}	86.58	75.75	0.58
	I_{grid}	87.94	86.56	0.66

III. DISCUSSION

γ_C values have been calculated using Forbes method [1] where it is defined as the ratio between the cathode-anode macroscopic distance, 5 µm in this case, and the extracted value for ζ_C that can be found in table I for each case. This explains the low values for γ_C. The values of A_f^{SN} describe the area that can be extracted from any type of the field emission analysis plots but not the real geometrical area, which still cannot be calculated accurately from any type of the analysis plots. A_f^{SN} can be used to describe the status of the emitter shape as a first approximation.

In this study, the tested voltage range for the <111> sample is 700-900 V while it is 810-950 V for the <100> sample. The current-voltage characteristics shows nearly exact behavior for the grid and the total cathode currents in both cases, and not all the samples are operating properly. The extracted results show very small values for the characterization parameters for the case of the <111> sample in comparison with the obtained results from the tips of the <100> sample. As can be seen from table I, lower values of ζ_C and A_f^{SN} can be relaed to the noisy behaviour and the need for higher voltage values for the field emission process to start.

IV. CONCLUSIONS

For both samples, not all cathodes are operating properly. The most likely reason behind this can be found from the values of the field enhancement factor in Table I since the variation in the radius and height of the individual tips strongly affects their emission behavior. Also, the existence of the safety resistance can affect the emission process by limiting the current feed from the power source for the parallel individual tips.

In conclusion, <100> samples presented better behavior as an electron source than the <111> samples since they provided better current-voltage characteristics and have better characterization parameters values.

ACKNOWLEDGMENT

The research described in this paper was financially supported by the Ministry of the Interior of the Czech Republic (project. No. VI20192022147) and Bayerische Forschungsstiftung (SI-FE-X, project No. AZ-1396-19).

REFERENCES

[1] M. M. Allaham, R. G. Forbes, and M. S. Mousa, "Applying the Field Emission Orthodoxy Test to Murphy-Good Plots" Jordan J. Phys., vol. 13 no. 2, pp. 101-111, 2020.
[2] M. M. Allaham, R. G. FORBES, A. Knápek, and M. S. MOUSA. Field emission analysis software [online]. Mu'tah University [cit. 2021-6-11]. Available from: https://fieldemissionanalysis.weebly.com.
[3] R. Ławrowski, M. Hausladen, and R. Schreiner, "Individually addressable silicon field emission cathode fabricated by laser micromaching," in 2020 7th International Vacuum Electronics Workshop (IVEW) & 13th International Vacuum Electron Sources Conference (IVeSC), May 2020.
[4] J. A. Dillon, and H. E. Farnsworth, "Work Function and Sorption Properties of Silicon Crystals" J. Appl. Phys. vol. 29, pp. 1195-1202, 1958.

Gap in pagination due to unavailable paper.

Page 142

Confined Electron Laser

Arya Fallahi[1,2*], Niels Kuster[1,2], Lukas Novotny[1]

[1]*ETH Zurich, Zurich, Switzerland*
[2]*IT'IS Foundation, Zurich, Switzerland*
*Contact: afallahi@itis.ethz.ch

Abstract— **Free-electron laser (FEL) is currently the only viable solution for producing strong THz radiation pulses and coherent beams in a hard x-ray regime, one of the ground-breaking inventions in the twentieth century. Nevertheless, the large size and high operational costs of an FEL has limited its use in research experiments and for advancing the technologies in the THz and X-ray radiation spectrum. In this contribution, we introduce the confined electron laser (CEL) concept for producing radiation with similarities to FEL outputs, but with less limitations. By confining the electrons using the gradient forces inside an optical cavity, electron beams with high charge densities and low divergence can be realized. This CEL concept can result in orders of magnitude improvements in the source efficiency and coherence of the radiation for compact setups compared with the FEL concept.**

I. Introduction

A free-electron laser (FEL) facility offers peak brightness many orders of magnitude greater than that of third-generation light sources, as well as pulse lengths of 100 fs or shorter, with fully coherent characteristics [1]. In a free-electron laser, relativistic electrons traverse a static undulator and follow the trajectory of a wiggling motion. After long interaction lengths, the radiated electromagnetic wave interacts with the bunch and the micro-bunching phenomenon occurs, leading to a periodic modulation of charge density inside the bunch with the periodicity equal to the radiation wavelength. This effect results in a coherent radiation scaling with the square of the bunch numbers. By adjusting the energy of the electron beam and the undulator period, coherent radiation in x-ray wavelengths can be generated. Coherent x-rays offer unprecedented potential for enabling biologists, chemists and materials scientists to study various evolutions and interactions with nanometer and sub-nanometer resolutions [2].

Besides numerous investigations and studies on x-ray FELs, research efforts have focused on building compact x-ray sources, where novel schemes for generating x-ray radiation in a so-called tabletop setup are examined and verified. These efforts can be categorized into two categories: (i) research on compact accelerators and (ii) compact undulator studies. There are currently several pathways towards making miniaturized accelerator modules. In another approach, compact undulators like optical undulators, where the oscillations in an electromagnetic wave realize the wiggling motion of electrons, are used. Sources based on optical undulators are typically referred to as Inverse Compton Scattering (ICS) or Thomson Scattering (TS) sources. Several studies have predicted the

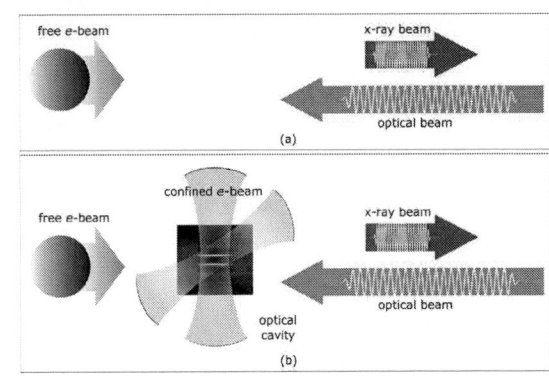

Fig. 2: *Schematic comparison of (a) FEL concept and (b) CEL concept.*

possibility of achieving FEL-like performance using optical undulators [3]–[5]. However, until now, no operational FEL based on optical undulator exists. For the coherent gain to be triggered, certain conditions, including (a) small space-charge forces to prevent expansion before micro-bunching, (b) small beam divergence to maintain co-propagation of e-beam and radiated light for the required interaction length, (c) small electromagnetic recoil due to photon emission, and (d) sufficiently long interaction length, must be met. Fulfilling these conditions is the main challenge in realizing a FEL interaction and reaching to the radiation saturation regime.

This contribution hypothesizes that the challenges originate from the free nature of the electron beams in the FEL mechanism (Fig. 1a). Through full-wave simulation of optical undulator radiation, we first show that the strong space-charge forces in the e-beams cause longitudinal expansion of the electron bunch. This expansion leads to continuous change in the charge density and consequently prevents the process of coherent gain. Secondly, it is shown that by adding the fields of an optical cavity that transversally confines the electron beam to field nodes (Fig. 1b), the coherent gain can be triggered. This occurs due to realization of electron microbeams with high charge density and zero divergence inside the optical cavity fields [6].

II. Results

We used the code MITHRA 2.0 to simulate the interaction through a Finite-Difference Time-Domain / Particle-in-Cell (FDTD/PIC) algorithm [7]. The selected interaction parameters, as shown in Table 1, were based on the state-of-the-art electron gun and laser technology. The results of the simulation are shown in Fig. 2a. The radiated power in front of the bunch is plotted versus the propagation length. The simulation results for a laser beam scattered off a free electron beam with and without space-charge effects demonstrate that the Coulomb repulsion between the electrons in the low energy (5 MeV) beam precludes the coherent gain. In other words, the space-charge effect that is typically neglected in FEL operation due to ultra-relativistic energies is not negligible in an ICS scheme.

Table 1: Source Parameters

Parameter	Value
Current profile	Uniform
Bunch RMS size ($\sigma_x = \sigma_y$)	5 μm
Bunch length (σ_z)	20 μm
Bunch charge (Q)	0.64 pC
Bunch energy (E)	5.12 MeV
Bunch current (I)	9.6 A
Longitudinal momentum spread ($\sigma_{\gamma\beta_z}$)	2.5×10^{-3}
Normalized emittance (ϵ_n)	50 nm-rad
Laser wavelength (λ_0)	10 μm
Laser strength parameter (a_0)	1.0
Pulse duration (cT)	18 mm
Laser pulse type	Flat-top
Radiation wavelength (λ_x)	37.6 nm
Cavity wavelength (λ_c)	11.8 μm
Cavity beam size ($w_{0x} \times w_{0y}$)	7.4×0.1 mm²
Cavity beam strength parameter (a_{0c})	0.07
Beam focal position ($c\delta t$)	8.4 mm

The radiaton was enhanced by a factor of 300 for the same simulation with the cavity fields. Analagous to the conventional FEL operation, the radiation began with an initial incoherent radiation, originating from the shot noise in the bunch electron distribution, proceeded with a gain regime in the radiation, and ultimately, the radiated power was saturated. The radiation gain was thus triggered in the CEL scheme but not in the FEL scheme. The bunch profile at the maximum radiation instant for the two schemes were compared, as shown in Fig. 2b. High density microbeams produced in the cavity were microbunched, whereas the low-density electron beam did not experience any visible microbunching.

From FEL physics, the FEL parameter in an undulator can be written as

$$\rho_{FEL} = \left[\frac{\mu_0 \widehat{K}^2 e^2 \lambda_X^2 n_e \gamma}{32\pi^2 (1 + K^2/2)^2 m_e} \right]^{1/3} = C[n_e \gamma]^{1/3},$$

where μ_0, K, n_e, γ, λ_X, and m_e denote the vacuum permeability, undulator parameter, electron density, electron Lorentz factor, x-ray wavelength, and electron mass respectively. From this equation, a FEL can be realized with a small undulator period (i.e. small electron energy) and with the same efficiency as a long undulator period (i.e. large electron energy), by using higher charge densities. In the confined electron laser scheme,

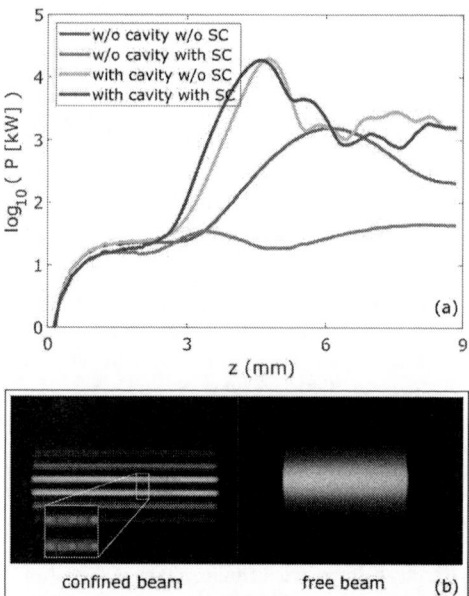

Fig. 2: (a) radiated power on logarithmic scale versus undulator length for free and trapped electrons with and without considering space-charge effect and (b) the electron

such high charge densities are produced using the gradient forces inside the optical cavity.

III. CONCLUSIONS

A novel confined electron laser device that can generate a compact coherent x-ray source by inverse Compton scattering of low-energy electrons transversely confined inside fields of an optical cavity is introduced. Transverse gradient forces push the particles towards regions with weak field values in the fields of an optical cavity, leading to high density microbeams with zero divergence. Numerical simulations show that this is a potentially better solution for microbunching and lasing than a free electron beam.

REFERENCES

[1] C. Pellegrini, A. Marinelli, and S. Reiche, "The physics of x-ray free-electron lasers," *Rev. Mod. Phys.*, vol. 88, no. 1, p. 15006, 2016.

[2] K. J. Gaffney and H. N. Chapman, "Imaging atomic structure and dynamics with ultrafast X-ray scattering," *Science (80-.).*, vol. 316, no. 5830, pp. 1444–1448, 2007.

[3] J. C. Gallardo, R. C. Fernow, R. Palmer, and C. Pellegrini, "Theory of a free-electron laser with a gaussian optical undulator," *IEEE J. Quantum Electron.*, vol. 24, no. 8, pp. 1557–1566, 1988.

[4] C. Chang *et al.*, "High-brightness X-ray free-electron laser with an optical undulator by pulse shaping," *Opt. Express*, vol. 21, no. 26, pp. 32013–32018, 2013.

[5] A. Bacci, C. Maroli, V. Petrillo, A. R. Rossi, L. Serafini, and P. Tomassini, "Compact X-ray free-electron laser based on an optical undulator," *Nucl. Instruments Methods Phys. Res. Sect. A Accel. Spectrometers, Detect. Assoc. Equip.*, vol. 587, no. 2–3, pp. 388–397, 2008.

[6] A. Fallahi, N. Kuster, and L. Novotny, "Transverse Confinement of Electron Beams in a 2D Optical Lattice for Compact Coherent X-Ray Sources," *arXiv Prepr. arXiv2104.11586*, 2021.

[7] A. Alba, A. Adelmann, and A. Fallahi, "OPAL-MITHRA: Self-consistent Software for Start-to-End Simulation of Undulator-based Facilities," in *33rd International Vacuum Nanoelectronics Conference, IVNC 2020*, 2020.

Single - Cycle THz Accelerating Structure with Wave Beam Focusing Lens

Sergey Antipov[1], Sergey Kuzikov[1]*, and Alexander Vikharev[2]

[1]*Euclid Techlabs LLC, Bolingbrook, IL, 60440*

[2]*Institute of Applied Physics, Russian Academy of Sciences, Nizhny Novgorod, 603950, Russia*

Contact: sergeykuzikov@gmail.com, phone +1-203-435-6400

Abstract— **Recently, gradients on the order of 1 GV/m level have been obtained in a form of a single cycle (~1 ps) THz pulses produced by conversion of a high peak power laser radiation in nonlinear crystals (~1 mJ, 1 ps, up to 3% conversion efficiency) [1]. Such high intensity radiation can be utilized for charged particle acceleration. However, these pulses are short in time (~1ps) and broadband, therefore a new accelerating structure type is required. In this paper we propose a novel structure based on focusing of THz radiation in accelerating cell and stacking such cells to achieve a long-range interaction required for an efficient acceleration process. We present an example in which a 100 microJoule THz pulse produces a 600 keV energy gain in 5mm long 10 cell accelerating structure for an ultra-relativistic electron. This design can be readily extended to non-relativistic particles. Such structure had been laser microfabricated and appropriate dimensions were achieved.**

I. INTRODUCTION

The progress of particle and accelerator physics relies on achievement of high gradient charged particle acceleration. The record gradients above GV/m are being produced in laser plasma wakefield acceleration (LPWA) and beam driven plasma wakefield acceleration (PWFA). Traditional technology reaches the level of 100 MV/m in normal conducting microwave accelerating cavities with frequency range of 1-10 GHz [1]. The further progress in normal conducting structure acceleration had been limited due to breakdown and thermal fatigue effects associated with pulsed heating [2-3]. Threshold values of surface fields strongly depend on the time of material exposure to these extreme fields. Experimental statistics [2] yield the following scaling law for surface electric E_s field as a function of exposure time τ (microwave pulse length): $E_s^p \tau = const, p = 5 - 6$. It was also discovered that a breakdown threshold correlates with pulsed heating of accelerating structure ΔT, which is defined by the surface magnetic field Hs: $\Delta T \sim H_s^2 \sqrt{\tau} = const$ [2]. These experimental dependencies strongly suggest that the gradient can be increased further if the pulse length is reduced. We present here an idea for a normal conducting accelerating structure that takes this idea to its ultimate limit. We propose to use an ultra-short pulse of a single cycle duration. In the case of short pulse accelerating structure must be extremely broadband, completely different from standard arrays of coupled resonators. A new approach is required for accelerating structure design to facilitate efficient energy transfer from a broadband pulse to a charged particle.

II. ACCELERATING STRUCTURE DESIGN

In the geometry presented here a picosecond pulse is fed symmetrically from two sides into an array of accelerating cells (Figure 1). Each accelerating gap has a dielectric delay line that is tuned to maintain synchronism between the THz pulse and electrons. The use of dielectrics of increased thickness along the length of the accelerating array ensures that electron passes each accelerating gap at the maximum of electric field. To increase the accelerating gradient in the gap it was offered to use dielectric lenses to focus the THz pulse into an accelerating gap [4] (Figure 2). Dielectric lens made of a dispersion-less material is a broadband element and does not produce noticeable losses and distortions of accelerating field. The accelerating gradient can be increased multiple times due to focusing of the THz pulse.

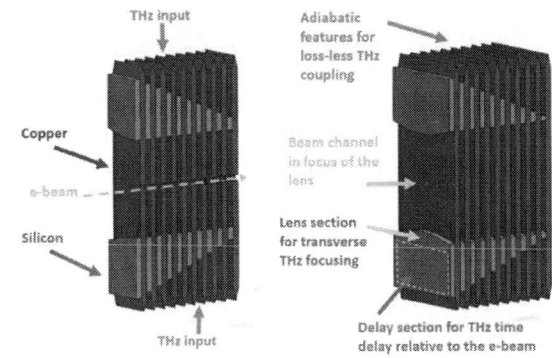

Fig. 1 Sketch of broad band THz structure based on dielectric delay sections and lenses.

In the case of ultra-relativistic electron beam, that moves with velocity virtually equal to speed of light the lengths of delay inserts decreases correspondingly to maintain synchronization between the particle and THz pulse: $\Delta = P/(\sqrt{\varepsilon} - 1)$, here ε - is dielectric permittivity of the delay material, p – is a period of the structure and Δ is the delay line length change. The Figure 3 demonstrates bunch acceleration basics and timing of cells. One can see the importance of dielectric delay lines for the acceleration process (time proceeds from left to right).

978-1-6654-2590-2/21 $31.00 © 2021 IEEE

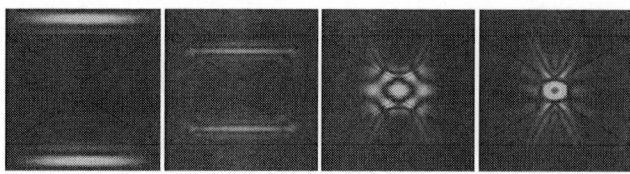

Fig.2. E-field distributions at the quartz lens while focusing the short THz pulse, from the time correspondent to THz pulse arrives to lens to the maximum of focusing.

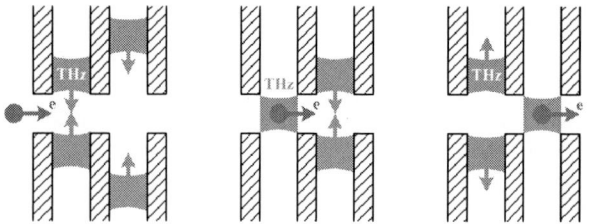

Fig.3. Timing of THz pulses in the structure.

For a fixed energy THz pulse the highest accelerating gradient can be achieved if this whole energy is focused into a single accelerating cell. When the same pulse has to be split to feed N accelerating cells the gradient in each cell decreases as $\sim\sqrt{N}$. However the total energy gain by a charged particle grows as $\sim\sqrt{N}$, being a product of gradient and structure length. For this reason we are considering an extended interaction of THz pulse with the charged particle in multi-cell accelerating structures with $N \geq 10$.

The number of accelerating cells is limited by the dispersion of the dielectric properties of the delay line material and the wide waveguide dispersion which may lead to THz pulse shape distortion. For a broad band THz pulse the choice of delay line dielectric is more important as the dielectric constant should not have much variation over the wide range of frequencies corresponding to the THz pulse. For a practical structure design presented below we had chosen a high resistivity silicon and quartz as delay material. Their frequency dispersion is minimal in the range of frequency of interest for this application [5] for both, index of refraction and power loss. A short, picosecond THz pulse propagating in 2 millimeter long silicon delay line, experiences a 1% power loss. Quartz is a bit more lossy having 4% power loss in the same conditions.

III. SIMULATION OF ACCELERATION

We simulated particle acceleration in the quartz-delay-focusing line multi-cell accelerating structure. Parameters of simulated structures are given in the Table 1. This simulation had been done in CST Microwave Studio [CST]. We assumed a 3 MeV bunch of electrons with 1 pC charge. The electron beam duration was 100fs and transverse size of 100um, which is typical for state of the art ultrafast electron microscope facilities [6]. The structure had 10 cells with proper delay lines. In our simulation we assumed a 100 uJ THz pulse delivered into accelerating cells. Considering all losses (reflection at the entrance, 0.5%, reflection at the delay line (11% for quartz), power attenuation in the delay line and lens (4% for quartz) the input energy of the pulse should be 116

microJoules. With total of 10uJ per cell maximum accelerating field reaches $3.85 \cdot 10^3$ kV/c. Total energy gain is 600keV. The timing between the THz pulse and the electron beam was adjusted so that the maximum field in the accelerating cell was produced when the beam was in the middle of the cell.

TABLE I
PARAMETERS OF THz ACCELERATING STRUCTURES

Parameters	Silicon	Quartz
Number of cells	10	10
Dielectric permittivity	11.9	3.75
Cell length	0.2 mm	
Beam pipe diameter	0.1 mm	
Focal length	0.2 mm	
Iris thickness	0.2 mm	
Width	3 mm	
Length	4 mm	

In the design and simulation of this accelerating structure we considered all possible losses. We show that an effective long-range interaction between a broadband THz pulse and electron beam can be achieved. The energy gain in such a multi-cell structure is almost an order of magnitude larger than what is currently a record for such acceleration process [7]. This approach can be used for manipulation of the ultrafast electron microscope beam: beam compression, energy chirping or modulation.

IV. CONCLUSIONS

The presented THz structure is based on an accelerating cell in which a short, broadband THz pulse is focused to achieve both: a high accelerating gradient and an appropriate field configuration for charged particle acceleration. The long-range interaction is achieved by stacking such accelerating cells together and timing the THz pulse arrival with respect to the electron beam travel by means of delay-lines. Such broadband accelerating structures can withstand highest accelerating gradients in due to minimal duration of the pulse.

ACKNOWLEDGMENT

Electromagnetic simulations of THz structures were supported by the Russian Science Foundation under Grant 19-42-04133.

REFERENCES

[1] W. Wuensch, "High-gradient X-band technology: from TeV colliders to light sources and more," CERN Courier, 23 March 2018

[2] D. P. Pritzkau, R. H. Siemann, "Experimental study of rf pulsed heating on oxygen free electronic copper," Phys. Rev. ST Accel. Beams, vol.5, p.112002, 2002

[3] S.V. Kuzikov et al., "Single Cycle Terahertz Acceleration Structures," NAPAC2019 Conference, Lansing, MI, 1–6 September, 2019

[4] Donfang Zhang et al., "Femtosecond phase control in high-field terahertz-driven ultrafast electron sources," Optica, vol.6, p.872, 2019

[5] Dongfang Zhang et al., "Cascaded Multicycle Terahertz-Driven Ultrafast Electron Acceleration and Manipulation," Phys. Review X, vol.10, p.011067, 2020

[6] S.V. Kuzikov et al., "Millimeter and submillimeter-wave, high-gradient accelerating structures," AIP Conf. Proc., vol.1812, p.060002, 2017

[7] Fulop, J. A. et al., "Efficient generation of THz pulses with 0.4 mJ energy. In CLEO: Science and Innovations," SW1F–5, Opt. Soc. of America, 2014

978-1-6654-2590-2/21 $31.00 © 2021 IEEE

Magnetron Sputtering Formation of Molybdenum-Copper Alloys for Fabrication of Millimeter-Band Planar Slow Wave Structures

A.V. Starodubov[1,2]*, D.A. Nozhkin[2], A.A. Serdobintsev[2], I.O. Kozhevnikov[2], A.M. Pavlov[2], V.V. Galushka[1,2], N.M. Ryskin[1,2], G. Ulisse[3], V. Krozer[3]

[1]Saratov Branch, Kotelnikov Institute of Radio Engineering and Electronics RAS, Saratov, Russia
[2]Saratov State University, Saratov, Russia
[3]Goethe University Frankfurt, Physics Institute, Frankfurt, Germany
*Contact: StarodubovAV@gmail.com, phone +7-917-212-4399

Abstract—**The goal of this work is an investigation of peculiarities of the molybdenum-copper alloys formation by magnetron sputtering to meet the requirements of low thermal expansion coefficient together with high electrical/thermal conductivity of the material for potential use in microfabrication of slow-wave structures for miniaturized millimeter-band vacuum electron devices. The study is focused on controlling the roughness and adhesion of the fabricated thin Cu-Mo alloy films. The roughness mainly determines the transmission losses at millimeter and THz frequencies, whereas the adhesion of the film is quite important for further processing, such as laser ablation.**

I. Introduction

The trend towards miniaturization of vacuum electron devices and shifting their operating wavelength to the millimeter and submillimeter (THz) bands [1], [2] puts forward special requirements to the properties of structural materials used for fabrication of key components, such as slow-wave structures (SWS) [3], [4]. Low thermal expansion coefficient together with high electrical/thermal conductivity of the material are among the key factors here [5], [6]. A common strategy in the field is utilization of special composite materials with laminated, fibrous or dispersed structure, based on various combinations of metals, alloys, and compounds. One of such promising composite materials for vacuum microelectronic devices is the copper-molybdenum (Cu-Mo) alloy [6]. This work is aimed at study of the Cu-Mo alloys formation by magnetron sputtering to meet the abovementioned requirements for potential use in microfabrication of planar slow-wave structures for miniaturized millimeter-band vacuum electron devices. Using the planar microstrip slow wave structures (SWS) on dielectric substrates is very promising due to their high slow-wave factor which allows for low-voltage operation, device size reduction and adoption of a high-current and high-aspect-ratio sheet electron beam [7]–[11]. The study is focused on controlling the roughness and adhesion of the fabricated thin Cu-Mo alloy films. The roughness mainly determines the transmission losses at millimeter and THz frequencies, whereas the adhesion of the film is quite important for further processing, such as laser ablation [10]–[12].

II. Results and Discussion

For the fabrication of thin Cu-Mo films, we use a simultaneous deposition by two magnetron sources, which allows for varying the alloy composition in a broad range. Quartz slides of 500 µm thick are used as substrates. The NexDep magnetron sputtering unit (Angstrom Engineering, Canada) with oil-free pumping was applied. We use the 99.997%, 76 mm oxygen free copper target (Girmet, Russia) and 99,95%, 76 mm molybdenum target (Girmet, Russia). The base pressure in the working chamber did not exceed 2×10^{-5} Torr, with the deposition pressure regulated at 5×10^{-3} Torr. Series of Cu-Mo alloy films on quartz slides were fabricated with several varied deposition process parameters: (1) substrate temperature (room and heated up to 200 °C), (2) sputtering duration, and (3) molybdenum and copper magnetron sources power ratio. The samples of Cu-Mo alloy films were deposited for 10 and 40 minutes which yielded film thickness in the range of 1.1-1.15 µm and 4.3-4.5 µm respectively. The thickness of fabricated films was measured by means of SEM (Mira II microscope, Tescan, Czech Republic) and by contact stylus profilometry technique with the help of the Dektak 150 Surface Profiler (Veeco, USA). The latter was also utilized for the source roughness measurements (arithmetic average value R_a and root-mean-square value R_q). The experiments show that the variation of the magnetron sources power ratio during the sputtering process does not significantly affect the surface roughness of the fabricated films. When the substrate was heated up to 200°C, the surface roughness of the film increased significantly. The results of the roughness study are shown in Fig. 1. The samples fabricated on the substrate heated up to 200°C have $R_a = 87.47$ nm and $R_q = 116.07$ nm while in the case of the room temperature substrate $R_a = 4.64$ nm and $R_q = 5.93$ nm. An increase in the thickness of the film by a factor of 4 leads to an increase in the surface roughness by a factor of 3.

The adhesion properties were studied in the case of the Mo-Cu thin films fabricated on the substrate heated up to 200°C. In this study, the strips with the width of 10, 20, 30, 40 and 50

978-1-6654-2590-2/21 $31.00 © 2021 IEEE

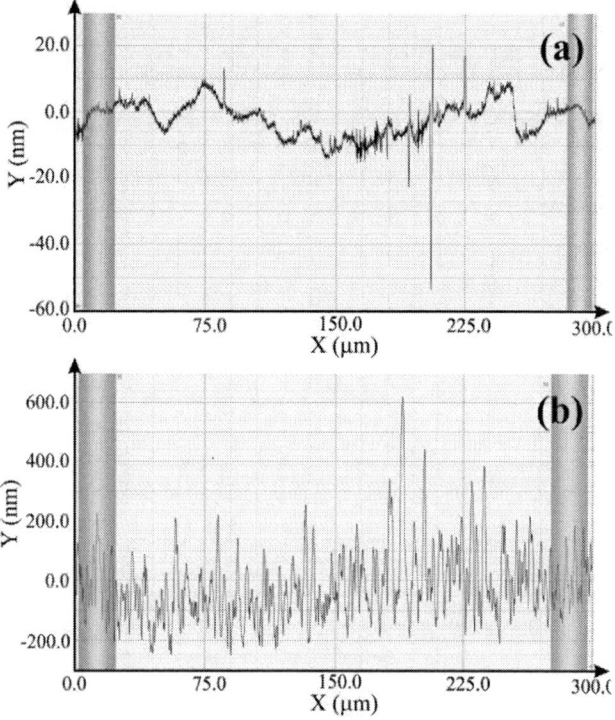

Fig. 1. Results of the roughness study of the Mo-Cu alloy samples fabricated in the case of the room temperature substrate (a) and heated up to 200°C (b).

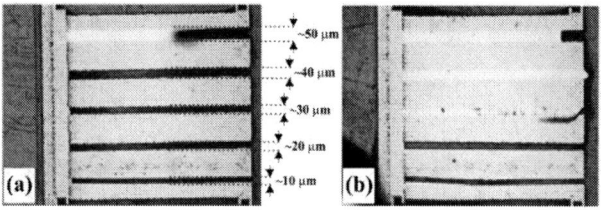

Fig. 2. Results of the fabrication of the strips with width of 10, 20, 30, 40 and 50 μm with the help of the nanosecond laser ablation of the Mo-Cu film: (a) photo of the sample just after laser micromachining. (b) photo of the sample after chemical treatment and polishing

μm were fabricated by using the nanosecond laser ablation technology [11]. The sizes of the strips are chosen according to the typical dimensions of the millimeter-band planar slow-wave structures [8], [9], [11], [13]. A commercially available high-precision CNC laser machine ("MiniMarker-2", Laser Center, Russia) equipped with a 1064-nm pulsed YAG:Nd fiber laser and galvanometric 2D scanning heads (Cambridge Technology, USA) was used. The pulse duration was 30 ns, pulse repetition rate was 100 kHz, pulse power was varied in 15-25 μW, and the scanning velocity was varied in 200-400 mm/s. In Fig. 2, the optical microscopic images of the fabricated structures before and after chemical treatment and polishing are presented.

Fig. 2 clearly shows that some of the fabricated strips can detach from the quartz substrate after the chemical treatment or even during the laser micromachining process. We suggest that one of the reasons for such weak adhesion of the Mo-Cu thin film is the absence of the adhesion sub-layer. Therefore, as a next step, we are going to fabricate another series of films made from the Mo-Cu alloys with the molybdenum adhesion sub-layer and repeat the treatment by laser micromachining and chemical polishing.

ACKNOWLEDGMENT

This work is supported by the grant of Russian Foundation for Basic Research # 20-57-12001 and DFG # 430109039

REFERENCES

[1] T. S. Rappaport et al., "Wireless communications and applications above 100 GHz: opportunities and challenges for 6G and beyond," IEEE Access, vol. 7, pp. 78729–78757, 2019.

[2] T. S. Rappaport, Y. Xing, G. R. MacCartney, A. F. Molisch, E. Mellios, and J. Zhang, "Overview of millimeter wave communications for fifth-generation (5G) wireless networks-with a focus on propagation models," IEEE Transactions on Antennas and Propagation, vol. 65, no. 12. pp. 6213–6230, Dec-2017.

[3] D. Gamzina, N. C. Luhmann, and B. Ravani, "Thermo-mechanical stress in high-frequency vacuum electron devices," J. Infrared, Millimeter, Terahertz Waves, vol. 38, no. 1, pp. 47–61, Jan. 2017.

[4] D. Gamzina and B. Ravani, "Thermomechanical Fatigue in Sub-THz Vacuum Electron Devices," IEEE Trans. Electron Devices, vol. 63, no. 12, pp. 4948–4954, Dec. 2016.

[5] S. Gorbatyuk, A. Pashkov, and N. Chichenev, "Improved copper-molybdenum composite material production technology," Mater. Today Proc., vol. 11, pp. 31–35, 2019.

[6] A. Serdobintsev et al., "Molybdenum-copper alloys as a base material for microfabrication planar slow-wave structures of millimeter-band vacuum electron devices," in 2020 7th International Congress on Energy Fluxes and Radiation Effects (EFRE), 2020, pp. 809–812.

[7] G. V. Torgashov, R. A. Torgashov, V. N. Titov, A. G. Rozhnev, and N. M. Ryskin, "Meander-line slow-wave structure for high-power millimeter-band traveling-wave tubes with multiple sheet electron beam," IEEE Electron Device Lett., vol. 40, no. 12, pp. 1980–1983, 2019.

[8] W. Shaomeng and S. Aditya, "A microfabricated V-shaped microstrip meander-line slow-wave structure," pp. 2–3, 2017.

[9] G. Ulisse and V. Krozer, "W-band traveling wave tube amplifier based on planar slow wave structure," IEEE Electron Device Lett., vol. 38, no. 1, pp. 126–129, 2017.

[10] R. A. Torgashov et al., "Theoretical and experimental study of a compact planar slow-wave structure on a dielectric substrate for the W-band traveling-wave tube," Tech. Phys., vol. 65, no. 4, pp. 660–665, Apr. 2020.

[11] N. M. Ryskin et al., "Development of microfabricated planar slow-wave structures on dielectric substrates for miniaturized millimeter-band traveling-wave tubes," J. Vac. Sci. Technol. B, vol. 39, no. 1, p. 013204, Jan. 2021.

[12] N. M. Ryskin et al., "Planar microstrip slow-wave structure for low-voltage V-band traveling-wave tube with a sheet electron beam," IEEE Electron Device Lett., vol. 39, no. 5, pp. 757–760, May 2018.

[13] S. Wang, S. Aditya, X. Xia, Z. Ali, J. Miao, and Y. Zheng, "Ka-band symmetric V-shaped meander-line slow wave structure," IEEE Trans. Plasma Sci., vol. 47, no. 10, pp. 4650–4657, 2019.

A Facile Approach for Surface Quality Improvement of Mm-Band Planar Electromagnetic Structures Fabricated by Laser Ablation

A.V. Starodubov[1,2*], A.A. Serdobintsev[2], I.O. Kozhevnikov[2], A.M. Pavlov[2], V.V. Galushka[1,2], N.M. Ryskin[1,2]

[1]Saratov Branch, Kotelnikov Institute of Radio Engineering and Electronics RAS, 38 Zelenaya st., 410019 Saratov, Russia
[2] Saratov State University, 83 Astrakhanskaya st., 410004 Saratov, Russia
*Contact: StarodubovAV@gmail.com, phone +7-917-212-4399

Abstract—**Travelling-wave-tube (TWT) power amplifiers with planar microstrip slow wave structures (SWS) on dielectric substrates are very promising for operation at high frequencies due to their low-voltage operation, small size and relatively high output power. Recently, the interest to the planar SWSs has increased since their using can significantly decrease the complexity of microfabrication process. In our previous work, we reported successful fabrication of millimeter-band microstrip planar SWSs using the scalable and cost-effective laser ablation process with nanosecond pulse duration. However, the SWS samples fabricated by nanosecond laser ablation suffer from surface distortions and debris on the borders of the ablated zone. In order to improve the quality of resulting structures, here we propose deposition of a supplementary metallic (Al or Ti) layer on top of copper film prior to laser treatment. Such coating is aimed to prevent defects formation in the course of laser irradiation. In the proposed system, debris and distortions from Cu layer irradiation, naturally sediment on the sacrificial layer instead of sticking to the copper layer itself. Sacrificial aluminium coating can later be easily chemically removed together with sedimented debris leaving a clean underlying patterned Cu surface. In contrast, titanium layer prevents sticking of copper debris due to the higher melting temperature of Ti, thus preventing attachment of those to the surface in the first place.**

I. INTRODUCTION

Miniaturized millimeter- and THz-band travelling-wave-tube (TWT) power amplifiers are of great interest for such applications as high-data-rate wireless communications, security systems, and high-resolution radar [1]–[4]. In particular, the microstrip planar slow wave structures (SWS) on dielectric substrates are very promising due to their high slow-wave factor and high-current and high-aspect-ratio sheet electron beam adoption capabilities [2], [5]–[8]. Commonly used technologies for microfabrication of such miniature structures are photolithography, deep reactive ion etching, electrical discharge micromachining, and additive manufacturing (3-D printing, selective laser sintering, selective laser melting). In our previous work [9]–[12], we reported the fast, scalable and cost-effective technology for fabrication of millimeter-band microstrip planar SWSs using laser ablation with nanosecond pulse duration. Unfortunately,

the patterns fabricated by nanosecond laser ablation suffer from surface distortions and debris on the borders of the ablated zone [13]. These defects significantly increase the surface roughness of the fabricated structures, which is one of the major parameters determining the transmission losses of the SWSs, especially in the millimeter band. In this work, we consider several ways to improve the quality of the structures.

We suggest covering the surface of the copper film by a thin metallic (aluminium or titanium) film prior to the laser patterning. These metallic coverages can prevent the formation of defects during the laser irradiation in two different ways. The sacrificial aluminium coating can be chemically removed from the target structure together with debris and surface distortions, while the titanium coating prevents sticking and bonding of the debris with the functional copper surface after the laser pattering due to different melting temperatures of the copper and titanium.

II. RESULTS AND DISCUSSION

To test the proposed approaches, were carry out several preliminary experiments. Glass slides of 1.0-mm thickness were used as substrates. Two samples of copper coatings were deposited onto the glass slides by NexDep magnetron sputtering unit (Angstrom Engineering, Canada) with oil-free pumping and 99.997%, 76-mm oxygen free copper target (Girmet, Russia). To improve the adhesion of the copper films, we used the titanium film with around 30-nm thickness as an adhesion sublayer. The base pressure in the working chamber did not exceed 2×10^{-5} Torr, with deposition pressure regulated at 5×10^{-3} Torr. The deposition time was 60 minutes for all the samples, yielding in copper films thickness in the range of 5 μm as measured by the means of scanning electron microscopy (SEM, Mira II microscope, Tescan, Czech Republic) and stylus profilometry (Dektak 150, Veeco, USA). As the next step, one of the samples was coated with aluminium by magnetron sputtering also. The other sample was coated with titanium in the same way. The thickness of the aluminium and titanium layers was in the range of 140-150 nm. After that, the prepared samples were micromachined by the nanosecond laser ablation. We used a commercially

978-1-6654-2590-2/21 $31.00 © 2021 IEEE

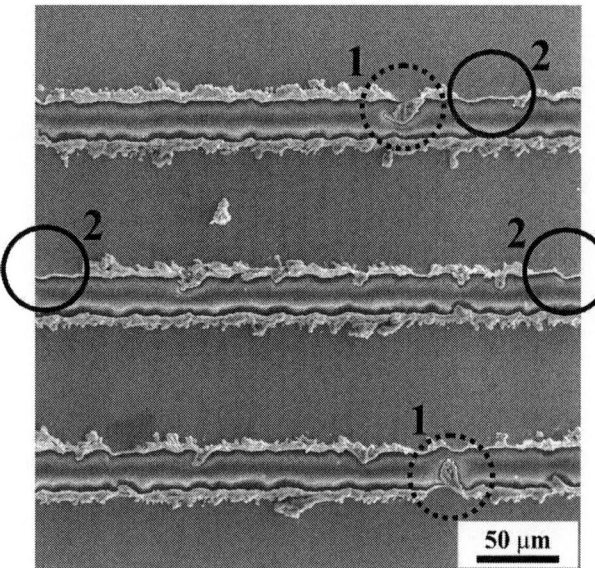

Fig. 1. SEM image of the surface of the copper thin film after the laser micromachining with nanosecond pulses. The copper surface was precoated by an aluminium layer. After the laser micromachining, the sacrificial aluminium layer was chemically removed with the help of alkali. Symbol "1" denotes areas where the process of debris removal begins, symbol "2" denotes areas without the debris on the borders of the ablated zone.

available high-precision CNC laser machine ("MiniMarker-2", Laser Center, Russia) equipped with 1064-nm pulsed YAG:Nd fiber laser and galvanometric 2D scanning heads (Cambridge Technology, USA). The pulse duration was 8 ns, pulse repetition rate was 100 kHz, pulse power was 60 µW, and the scanning velocity was 400 mm/s. A single strip was used as a pattern for micromachining. After laser micromachining, the aluminium-coated sample was chemically treated with alkali in order to remove the coating together with the debris and surface distortions from the target structure. Fig. 1 illustrates the obtained results. This figure demonstrates that the proposed approach allows the removing of the debris from the border of the ablation area. However, most of the debris cannot be completely removed. We assume that one of the reason is that the thickness of the aluminium layer of 140-150 nm is not enough to prevent the sticking and bonding of the debris to the border of the ablation zone. Therefore, we are going to increase the aluminium layer up to several µm and repeat the experiments.

Almost the same results were obtained during the laser micromachining of a thin copper film previously coated with a titanium layer. In that case, we suggest covering the copper film by the metal with a melting temperature much higher than that of the titanium, for example, by the tungsten having the highest proven melting and boiling points. Such a significant difference in melting points will also prevent the sticking and bonding of the debris to the border of the ablation zone.

III. CONCLUSIONS

The preliminary results presented in this paper demonstrate a successful local minimization of the debris volume, resulting in decreased surface roughness at the border zone of the ablation area using both aluminium and titanium coatings. At the next step, to increase the surface roughness uniformity, we are going to increase the thickness of the sacrificial aluminium layer in the first approach and replace the titanium with tungsten (which attains even higher melting temperature) in the second one.

We believe that the proposed approaches will significantly improve the quality of planar electromagnetic structures microfabricated by a scalable and cost-effective nanosecond laser ablation process.

ACKNOWLEDGMENT

This work is supported by the grant of Russian Foundation for Basic Research # 20-07-00929.

REFERENCES

[1] R. A. Lewis, "A review of terahertz sources," *J. Phys. D. Appl. Phys.*, vol. 47, no. 37, p. 374001, Sep. 2014.

[2] A. D. Grigoriev, V. A. Ivanov, and S. I. Molokovsky, *Microwave Electronics*, vol. 61. Cham: Springer International Publishing, 2018.

[3] T. S. Rappaport, Y. Xing, G. R. MacCartney, A. F. Molisch, E. Mellios, and J. Zhang, "Overview of millimeter wave communications for fifth-generation (5G) wireless networks-with a focus on propagation models," *IEEE Transactions on Antennas and Propagation*, vol. 65, no. 12. pp. 6213–6230, Dec-2017.

[4] T. S. Rappaport *et al.*, "Wireless communications and applications above 100 GHz: opportunities and challenges for 6G and beyond," *IEEE Access*, vol. 7, pp. 78729–78757, 2019.

[5] S. Wang, S. Aditya, X. Xia, Z. Ali, and J. Miao, "On-wafer microstrip meander-line slow-wave structure at Ka-band," *IEEE Trans. Electron Devices*, vol. 65, no. 6, pp. 2142–2148, 2018.

[6] S. Wang, S. Aditya, X. Xia, Z. Ali, J. Miao, and Y. Zheng, "Ka-band symmetric V-shaped meander-line slow wave structure," *IEEE Trans. Plasma Sci.*, vol. 47, no. 10, pp. 4650–4657, 2019.

[7] G. Ulisse *et al.*, "Fabrication and measurements of a planar slow wave structure operating in V-band," in *2019 International Vacuum Electronics Conference (IVEC)*, 2019, pp. 1–2.

[8] M. Sumathy, D. Augustin, S. K. Datta, L. Christie, and L. Kumar, "Design and RF characterization of W-band meander-line and folded-waveguide slow-wave structures for TWTs," *IEEE Trans. Electron Devices*, vol. 60, no. 5, pp. 1769–1775, 2013.

[9] A. Starodubov *et al.*, "Technological approaches to the microfabrication of planar slow-wave structures for millimeter- and THz-band vacuum electron devices," in *2020 International Conference on Actual Problems of Electron Devices Engineering (APEDE)*, 2020, pp. 256–261.

[10] N. M. Ryskin *et al.*, "Planar microstrip slow-wave structure for low-voltage V-band traveling-wave tube with a sheet electron beam," *IEEE Electron Device Lett.*, vol. 39, no. 5, pp. 757–760, May 2018.

[11] R. A. Torgashov *et al.*, "Theoretical and experimental study of a compact planar slow-wave structure on a dielectric substrate for the W-band traveling-wave tube," *Tech. Phys.*, vol. 65, no. 4, pp. 660–665, Apr. 2020.

[12] N. M. Ryskin *et al.*, "Development of microfabricated planar slow-wave structures on dielectric substrates for miniaturized millimeter-band traveling-wave tubes," *J. Vac. Sci. Technol. B*, vol. 39, no. 1, p. 013204, Jan. 2021.

[13] A. V. Starodubov *et al.*, "Comparison of nanoseconds and picoseconds laser ablation for microfabrication of planar slow-wave structures for D-band vacuum electronic devices with sheet electron beam," in *Saratov Fall Meeting 2019: Laser Physics, Photonic Technologies, and Molecular Modeling*, 2020, vol. 1145803, no. April, p. 41.

978-1-6654-2590-2/21 $31.00 © 2021 IEEE

Analyses of field electron emission Molybdenum current-voltage data using Fowler-Nordheim and Murphy-Good plots

Mohammad M. Allaham[1,2*], Marwan S. Mousa[3], Daniel Burda[1], Mohammad H. AlSa'eed[3], Sabreen Y. AlJrawen[3]
and Alexandr Knápek[1]

[1] *Institute of Scientific Instruments of CAS, Královopolská 147, 612 64 Brno, Czech Republic*
[2] *Central European Institute of Technology, Brno University of Technology, Purkyňova 123, 612 00 Brno, Czech Republic*
[3] *Physics Department, Mu'tah University, Al-Karak 6170, Jordan*
*Contact: allaham@isibrno.cz, Phone +420-735-039950

Abstract— **Field electron emission theory and experiments include testing and analyzing the measured current-voltage characteristics using the so-called Fowler-Nordheim or Murphy-Good plots. Although Fowler-Nordheim plots are theoretically predicted to be slightly curved and Murphy-Good plots are predicted to be almost-exactly straight, still they provide the same results when applying the field-emission orthodoxy test to practical experimental data, and near results when extracting the emitter characterization parameters. This study is to compare the analysis results that will be obtained when applying the two methods to experimental data obtained from Molybdenum single field emitters, mounted in a traditional field emission microscope, and operated in high vacuum conditions (~10^{-6} Pa).**

I. INTRODUCTION

In field electron emission (FE) theory, Fowler-Nordheim (FN), and Murphy-Good (MG) analysis plots; are methods to present the field emission current-voltage $I_m(V_m)$ data in a nearly linear form. MG plots have several advantages to be used such as: the absence of the correction factors in its mathematics and having a very-nearly straight-line curve [1]. This allows to easily test and analyze the obtained FE-$I_m(V_m)$ data using the field emission orthodoxy test and then to use each plot parameters to extract the characterization parameters of the used emitter. Both types of plots can be presented in a general form of $\ln(Y/X^\kappa)$ vs X^{-1}. With Y is the measured total emission current I_m, X is the measured voltage V_m.

The orthodoxy test is a quantitative test that can be applied to both analysis plots and to any geometrical emitter shape. It is based on extracting a specific and important factor in the theory of field emission. This factor is the local scaled field at

a characteristic point on the apex of the emitter tip, *the characteristic scaled field* $f_C = c_S^2 \phi^{-2} F_C = c_S^2 \phi^{-2} \zeta_C^{-1} V_m$, F_C is the *characteristic electrostatic field* at the same characteristic point on the emitter tip apex, and ζ_C [$\equiv V_m/F_C$] is the related *characteristic voltage conversion length (VCL)*. The extracted f_C values are then compared to a set of data that forms the orthodoxy test criteria as listed in [2]. The test provides three results: (1) Pass; the data is reasonable, (2) Fail; the data is unreasonable, and the extracted results are likely to be spurious and (3) Inconclusive; the data need more study and analysis. If the tested data pass the orthodoxy test, then the characterization parameters of the emitter can be validly extracted, mainly: the formal emission area from a Schottky-Nordheim barrier A_f^{SN}, and ζ_C [3].

In what follows, S_{FN}^{fit} and S_{MG}^{fit} are the slopes of the fitted to the curve lines for both types of plots, R_{FN}^{fit} and R_{MG}^{fit} are the exponents of the same fitted line for both types of plots, the superscript *extr* refers to an extracted value, s_t and r_t are the slope and vertical axis intercept correction factors for a FN plot, $\eta(\phi) = bc_S^2\phi^{-1/2}$ and $\theta(\phi) = ac_S^{-4}\phi^3$ are scaled parameters with a and b as the first and second Fowler-Nordheim constants, c_S is the Schottky constant and ϕ is the local work function and $V_{mR} = c_S^{-2}\phi^2\zeta_C$ is the reference measured voltage that is required to pull down the top of the Schottky-Nordheim barrier to the Fermi level. Table I describes the comparison between FN and MG plots which includes: comparing the κ values, definitions of the theoretical slope and vertical axis intercept, mathematical assumptions of the extraction process for f_C, A_f^{SN}, and ζ_C from both types of plots.

TABLE I
Comparison of the theoretical assumptions of the extraction process from Fowler-Nordheim (FN) and Murphy-Good (MG) plots

Plot Type	k	*Slope*	*Vertical axis intercept*	f_C^{extr}	ζ_C^{extr}	$\{A_f^{SN}\}^{extr}$
FN	2	$\dfrac{-s_t\eta V_{mR}}{V_m}$	$\ln\left(A_f^{SN}\theta V_{mR}^{-2}\right)$	$\dfrac{s_t\eta}{\left|S_{FN}^{fit}\right|V_m^{-1}}$	$\dfrac{\left|S_{FN}^{fit}\right|}{s_t\,b\phi^{3/2}}$	$\dfrac{R_{FN}^{fit}\left(S_{FN}^{fit}\right)^2}{r_ts_t^2\,ab^2\phi^2}$
MG	$2-\eta/6$	$\dfrac{-\eta V_{mR}}{V_m}$	$\ln\left(A_f^{SN}\theta\exp(\eta)\,V_{mR}^{-\kappa}\right)$	$\dfrac{\eta}{\left|S_{MG}^{fit}\right|V_m^{-1}}$	$\dfrac{\left|S_{MG}^{fit}\right|}{b\phi^{3/2}}$	$\dfrac{R_{MG}^{fit}\left|S_{MG}^{fit}\right|^\kappa}{\exp\{\eta\}\eta^{\kappa-2}\,ab^2\phi^2}$

978-1-6654-2590-2/21 $31.00 © 2021 IEEE

II. METHODOLOGY

Molybdenum single field emission tips (ϕ=4.45 eV) are prepared by the electrochemical polishing technique as explained in [4]. The samples are then mounted to a traditional field emission microscope FEM which is connected to a DC high voltage power source and operated under high vacuum level (~10^{-6} Pa). This value of vacuum is excellent for this study to be able to achieve both, Pass and Fail, results for the purpose of comparison. To obtain the FEM emission patterns, the phosphorus screen of the FEM is connected to a picoamperemeter. The $I_m(V_m)$ characteristics is obtained, tested, and are being analyzed using both FN and MG plots.

the FE orthodoxy test to the FN and MG plots and to extract the corresponding characterization parameters [5].

III. RESULTS AND DISCUSSION

Fig. 1-a presents the logarithmic scale for the obtained current-voltage characteristics in comparison to the results of a simulation carried for the extracted parameters of the tested samples. Fig. 1-b presents the corresponding FN and MG plots. Each of the two analysis plots is split into two curves (lines), to describe the pass and fail regions for each plot type.

Table II presents the results of applying the FE orthodoxy test and the extraction process as achieved from [4]. In this presentation, A_f^{SN} and ζ_C are extracted only for the curves that

pass the FE orthodoxy test. For this, these parameters have been found only for L_1 and L_3 in Fig. 1-b. The curves L_2 and L_4 from the same figure have failed the FE orthodoxy test. So, no further discussion for their results will be carried on.

TABLE II
The extraction results for each of the curves in Fig. 1 analysis plots

Curve	S^{fit} Np.V	f_C^{extr}	Test result	ζ_C^{extr} nm	$\{A_f^{SN}\}^{extr}$ nm^2
L_1	-8380.00	0.22-0.28	Pass	130.68	5087.84
L_2	-1217.39	1.99-3.88	Fail	-	-
L_3	-8020.00	0.22-0.28	Pass	131.65	5598.48
L_4	-663.04	3.48-6.86	Fail	-	-

IV. CONCLUSIONS

The shape and the ability to use MG plots is related to the knowledge of the nearly exact value of the local work function value for the used emitter, since the power κ is a function of ϕ for the case of this type of plots. this applies to the extraction process from both types of plots also.

From Table 2, the extracted f_C values are exactly extracted equal from the FN and MG plots, this means that both types of plots provide precise and exact FE orthodox test results. The values of ζ_C are extracted also nearly equal with slight difference which is related to the presence of the slope correction factor s_t in the theoretical slope of FN plots. In contrast, there is large difference in the extracted values of A_f^{SN} between the two types of plots which is related to the presence of two correction factors in the theory of FN plots. From this, using MG plots is more advised for the case of extracting A_f^{SN}.

The kinked behavior of the analysis plots is believed to be related to the slightly change in the emission surface geometry or its chemical composition because of the high gas ions bombardment to the surface or even its reaction with the residual gases during the experiment; since the experiment was not set in relative high-pressure value which allows fast creation of the contamination layers on the surface of the tip [6]. this will affect the emission process by changing the behavior of the local electrostatic field at the surface of the tip.

ACKNOWLEDGMENT

The research described in this paper was financially supported by the Ministry of the Interior of the Czech Republic (project. No. VI20192022147)

REFERENCES

[1] M. M. Allaham, M. S. Mousa, and R. G. Forbes,"Comparing the performance of Fowler-Nordheim plots and Murphy-Good plots" in 2020 33rd International Vacuum Nanoelectronics Conference (IVNC) (IEEE, 2020), pp. 1–2.

[2] M. M. Allaham, R. G. Forbes, A. Knápek, and M. S. Mousa,"Implementation of the orthodoxy test as a validity check on experimental field emission data" J. Electr. Eng. Slovak 71, 37 (2020).

[3] R. G. Forbes,"The Murphy-Good plot: A better method of analysing field emission data" R. Soc. Open Sci. 6, 190912 (2019).

[4] M. Madant, M. Al Share, M. M. Allaham and M. S. Mousa, "Information extraction from Murphy-Good plots of tungsten field electron emitters" J. Vac. Sci. Technol. B 39, 024001 (2021)

[5] M. M. Allaham, R. G. FORBES, A. Knápek, and M. S. MOUSA. Field emission analysis software [online]. Mu'tah University [cit. 2021-6-11]. Available from: https://fieldemissionanalysis.weebly.com.

[6] W. Umrath, "Fundamentals of Vacuum Technology", Cologne 2007, p. 199.

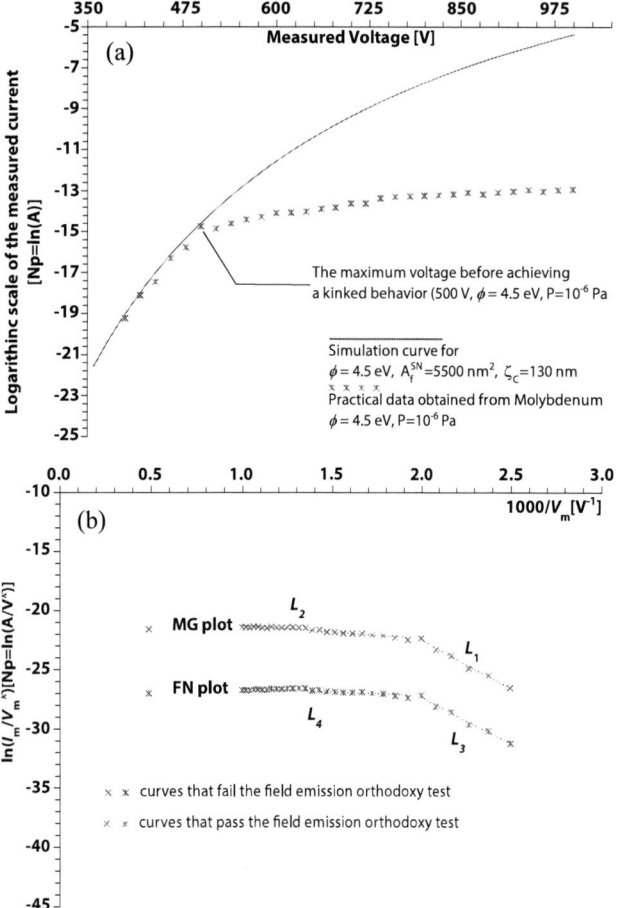

Fig. 1 The obtained data from Molybdenum sample. (a) The current-voltage characteristics and (b) the corresponding FN and MG plots.

Universality of Characteristic Field Enhancement Factor from Arched Carbon Nanofibers

Thiago A. de Assis

Institute of Physics, Federal University of Bahia,
Barão de Jeremoabo street s/n, 40170-115, Salvador, Ba, Brazil
E-mail: thiagoaa@ufba.br

Fernando F. Dall'Agnol

Department of Exact Sciences and Education (CEE), Universidade Federal de Santa Catarina
Campus Blumenau, Rua João Pessoa, 2514, Velha, Blumenau 89036-004, SC, Brazil
E-mail: fernando.dallagnol@ufsc.br

Marc Cahay

Spintronics and Vacuum Nanoelectronics Laboratory, University of Cincinnati
Cincinnati, OH 45221, USA
E-mail: cahaymm@ucmail.uc.edu

Abstract— In this contribution, we perform electrostatic numerical simulations of conducting semicircular arched fibers and predict a scaling law of their characteristic field enhancement factor (FEF) while varying their geometrical parameters. Conditions for maximum FEF inferred from the isolated emitter allow us to analyze arrays of arched fibers, as a function of the lattice parameters.

Keywords—Field enhencement factor; arched fibers;carbon nanofibers.

I. Introduction/Background

Arched carbon nanotubes (CNTs) and nanofibers (CNFs) has been subject of recent experimental research in many fields of the science, including field electron emission (FE) [1-4]. Some advantages of these architectures include low-power consumption, good field emission uniformity and stable operation, necessary conditions for possible technological FE applications. For example, a synthesis of CNT FE device allows thin multi-walled CNTs to stick out from arched structures in a stable array [1]. In these structures, FE can occur at their apexes, resulting in a low applied turn-on field, due to a high field enhancement factor (FEF) at that location.

In the presence of a uniform applied macroscopic field, F_M, a conducting protrusion tends to develop a high local field, $F(r)$, near its apex. At any point r, a local FEF can be defined by [5, 6]

$$\gamma(r) = \frac{F(r)}{F_M}.$$

Recently, we demonstrated that it is possible to achieve a very large local FEFs at the top of arched CNFs as well, due to the presence of CNT fibrils near their apexes [2–4]. Our simulations were used to analyze FE characteristics of several looped CNT fibers and the role of each stage for producing FE

characteristics in agreement with experimental results. We showed that the modeling of a combined two-stage structure (looped CNT fiber + fibrils) leads to apex FEF values in excellent agreement with an orthodoxy theory analysis of FE experiments performed on these fibers [7-9].

Strain-engineered growth of aligned CNTs have been the subject of recent investigations which show promise for next-generation electronic devices [10,11]. Therefore, techniques to grow arched CNFs in a precisely controlled way show strong potential for FE applications. In particular, homogeneous arrays of CNT arched field emitters, with controllable pitch and height, should find application as precisely controlled uniform large area field emitters (LAFEs).

In this work, we report how to theoretically predict the local FEF close to the tip of growing arched conducting protrusions. Moreover, we report universal features that allow one to predict the maximal FEF achieved during the growth of arched structures, for any curvature of the tip, providing a theoretical support for its optimization.

II. Simulation Procedures

We used commercial simulator COMSOL based on Finite Elements Method to calculate the potential and local electrostatic field distributions. Figure 1(a) shows the top view of the Computer Aided Design (CAD) of our simulation domain. It consists of a hollow arched-tube in a unit cell that represents the fiber in an infinite array. The simulation domain excludes the interior of the tube because we assume that the fiber is a perfect conductor, so that the electric potential is grounded ($\Phi = 0$) everywhere inside and at the surface of the tube. The bottom surface is assumed to be grounded. The top of the tube is truncated and the rim of the top has curvature R_{rim}, at which we extract the characteristic FEF. The boundary

978-1-6654-2590-2/21 $31.00 © 2021 IEEE

conditions (BCs) on opposite side-walls impose that they have the same electric potential distribution, i.e., $\Phi_{left}(x, y, z) = \Phi_{right}(x + L_x, y, z)$ and $\Phi_{front}(x, y, z) = \Phi_{back}(x, y + L_y, z)$. These BCs account for the periodicity of the array. The front and back walls are kept parallel, but the left and right hand side walls are tilted θ with respect to the y-axis. The columns in the array can be made misaligned by an angle θ. The top boundary has a uniform surface charge density $\sigma = \varepsilon_0 F_M$ (ε_0 being the permittivity of vacuum) that imposes a uniform electric field of magnitude F_M into the simulation domain.

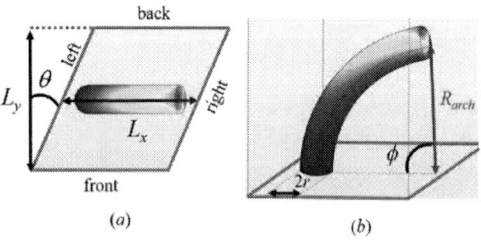

Fig.1: (a) Top view of our simulation domain containing an arched hollow tube that represents an arched CNF emitter. (b) Side view of the same emitter. In the color map, red (blue) color indicates a higher (lower) local FEF. Parameters used are also shown.

Figure 1(b) shows a side view of the emitter. To simulate an isolated emitter, we do L_x and L_y sufficiently large, so that the electrostatic influence of neighboring emitters can be considered negligible. To simulate an isolated emitter L_x and L_y obey the Minimum Domain Size (MDS) criterion [12]. The characteristic FEF can migrate to the body of the fiber if the arch becomes almost full ($\psi \equiv \phi/2\pi \to 1$).

III. RESULTS AND DISCUSSION FOR A CONDUCTING SINGLE TIP ARCHED FIBER

Figure 2 shows the maximum local FEF (characteristic FEF), $\gamma_C \equiv \max\{\gamma\}$ at the top rim of the arched fiber, as a function of ψ, for various curvature radius of the rim, α, and ratios R_{arch}/r. The results are presented for $r = 50$ nm.

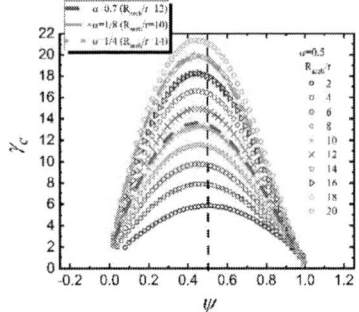

Fig.2: Maximum local FEF ($\gamma_C \equiv \max\{\gamma\}$) at the cap of the arched fiber, as a function of ψ, for various α and ratios R_{arch}/r. Vertical dashed lines represent $\psi = 0.5$. Results presented for a square array ($\theta = 0\ rad$).

The results shown in Fig. 2 clearly indicate that the optimal FEF as a function of ψ, for $R_{arch}/r \gtrsim 2$, is reached before the apex of the circle arch (at $\psi^* \approx 0.46$). At this optimal value, we define $\gamma_{max} \equiv \max\{\gamma_C(\psi)\}$. Since the circle arc has a constant curvature, given by the inverse of the radius, this result reinforces the non-trivial nature of the task of optimizing γ_C during surface growth, somewhat qualitatively consistent with an analytical treatment developed based on conformal mapping [13].

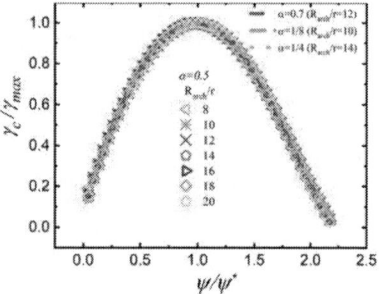

Fig.3: Collapse of the curves shown in Fig. 2 by replacing the variables $\gamma_c \to \gamma_c/\gamma_{max}$ and $\psi \to \psi/\psi^*$ (see text for more details). Results presented for a square array ($\theta = 0\ rad$).

Finally, our results show a universal behavior in the sense that we obtained an excellent collapse of all nearly parabolic curves shown in Fig. 2, by normalizing the vertical variable to γ_C/γ_{max} and replacing the horizontal variable by ψ/ψ^*, where ψ^* corresponds to the value of ψ when $\gamma_C=\gamma_{max}$ (see Fig.3).

REFERENCES

[1] D. N. Futaba, H. Kimura, B. Zhao, T. Yamada, H. Ku-rachi, S. Uemura, and K. Hata, Carbon **50**, 2796 (2012).

[2] P. Zhang, J. Park, S. B. Fairchild, N. P. Lockwood, Y. Y. Lau, J. Ferguson, and T. Back, Applied Sciences **8** (2018)..

[3] T. Y. Posos, S. B. Fairchild, J. Park, and S. V. Baryshev, Journal of Vacuum Science & Technology B **38**, 024006 (2020).

[4] F. F. Dall'Agnol, T. A. de Assis, S. B. Fairchild, J. Ludwick, G. Tripathi, and M. Cahay, Applied Physics Letters **117**, 253101 (2020).

[5] R. G. Forbes, C. Edgcombe, and U. Valdrè, Ultramicroscopy **95**, 57 (2003), iFES (2001).

[6] F. F. Dall'Agnol, S. V. Filippov, E. O. Popov, A. G. Kolosko, and T. A. de Assis, Journal of Vacuum Science& Technology B **39**, 032801 (2021).

[7] R. G. Forbes, Proceedings of the Royal Society of London A: Mathematical, Physical and Engineering Sciences **469**, 20130271 (2013).

[8] R. G. Forbes, Nanotechnology **23**, 095706 (2012).

[9] R. G. Forbes, J. H. B. Deane, A. Fischer, and M. S. Mousa, Jordan Journal of Physics **8**, 125 (2015).

[10] M. De Volder, S. Park, S. Tawfick, and A. J. Hart, Nature Communications **5**, 4512 (2014).

[11] S. Kim, Y. Jiang, K. L. Thompson Towell, M. S. H.Boutilier, N. Nayakanti, C. Cao, C. Chen, C. Jacob, H. Zhao, K. T. Turner, and A. J. Hart, Science Advances **5**, 1 (2019).

[12] T. A. de Assis, F. F. Dall'Agnol, Journal of Vacuum Science & Technology B **37**, 022902 (2019).

[13] E. M. C. Neto, T. A. de Assis, C. M. C. de Castilho, R. F. S. Andrade, submitted to Journal of Applied Physics (2021).

Using the parameter "formal area efficiency" (α_f^{SN}) to analyze current-voltage measurements on large-area field electron emitters

Richard G. Forbes

University of Surrey, Advanced Technology Institute & Dept. of Electrical and Electronic Engineering,
Guildford, Surrey GU2 7XH, UK
Permanent e-mail alias: r.forbes@ trinity.cantab.net

Abstract—**An effective methodology is indicated for using the parameter "formal area efficiency" to analyze field electron emission (FE) current-voltage measurements taken from large area field electron emitters (LAFEs). As part of this, difficulties with existing literature methods are systematically indicated.**

Keywords—Field electron emission, current-voltage characteristics, Fowler-Nordheim plots. Murphy-Good plots, formal emission area, formal area efficiency.

I. INTRODUCTION

Over the last twenty years or so, there has been much interest in the development of materials for large-area field electron sources (LAFEs), which have many potential uses.

In LAFE technological literature, measured current-voltage data have often been interpreted by using the *elementary* field electron emission (FE) equation for local emission current density (LECD) J_L. This equation, introduced about 20 years ago, is a partly corrected and simplified version of the (defective) equation developed by Fowler and Nordheim (FN) in 1928. It is convenient here to write the elementary FE equation in a form close to that often found in the literature:

$$J_L{}^{el} = a\phi^{-1}(\gamma F_M)^2 \exp[-b\phi^{3/2}/\gamma F_M] , \quad (1)$$

where a and b are the FN constants [1], ϕ is the relevant local work function, F_M is (the magnitude of) an applied macroscopic field, and γ is the related local field enhancement factor (FEF) (often denoted by β in the literature).

I have previously noted several difficulties with using this equation for data interpretation. This Poster brings these comments together, and also shows that (for so-called *ideal* emitters) it could be useful to characterize LAFEs by using both an apex FEF *and* a parameter (α_f^{SN}) called here the "formal area efficiency for the Schottky-Nordheim barrier". This parameter α_f^{SN} is a measure of what *fraction* of the LAFE is actually emitting electrons, and could be useful in research and development. The difficulties with the literature and with using (1), and remedies, are summarized below.

A necessary initial question is whether a given FE system (emitter plus measurement circuit) is *ideal*. An ideal system does not change physically with time or current level, and the measured current-voltage $[I_m(V_m)]$ characteristics are determined by emission theory and the system geometry alone. Due to the possibility of "complications", many real systems

are not ideal. However, even with ideal systems, difficulties exist both with the data-analysis methods used in the literature, and with the provision of accurate theoretical information.

A second general point is that nearly all FE $[I_m(V_m)]$ data analysis is carried out on the basis of an underlying Sommerfeld-type physical model introduced in the early 20th Century. This disregards atoms and atomic-level wave-functions, and assumes that the emitter has a smooth planar surface of large lateral extent. It has been called *smooth planar metal-like emitter (SPME) methodology*. FE theory is now well beyond this, but a 21st-Century version of SPME methods (used here) still seems a good basis for $[I_m(V_m)]$ data analysis.

II. EMISSION-THEORY-RELATED DIFFICULTIES

(1) It is common practice to cite the original 1928 FN paper [2] as the source of (1). Taken by itself, this 1928 paper is not a helpful citation for non-experts, because the paper contains many technical errors [1] and is well out of date. In fact, the discrepancy between theoretical equation (22) in [2] and experiment is typically a factor around 10^{15} to 10^{20}. It is unhelpful to non-experts that this discrepancy is usually not mentioned when the paper is cited (over 7000 times, so far). The related numerical error was corrected in 1929 in [3].

The actual sources of the simplified equation (1) appear to be two much later papers, published in 1999 [4] and in 2004 [5]. There are many descriptions of modern ("21st Century") FE theory (e.g., [6–8]) that could be used to supplement [2].

(2) Equation (1) assumes the tunneling barrier is exactly triangular. It is better physics [9] to assume that the tunneling barrier is the *Schottky-Nordheim* (SN) ("planar image-rounded") barrier used in Murphy-Good (MG) (1956) FE theory [10]. This causes correction factors, here written v_F and $t_F{}^{-2}$, to appear in the LECD equation, which I write in the form

$$J_L{}^{MG} = t_F{}^{-2}J_{kL}{}^{SN} \equiv t_F{}^{-2}a\phi^{-1}(\gamma F_M)^2\exp[-v_F b\phi^{3/2}/\gamma F_M] , \quad (2)$$

where the *kernel current density for the SN barrier*, $J_{kL}{}^{SN}$ is defined by (2).

Paper [9] shows that $J_{kL}{}^{SN}$ is greater than $J_L{}^{el}$ by typically $100\times$ to $500\times$. When research is funded in order to develop high-current-density emitters, there is no merit in using an equation that under-predicts current densities by a large factor.

(3) Equations (1) and (2) both disregard the infuence that atomic wavefunctions have on transmission probabilities. The

978-1-6654-2590-2/21 $31.00 © 2021 IEEE

size of this effect is not well known, so I now eplace t_F^{-2} by a *knowledge uncertainty factor* λ and call the resulting equation the *Extended* Murphy-Good (EMG) FE equation for LECD:

$$J_L^{EMG} = \lambda J_{kL}^{SN}. \qquad (3)$$

My best guess is that λ varies with field, with $0.005 < \lambda < 14$.

(4) Real emitters are usually post or needle shaped. By making the *planar transmission approximation*, and integrating over the emitter surface, we can write an equation for the EMG-theory emission current I_e^{EMG} in the form

$$I_e^{EMG} = \lambda A_n^{SN} J_{ka}^{SN} \equiv A_f^{SN} J_{ka}^{SN}, \qquad (4)$$

where J_{ka}^{SN} is the apex value of J_{kl}^{SN}, and the *notional emission area* is defined via (4). The *formal emission area* A_f^{SN} is then formally obtained as $A_f^{SN} \equiv \lambda A_n^{SN}$. The "area" derived from a FN or related plot is a *formal* emission area, (not a notional emission area) and the value obtained depends on the assumption made about the tunnelling barrier form.

(5) The EMG-theory *macroscopic* (or "LAFE-average") emission current density J_M^{EMG} is obtained from I_e^{EMG} by dividing by the LAFE *macroscopic area* (or "footprint") A_M:

$$J_M^{EMG} = I_e^{EMG}/A_M = (A_f^{SN}/A_M)J_{ka}^{SN} \equiv \alpha_f^{SN} J_{ka}^{SN}, \qquad (5)$$

where the *formal area efficiency* $\alpha_f^{SN} \equiv (A_f^{SN}/A_M)$.

Technological FE literature often fails to distinguish between local and macroscopic current density, and equations for "current density" often omit any parameter representing area efficiency. This often leads to large apparent discrepancies (by a factor of 10^5 or more) between theory and experiment.

Terms relating to "current density" need careful definition, and α_f^{SN} needs to be included in equations for macroscopic current density. With ideal systems, the value of α_f^{SN} needs to be extracted, due to its technological importance. (A LAFE where 10^{-5} of the footprint area A_M is emitting is likely to be a better emitter than one where only 10^{-9} of A_M is emitting.)

III. CURRENT-VOLTAGE DATA-ANALYSIS FOR IDEAL SYSTEMS

With a few exceptions, the method of analysis used for measured FE current-voltage data is to make a Fowler-Nordheim (FN) plot, i.e., a plot of $\ln\{Y/X^2\}$ vs $1/X$, where X and Y are the chosen independent and dependent variables.

(7) Often in FE literature, experimental current-voltage data is "pre-converted" to give the macroscopic current density J_M as a function of the (apparent) macroscopic field F_M^{app}. This procedure is dangerous, because for non-ideal systems the conversion of V_m to F_M^{app} can be defective [11], and the outcome may be spurious values of characterization parameters (see Section IV). Best scientific and technological practice is that data-analysis plots should be made using the raw experimental $I_m(V_m)$ data (before "pre-conversion").

(8) Murphy-Good FE theory predicts that a FN plot should be very slightly curved. Thus, for the last 60 years or so, FE experimentalists have been analyzing their data by fitting a straight line to data points that theoretically lie on a curve (and are well away from the vertical axis at $1/X=0$). The slope of the fitted line and (far less frequently) the intercept are then used to provide emitter characterization parameters. Several

different approximations have been used for v_F in (2), and several technically different methods have been used to extract "area" values. As a result, the reliability of area extraction is poor, even for ideal FE systems; [12] implies that uncertainty by 50% might be plausible. This is unfortunate, because recent developments in FE theory [13] suggest important information about theory is contained in the parameter "formal area".

Recently [14], a new form of data plot, the so-called *Murphy-Good (MG) plot*, has been proposed. This is predicted to be "very nearly straight". Simulations suggest that values of A_f^{SN} can be self-consistently extracted to a precision of a few percent. It is suggested that use of MG plots should replace use of FN plots. Values of α_f^{SN} can then be found as A_f^{SN}/A_M.

IV. DATA ANALYSIS FOR NON-IDEAL SYSTEMS

As already noted, analysis using FN or MG plots yields demonstrably valid characterization parameters *only if* the FE system is ideal. In 2013, a so-called *orthodoxy test* was introduced [15] that can be applied to a FN plot to check if the system is ideal. This measures the local fields apparently used in the experiments and compares these with the field range in which emitters are known to normally operate. A small survey [15] suggested that around 40% of experimental papers were reporting spuriously high field enhancement factors taken from non-ideal systems. Best practice would be to always apply the orthodoxy test before publishing results or using published data. This test also works with MG plots [16].

With non-orthodox systems it is sometimes possible to estimate FEF values by *phenomenological adjustment* [17].

REFERENCES

[1] R. G. Forbes and J. H. B. Deane, Proc. R. Soc. Lond. A 467, 2927 (2011). See electronic supplementary information.

[2] R. H. Fowler and L. Nordheim, Proc. R. Soc. Lond. A 119, 173 (1928).

[3] T. E. Stern, B. S. Gossling and R. H. Fowler, Proc. R. Soc. Lond. A 124, 699 (1929).

[4] R. G. Forbes, J. Vac. Sci. Technol. B 17, 526 (1999).

[5] N. de Jonge and J.-M. Bonard, Phil. Trans. R. Soc. Lond. A 362, 2239 (2004).

[6] S.-D. Liang, Quantum Tunneling and Field Electron Emission Theories. New Jersey: World Scientific, 2014.

[7] K. L. Jensen, Introduction to the Physics of Electron Emission. Chichester, UK: Wiley, 2018.

[8] G. Gaertner, W. Knapp and R. G. Forbes, Eds. Modern Developments in Vacuum Electron Sources. Switzerland: Springer Nature, 2020.

[9] R. G. Forbes, J. Appl. Phys. 126, 210901 (2019).

[10] E. L. Murphy and R. H. Good, Phys. Rev. 102, 1464 (1956).

[11] R. G. Forbes, J. Vac. Sci. Technol. B 37, 051802 (2019).

[12] R. G. Forbes and J. H. B. Deane, 30th IVNC, Regensburg, July 2017. [Technical Digest, p. 234.] doi:10.13140/RG.2.2.33297.74083 .

[13] R. G. Forbes, E. O. Popov, A. G. Kolosko, and S. V. Filippov, Roy. Soc. Open Sci. 8, 201986 (2021).

[14] R. G. Forbes, Roy. Soc. Open Sci. 6, 190912 (2019).

[15] R. G. Forbes. Proc. R. Soc. Lond. A. 469, 20130271 (2013).

[16] M. M. Allaham, R. G. Forbes and M. S. Mousa, Jordan J. Phys. 12, 101 (2020).

[17] R. G. Forbes, J. H. B. Deane, A. Fischer and M. S. Mousa, Jordan J. Phys. 8, 125 (2015).

A Tutorial Commentary on the Schottky Constant

Richard G. Forbes

University of Surrey, Advanced Technology Institute & Dept. of Electrical and Electronic Engineering,
Guildford, Surrey GU2 7XH, UK
Permanent e-mail alias: r.forbes@ trinity.cantab.net

Abstract—The Schottky constant plays a central role in modern theories of field electron emission, thermal electron emission, and ionic field evaporation, particularly since the 1970s reforms in the international system of measurement. However, it is not widely recognised as a useful universal constant. This Poster provides a brief "tutorial" introduction to the Schottky constant, primarily for those not familiar with it. It provides a definition, and proof of relevant equations. It indicates the main contexts in which the Schottky constant is used, and demonstrates that the Schottky constant is a "property of the world" that is represented by technically different physical quantities in different equation systems.

Keywords.—Schottky constant, field electron emission, thermal electron emission, field evaporation theory.

I. INTRODUCTION

An electron in a piece of condensed matter is held into the material by a surface barrier. In simple basic models, the force preventing electron escape is attributed to an image attraction between the electron and the material surface. This gives rise to an energy barrier of zero-field height H that prevents classical escape. Applying a classical electrostatic field of appropriate polarity and magnitude F lowers this barrier, and can reduce it to zero if the field magnitude is sufficiently large.

Image effects in conducting spheres, and the existence of a field-lowered barrier, were discussed by Kelvin in 1849 [1] and by Maxwell in 1873 and 1891 [2]. In 1903, when discussing [3, p. 386] spark discharges observed by Earhart [4] between closely spaced planar surfaces, J.J. Thomson suggested the cause might be electron emission due to barrier lowering. However, relevant planar-geometry equations were first clearly formulated by Schottky in 1914 [5]. Hence the barrier-lowering effect is known as the *Schottky effect*.

Schottky's original treatment used "electric potentials", and the Gaussian equation system. Modern treatments discuss the component of electron total energy in the direction (z) normal to the material surface, and use the modern International System of Quantities (ISQ) [6], which puts the vacuum electric permittivity ε_0 in Coulomb's Law. For a good conductor, such as a metal, the energy barrier is described by an energy-like quantity (the so-called *electron motive energy*) $M^{SN}(z)$ [7] given (for a classical planar surface) by

$$M^{SN}(z) = H - eFz - e^2/16\pi\varepsilon_0 z , \qquad (1)$$

where e is the *elementary positive charge*. This barrier is often called the *Schottky-Nordheim (SN) barrier*.

II. SCHOTTKY REDUCTION AND THE SCHOTTKY CONSTANT

With a barrier-reducing field present, the maximum height M_{max} of a SN barrier is less than its zero-field height H by an energy Δ_S found as follows. The barrier maximum occurs for

$$dM^{SN}/dz = -eF + e^2/16\pi\varepsilon_0 z^2 = 0 , \qquad (2)$$

or:
$$eFz = e^2/16\pi\varepsilon_0 z , \qquad (3)$$

or:
$$z = (e/16\pi\varepsilon_0)^{1/2} F^{-1/2} . \qquad (4)$$

Hence, from (1) & (3):

$$M_{max} = H - 2eFz . \qquad (5)$$

Hence:
$$\Delta_S = H - M_{max} = 2eFz . \qquad (6)$$

Hence, using (4):
$$\Delta_S = c F^{1/2} , \qquad (7)$$

where
$$c \equiv (e^3/4\pi\varepsilon_0)^{1/2} . \qquad (8)$$

The energy Δ_S is called the *Schottky reduction*. For SN barriers, Δ_S depends only on the applied field and is proportional to $F^{1/2}$. The coefficient of proportionality c (sometimes written c_S) is an universal constant defined as above and now often called the *Schottky constant*. In the customary units often used in field emission, c has the value

$c \cong 1.199985$ eV $(V/nm)^{-1/2} \cong 3.794686\times10^{-5}$ eV $(V/m)^{-1/2}$. (9)

For those unfamiliar with it, this derivation can seem unexpectedly "tricky".

The numerical value of c was first given (but in cm-based units) in eq. (6) of Schottky's 1914 paper [5], but the name "Schottky constant" is much more recent, and is still not widely recognised.

The zero-barrier field (for the SN barrier), $F_0^{SN}(H)$, is the field that reduces a SN barrier of zero-field height H to zero (i.e., makes M_{max} zero), and is found by setting $\Delta_S = H$ in (7):

$$F_0^{SN}(H) = c^{-2}H^2 \cong (0.6944615 \text{ V/nm}) (H/\text{eV})^2 . \quad (10)$$

The numerical value of c^{-2} (for fields in V/cm) was first given by Schottky in his 1923 paper [8] [see Table 1, on p. 83 – the value given in eq. (5) is incorrect], where unsuccessfully tried to explain field electron emission (FE) as an effect occurring at field $F_0^{SN}(\phi)$, where ϕ is the relevant local work function.

III. GENERALIZATION TO OTHER BARRIER FORMS

The algebraic results just discussed refer specifically to the SN barrier. However, physically analogous effects occur with any barrier of the mathematical form:

$$M(z) = H - AFz - B/z , \qquad (11)$$

978-1-6654-2590-2/21 $31.00 © 2021 IEEE   157

where A and B are physical constants, and with any barrier of geometrically similar form. It seems reasonable to let the terms "Schottky reduction" and "the Schottky effect" also apply to these physically analogous effects, but to restrict the name "Schottky constant" to the parameter c as just defined for the SN barrier. However, for barriers of form (11) there will be, in each case, a formula similar to (7), with a specific barrier reduction constant equal to a constant times c. An example is eq. (15): here the multiplying constant is $n^{3/2}$.

IV. Common Applications of the Schottky Constant

A. Classical Schottky Emission

The *classical Richardson-Schottky equation* for emission current density J is conveniently written

$$J = A_0 T^2 \exp[cF^{1/2}/k_B T] \exp[-\phi/k_B T] , \qquad (12)$$

where $k_B T$ is the Boltzmann factor, and A_0 is the universal theoretical Richardson constant.

B. Murphy-Good (MG) Field Electron Emission (FE) Theory

Murphy-Good 1956 FE theory corrected errors in the original 1928 treatment of Fowler and Nordheim. A modern version [9] of MG theory uses a parameter f, the *scaled field for a SN barrier of zero-field height* ϕ, defined by

$$f \equiv F/F_0^{SN}(\phi) \equiv c^2 \phi^{-2} F . \qquad (13)$$

Older versions used the *Nordheim parameter y,* given by

$$y \equiv +\sqrt{f} = c\phi^{-1}F^{1/2} . \qquad (14)$$

Nowadays, use of f, rather than y, is normally to be preferred [7], particularly when discussing current-voltage characteristics. (Thus, in MG theory, it is best to express the correction factor "v_F" in the exponent as a function of f.)

C. Positive Ion Field Evaporation

Positive-ion field evaporation is used in field ion microcopy and atom probe microscopy [10]. A parameter of some interest is the so-called *zero-barrier escape field F_n^e* at which the activation energy for an ion of charge ne to escape locally from the surface becomes zero. The so-called *basic thermodynamic formula* [10] is an approximation like (10), in which the escape field is given by

$$F_n^e \approx n^{-3} c^{-2} K_n^2 , \qquad (15)$$

where the so-called *thermodynamic term K_n* is the activation energy needed for local ion escape when the field is zero. In this context, K_n is a parameter analogous to the zero-field H.

D. Replacement of dimensionally inconsistent formulas

When the SI system of units was introduced in the 1970s, there was also a change in the official method of writing scientific equations. Prior to 1970, symbols in equations were allowed to represent pure numbers. In the reformed system, which has *quantity calculus* as its mathematical basis, symbols represent quantities (numerical value × units), and equations are expected to be dimensionally consistent.

In FE, particularly in the definition of the Nordheim parameter y, pre-reform 1960s-style formulae commonly appear in current literature. Thus, instead of (14) one sees

$$y = 3.79 \times 10^{-4}\, \phi^{-1} F^{1/2} . \qquad (16)$$

This equation is dimensionally inconsistent, and will generate a correct result only if you are told (or can reliably guess) what units the field F is measured in. Quite apart from the issue of whether Editors of modern journals do or should require equations written in their journals to be dimensionally consistent, it is far clearer is to use (14) as the equation, and give the value of c separately, using your chosen field units.

V. Schottky Constant in Other Equation Systems

Equation (8) is an ISQ-system equation, and the quantities it contains are defined within this system. Equation (8) can be generalized to become a statement about *physical properties* of the world, written:

Schottky-reduction =
    Schottky-constant × (Field-quantity)$^{1/2}$ (17)

Table 1 compares the forms this relationship takes in the ISQ, Gaussian and Hartree equation systems, all of which have been used in FE. The subscripts "ISQ", "s", and "H" label quantities belonging to these systems, respectively.

TABLE I. SCHOTTKY CONSTANT IN DIFFERENT EQUATION SYSTEMS.		
Equation system	**Schottky–constant definition**	**Schottky-reduction formula**
ISQ	$(e_{ISQ}^3/4\pi\varepsilon_0)^{1/2}$	$(e_{ISQ}^3/4\pi\varepsilon_0)^{1/2} F_{ISQ}^{1/2}$
Gaussian	$e_s^{3/2}$	$e_s^{3/2} F_s^{1/2}$
Hartree	1	$F_H^{1/2}$

An intention of the 1970s reforms in the international system of measurement, made by international and national standards authorities ultimately on behalf of Governments, was for the ISQ to eventually become the primary equation system for scientific communication. In FE, this 50-year old imperative is mostly implemented, but not yet completely.

References

[1] W. Thomson, Camb. Dublin Math J. 4, 276 (1849).

[2] J. C. Maxwell, A Treatise on Electricity and Magnetism. Oxford: Clarendon, 1st ed. 1873; 3rd ed. 1891.

[3] J. J. Thomson, Conduction of Electricity through Gases (1st ed.) Cambridge Univ. Press, 1903.

[4] R. F. Earhart, Philos. Mag., Series 6, 1:1, 147 (1901).

[5] W. Schottky, Physik. Zeitschr. 15, 872 (1914).

[6] International Standards Organisation (ISO), International Standard ISO 800000-1:2009. Quantities and Units, Part I: General. ISO, Geneva, 2009, corrected 2011.

[7] R. G. Forbes, Chapter 9 (pp. 387–447) in: Modern Developments in Vacuum Electron Sources, G. Gaertner, W. Knapp and R. G. Forbes, Eds. Switzerland: Springer Nature, 2020.

[8] W. Schottky, Z. Phys., 14, 63 (1923).

[9] R. G. Forbes and J. H. B. Deane, Proc. R. Soc. Lond. A 463, 2907 (2007).

[10] M. K. Miller and R. G. Forbes, Atom Probe Tomography: The Local Electrode Atom Probe. New York: Springer, 2015.

Correction of Conceptual Error in Feynman's Textbook Treatment of Pointed-Conductor Electrostatics

Richard G. Forbes

University of Surrey, Advanced Technology Institute & Dept. of Electrical and Electronic Engineering,
Guildford, Surrey GU2 7XH, UK
Permanent e-mail alias: r.forbes@ trinity.cantab.net

Abstract—In the well-known textbook based on his Lectures, Feynman develops a theoretical "two-sphere" model and formula that aim to provide a basic explanation of why the electrostatic field is significantly higher above a sharp point protruding from a bulk electrical conductor. However, in the limit of infinite radius for the larger sphere, the Feynman formula for field enhancement factor (FEF) does not go to the correct limit as known from recent field emitter electrostatics. The conceptual oversight in Feynman's argument is that he has failed to take into account that the charges on one sphere will influence the electrostatic potential at the surface of the other sphere. When the two-sphere situation is analyzed correctly, the usual approximate formula "FEF≈(height/apex-radius)" is obtained. This re-analysis supports the view that this formula is a general formula of approximate electrostatics (in the absence of current flow) and deserves to be known more widely.

Keywords—*Zero-current electrostatics, pointed conductors, electrostatic potentials, field enhancement.*

I. INTRODUCTION

It has been well known for a very long time that when a conductor is electrically charged, then the electrostatic field is highest in magnitude over the sharpest parts of the conductor. However, leaving aside the large body of literature developed in the context of field enhancement in field emission in recent years, it is difficult to find a simple physical explanation of this sharp-point effect in the general physics literature. For example, slightly complicated explanations have been given by Robin [1] and Starling [2], (and a modified version of the Robin approach is now used in atom-probe-microscopy field-evaporation simulations [3]), but these explanations do not yield simple formulae.

An apparently simple discussion has been given by Feynman [4, Vol. II, p. 6.13], who uses a system comprising a large sphere (A) of radius a and a small sphere (B) of radius b to represent a pointed conductor, and argues as follows. If these spheres carry charges of Q and q respectively, then the electrostatic potentials at the sphere surfaces will be $Q/4\pi\varepsilon_0 a$ and $q/4\pi\varepsilon_0 b$ respectively. It follows that, if the spheres are electrically connected (say by a thin wire), then the two surface potentials will be equal, and hence that $q/Q = b/a$. However, the ratio F_B/F_A of the electrostatic fields at the sphere surfaces (and hence the field enhancement factor γ) is

$$\gamma \equiv F_B/F_A = (q/4\pi\varepsilon_0 b^2)/(Q/4\pi\varepsilon_0 a^2)$$
$$= (q/Q)(a^2/b^2) = a/b \quad (>1) \quad (1)$$

Feynman then argues that the small sphere can represent the tip of a pointed conductor and the large sphere can represent the body of the conductor, and that formula (1) shows that the field will be higher above the "point".

This argument is superficially persuasive, but details are clearly not physically correct, as can be seen if we apply it to the well known (in field emission) case of a cylindrical post with a hemispherical apex, standing on a flat plane of large lateral extent. In the limit that a becomes very large, formula (1) tends to the result $\gamma = \infty$, rather than to the correct result (well known in field emission, e.g. [5]) that $\gamma \sim h/b$, where h is the post height.

Thus, there is clearly some sort of conceptual error in Feynmnn's approach, although at this stage it may not be immediately obvious what this is. That Feynman's argument is unsound has previously been noted by Fricker [6], who proposed instead a system involving touching spheres of progressively decreasing radii. However, this is a cumbersome system to analyze.

II. AN ALTERNATIVE TWO-SPHERE APPROACH.

A better approach is to apply to Feynman's two-sphere system a method analogous to the "Floating Sphere at Emitter Plane Potential (FSEPP)" model used to discuss field enhancement in field emission contexts (e.g., [7]).

The physical thinking and procedure behind applying this approach to the two-sphere system is as follows.

(1) The theory of electricity in metallic conductors (and related conducting materials) is the theory of the physics and chemistry of electron behaviour in these materials.

(2) The fundamental principle that lies behind metal-conductor electrostatics (in the absence of current flow) is that the conductor must be *in static electrical equilibrium*. At the electron level, this requires that the Fermi level (as defined by statistical mechanics—see Appendix) be the same everywhere. With a classical-conductor model that takes the local work function to be the same everywhere on the surfaces of the conductors involved, this means that the electrostatic potentials immediately outside the sphere surfaces must be the same everywhere on the sphere and the same for both spheres.

978-1-6654-2590-2/21 $31.00 © 2021 IEEE

(3) In the two-sphere case, two characteristic points are selected, one on each sphere, and an arrangement of point charges and (if necesssary) point dipoles, and their values, is chosen such that the electrostatic potentials at the two characteristic points become equal.

(4) The ratio of electrostatic fields at the two characteristic points is then calculated using this charge distribution.

A preliminary investigation shows that the characteristic points are best taken on the line joining the sphere centres and that—providing the sphere surfaces are well separated—a lowest-order treatment needs to consider only point charges at the sphere centres. The characteristic points are taken at the "apexes" of the spheres. Using k to denote $1/4\pi\varepsilon_0$, the condition for equality of electrostatic potentials at the characteristic points is

$$\frac{Qk}{(a+h)} - \frac{Qk}{a} + \frac{qk}{b} - \frac{qk}{(h-b)} = 0, \qquad (2)$$

where h is the distance between the sphere apexes.

The approximations $a \gg h$ and $h \gg b$ yield $q/Q \approx hb/a^2$, and then yield

$$\gamma = (qk/b^2)/(Qk/a^2) = (q/Q)(a^2/b^2)$$
$$\approx (hb/a^2)(a^2/b^2) = h/b, \qquad (3)$$

which, obviously, is the usual FSEPP-model formula. This is sometimes known as the *conducting-post formula*.

Feynman's argument has failed to take into account that the charges on one sphere will influence the electrostatic potential at the surface of the other sphere.

The best approach for problems of this kind is to consider each placed charge or dipole separately, and determine the difference in potential that this charge or dipole causes between the characteristic points. The sum of all the differences is then set equal to zero. Any terms relating to the reference zeros for the potentials due to the individual charges or dipoles automatically cancel, so for each individual point charge and dipole any convenient choice of reference zero can be made.

Feynman's approach would apply if $h \gg a$, but this condition does not provide a good model for a pointed conductor. One can debate whether the error was to choose a poor model or to apply electrostatics inappropriately.

III. Discussion

Result (3) is better written as h/r_a, where r_a is the apex radius of the post, or in the more specific form

$$\gamma = \alpha\, h/r_a, \qquad (4)$$

where α is an adjustment factor related to the post shape (or, more generally, to the protrusion *and* substrate shapes). It is best to determine values and functional dependences of α numerically (e.g., [8]). (For the HCP model, $\alpha \sim 0.7$.)

The derivation of the conducting-post formula given here is slightly more general than that usually given in field emission contexts, and helps to establish that this formula is a *general* approximate result of electrostatics, for a sytem in static electrical equilibrium (i.e., in the absence of current flow).

The conducting-post formula and the general result (4) deserve to be better known in the wider physics community than they currently seem to be, not least because they help to explain the physics of many poorly understood situations involving the Earth's electrostatic field [9]. Amongst others, these situations include lightning rods (where there seems to be limited appreciation that both rod height and apex radius are important), and the electrostatics of trees and mountains.

Appendix: The Definition of Fermi Level

In the context of this paper, the *Fermi level* is defined as the value E_F of the electron total energy E at which the Fermi-Dirac distribution function has the exact value ½. If an electron state exists at the Fermi level, and system is in local thermodynamic equilibrium, then the occupation probability of this state would be exactly ½. Particularly when different systems are interacting, the Fermi level must be measured relative to a common reference level. Often the Fermi level of the local laboratory "Earth" (or "Ground") is a convenient reference level. When two systems are in static electrical equilibrium, their Fermi levels (measured relative to a common external level) will be equal.

References

[1] G. Robin, Ann. Sci. de l'É.N.S., 3rd series, vol. 3 (1886), pp. 3–58 (supplement).

[2] S. G. Starling, Electricity and Magnetism, 6th ed. London: Longmans, 1937; see p. 135.

[3] N. Rolland, F. Vurpillot, S. Duguay and D. Blavette, Microsc. Microanal. 21, 1649 (2015).

[4] R. P. Feynman, R. B. Leighton and M. Sands, The Feynman Lectures on Physics. Reading, Mass: Addison-Wesley, 1964.

[5] R. G. Forbes, C. J. Edgcombe and U. Valdré, Ultramicroscopy 95, 57 (2003).

[6] H .S. Fricker, Phys. Educ. 24, 157 (1989).

[7] R. G. Forbes, J. Appl. Phys. 120, 054302 (2016).

[8] F. F. Dall'Agnol, S. V. Filippov, E. O. Popov, A. G. Kolosko and T. A. de Assis, J. Vac. Sci. Technol. B 39, 032801 (2021).

[9] R. G. Forbes "A 'nearly semi-quantitative' explanation of electrical breakdown effects reported by Julius Caesar and Pliny the Elder", 8th Internat. Workshop on Mechanisms of Vacuum Arcs (MeVARC 2019), Padova, September 2019.

Features of the field enhancement factor on blade-type emitters

S.V. Filippov *, E.O. Popov, A.G. Kolosko
Div. of Plasma Physics, Atomic Physics and Astrophysics
Ioffe Institute
St Petersburg, Russia
*s.filippov@mail.ioffe.ru

F.F. Dall'Agnol
Department of Exact Sciences and Education, Federal
University of Santa Catarina, Campus of Blumenau
Blumenau, SC, Brazil

Abstract- **We analyze and compare the field enhancement factor(FEF) from blade-like emitters with the corresponding post-like counterpart. We attempt to improve mechanical and electrical properites as an advantejous tradeoff for the reduced FEF of blades. We discuss that the reduced FEF is mild and can be compensated by increase in the voltage to yield the same current.**

Keywords: field enhancement factor; blade-type emitter; knife-edge emitter; wedge-like emitter; HCP; field emission; emission area.

I. Introduction

Blade-type field emitters (BFE) are an interesting class of cathodes for applications as electron sources equivalent to their needle-like counterparts. BFEs find applications in time-of-flight mass spectrometers [1] and sources of terahertz radiation [2]. As examples, there are vertically oriented graphene sheets and reduced graphene oxide used in vacuum microelectronics [3, 4]. Blade-like emitters are sturdier, have better electrical conductance, and large emitting surface compared to needle-like emitters. On the other hand, it has a lower field enhancement factor (FEF) due to self-screening effect. Nevertheless, the lower FEF is not so severe that can be compensated by increasing the applied voltage to yield the same current.

Our study aims, first foremost, to analyze BFE under many different geometries to predict their emission properties. These predictions remain open for almost all BFEs. Secondly, we maximize the tradeoff between the BFE's good properties and their lower FEF.

Using the Finite Element Method, we calculated the electric field distribution at the top edge of the three BFEs, for which the needle counterpart we know: Hemisphere-on-Cylindrical-Post (HCP), HemiSphere-on-Orthogonal Cone (hSoC), semi-Ellipsoidal tip (Elli) (see Fig. 1 a). In this work, we mapped the maximum FEF (characteristic FEF) γ_C as a function of the geometrical parameters. In addition, we calculated the current-voltage characteristics to infer the optimal area efficacy.

II. Simulation Details

We used COMSOL v5.3 to solve the Laplace equation numerically. Here we studied three shapes of the BFE: HCP, hSoC, and Elli blades (see Fig. 1a-c). Elsewhere, we have shown that the apex FEFs of ellipsoidal, paraboloidal, and

hyperboloidal (with half-angle 5°) tips are very similar, so we ignored them here [5]. Symmetry allows us to simulate only one quadrant of the emitter in a simulation domain that respects the Minimum Domain Size criterion [6]. The emitter was considered as an ideal conductor with no electric field penetration; consequently, the internal part of the emitter was removed from the simulation domain. Figs 1a-c show the whole emitter for illustration purpose; however, only one quadrant is actually analyzed, due to symmetry. The colors indicate the relative field strength. Fig. 1d shows the first quadrant of the simulation domain, with the hollow emitter snapped to the corner.

The blades have height h, radius of curvature at the apex r_{apex}, and length l. The length is defined as the interspace between the corners plus $2r_{apex}$. The hSoC blade has an additional parameter θ, which defines the half-angle of the vertex (here $\theta = 10°$). In order to compare the field enhancement for different shapes of emitters, the apex aspect ratio $f_{apex}=h/r_{apex}$ is used instead of the usual aspect ratio $f=h/R_{base}$, where R_{base} is the radius of the base. All dimensions in our system are relative to r_{apex}. So, r_{apex} is a scale parameter and can be arbitrarily set to unity. However, to evaluate the notional area of emission we did r_{apex}= 50 nm. The FEF does not depend on the scale, but the current and the notional area is proportional to r_{apex}^2.

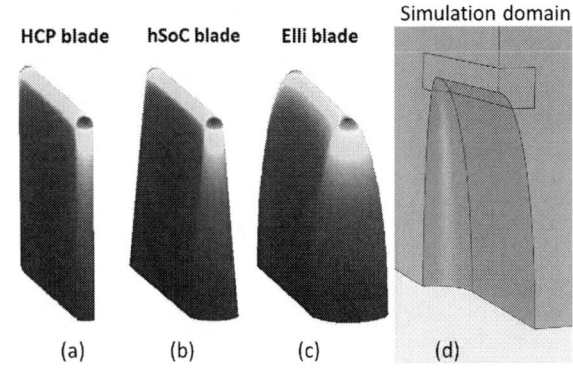

Fig. 1. Geometry of simulated blade-type emitters: (a), (b) and (c) show HCP, hSoC and Elli blades with r_{apex}=50 nm, f_{apex}=30, l/r_{apex}=50. The colors indicate the field strength on the surfaces. In (d) there is the simulation domain showing the interior of the emitter removed.

978-1-6654-2590-2/21 $31.00 © 2021 IEEE

III. Results and Discussion

Fig. 2 shows the FEF distribution along the apex edge of the BFEs with the same apex aspect ratio and length. The field enhancement is not uniform, as expected. The variation between the characteristic FEF (γ_C) and the FEF at the middle of the blade is less than 50% in all emitters.

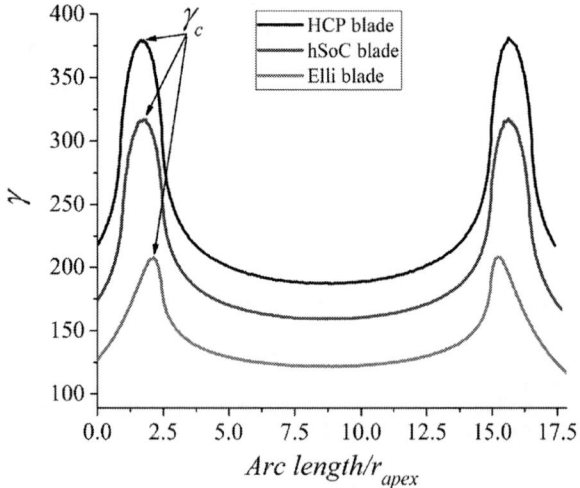

Fig.2. Field enhancement profile along the edge for HCP, hSoC, and Elli BFEs (f_{apex}=1001, l/r_{apex}=12.5]).

By increasing the blade's length, γ_C decreases about 1.6 times. When $l\to0$, apex FEF tends to γ_C of a tip emitter with the same geometric parameters (see Fig.3). The maximum FEF difference from the post counterpart is between 10 % (for $l\to0$) and 50% (for $l\to\infty$).

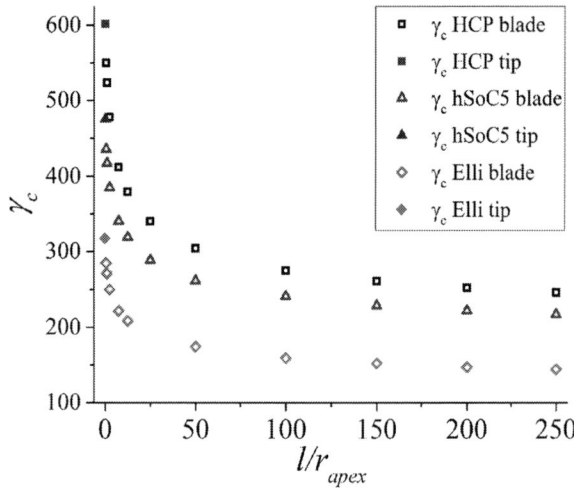

Fig.3. The $\gamma_C \times l$ (f_{apex}=1001 and $l \in$[0.02;10 μm]). Solid symbols correspond to characteristic FEF of needle-like emitter with the same geometric parameters.

From the field distribution shown in Fig.1 we derive the maximum current density, J_{kC}, the total current I, and the notional area of emission $A_n = I/J_{kC}$. We keep I=1 μA throughout.

The larger A_n should be one advantejous property that blades have over needles. However, Table I shows A_n larger only for Elli emitter. The values of A_n for the HCP and hSoC blades present a complicated and counterintuitive behavior. In these cases, A_n begins increasing between f_{lr}=l/r_{apex}=2 (needle) and $f_{lr}\cong2.1$ (not shown) then drops with increasing blade length. This is counterintuitive and it is being thoroughly analyzed as it is an interesting phenomenon on its own.

In any case, note that Fig. 1 and Fig.2 show that the field (therefore the J_{kC}) is very concentrated at the corners. The figures promptly instigate us to search for a profile of the edge, where the emission is better distributed, to increase the A_n. Hence, one could scale the area proportionally to the length of the blade.

Table I: Notional area of emission from blades and needle-like emitters.

	A_n (nm²)	
	Needle	Blade (f_{lr}=10)
Elli	1127	1904
hSoC	3023	3129
HCP	3735	3597

IV. Conclusion

We analyzed the apex FEF difference between blade emitters and their post counterpart. This difference is not larger than 50%, which can be compensated by increasing the applied field to yield the same current. As advantages, the blades are expected to have lower electrical resistance, mechanical firmness and the profile of the edge can be designed to uniformize the field distribution.

References

[1] G. Wen-Bin, D. Guo and G. Zhang, "Pulsed strip-shaped electron beam from cold cathode for time-of-flight mass spectrometer application," 2014 Tenth International Vacuum Electron Sources Conference (IVESC), 2014, pp. 1-1.

[2] Y. Zu, X. Yuan, X. Xu, M. T. Cole, Y. Zhang, H. Li, Y. Yin, B. Wang, and Y. Yan, "Design and Simulation of a Multi-Sheet Beam Terahertz Radiation Source Based on Carbon-Nanotube Cold Cathode," Nanomaterials, vol. 9, no. 12, p. 1768, 2019.

[3] J. Liu, B. Zeng, W. Wang, N. Li, J. Guo, Y. Fang, J. Deng, J. Li and Ch. Hao, "Graphene electron cannon: High-current edge emission from aligned graphene sheets", Appl. Phys. Lett., vol. 104, p. 023101, 2014.

[4] I.-K. Baek, R. Bhattacharya, J. S. Lee, S. Kim, D. Hong, M. A. Sattorov, S.-H. Min, Y. H. Kim and G.-S. Park, "Uniform high current and current density field emission from the chiseled edge of a vertically aligned graphene-based thin film", J. of Electromagnetic Waves and Applications vol. 31, no: 18, pp 2064-2073, 2017.

[5] S. V. Filippov, E. O. Popov, A. G. Kolosko and F. F. Dall'Agnol, "Modeling basic tip forms and its field emission," 2020 33rd International Vacuum Nanoelectronics Conference (IVNC), 2020, pp. 1-2.

[6] T. A. de Assis and F. F. Dall'Agnol, "Minimal domain size necessary to simulate the field enhancement factor numerically with specified precision," J. of Vac. Sci. Technol. B, vol. 37, p. 022902, 2019.

Comparison of the effective parameters of single-tip tungsten emitter using FN and MG-plots

Eugeni O. Popov*, Sergey V. Filippov**,

Anatoly G. Kolosko

Ioffe Institute
ul. Politekhnicheskaya 26, St.-Petersburg, 194021, Russia
*e.popov@mail.ioffe.ru, **f_s_v@list.ru

Alexandr Knápek

Institute of Scientific Instruments of the Czech Academy of
Sciences,
Královopolská 147, 612 64 Brno, Czech Republic

Abstract- **A comparison of the effective parameters obtained by processing the current-voltage characteristics in the Fowler-Nordheim and Merphy-Good coordinates for single-tip emitters was made. The tungsten tips were fabricated using of electrochemical sharpening method with sodium alkali. Statistical data of field enhancement factor (FEF) and emission area (EA) were obtained on the distribution of effective values obtained in real-time. An experimental technique is proposed for determining the shape of the tip based on the accumulation of statistical data in the coordinates $\ln(I_m/J_k)$ vs $\ln(f)$.**

Keywords: single-tip tungsten emitter, statistical data on effective parameters, FN plot, MG plot, real-time, k-power

I. Introduction

The modern theory of field emission poses several problems that require experimental support. The first challenge is to discover the benefits of the recently developed Murphy-Good coordinates (MG-plot) [1]. The result of plotting in these coordinates will be the values of FEF and EA, which in the idealized "plane" case do not depend on the applied voltage, that is, on the range of fields. It is important to compare the data of the plane case, 3D modeling of emitters of various shapes, as well as to obtain experimental values of the effective parameters.

The second task is to clarify the dependence of the degree of voltage in the preexponential factor on the shape of the tip [2]. Presumably single-tip metal systems will be able to give an unambiguous answer about the influence of the tip shape, in contrast to the plane case, on the functional dependence of the emission area on the applied voltage.

II. Calculation formulas and the emitter

The approximating expressions proposed in [3] for special functions of field emission made it possible to transfer the functional dependence of v on the field into a preexponential factor. And the adoption of a slightly varying function τ as a unit (or, more precisely, $\tau^2 \approx 1.1$), made it possible to introduce the concept of *kernal* current density J_k. As a result, the calculation formulas for the emission current and the processing of the graph in MG-plot will look like (here τ^2 is taken as a constant):

$$I = A_n \tau^{-2} J_k = A_n \tau^{-2} a_{FN} \varphi^{-1} (\alpha U)^{2-\eta/6} F_R^{\eta/6} \exp(\eta) *$$
$$* \exp(-b_{FN} \varphi^{3/2}/(\alpha U)) \tag{1}$$

$$\ln(I/U^{2-\eta/6}) = \ln(R^{Fit}) + S^{Fit}(1/U) \tag{2}$$
$$S^{Fit} = -b_{FN} \varphi^{3/2}/\alpha \tag{3}$$
$$R^{Fit} = A_n \tau^{-2} a_{FN} \varphi^{-1} F_R^{\eta/6} \exp(\eta) \alpha^{2-\eta/6} \tag{4}$$
$$\eta = b_{FN} \varphi^{3/2}/F_R = b_{FN} c_S^2 \varphi^{-1/2} \tag{5}$$

where A_n - notional emission area, a_{FN}, b_{FN} - Fowler-Nordheim constants, c_S - Schottky constant, $F_R = \varphi^2 c_S^{-2}$ - reference field or barrier removal field, $F = \alpha U$ - emission field, $\alpha = \zeta^{-1}$, ζ - characteristic length.

The standard approach is to obtain effective values from the FN-plot, which we carry out using the Spindt formula [4] or the Elinson-Schrednik formula [5] (where $\tau^2 \approx 1.1$):

$$I = A_n \tau^{-2} a_{FN} \varphi^{-1} (\alpha U)^2 \exp(1.03\eta) *$$
$$* \exp(-0.95 b_{FN} \varphi^{3/2}/(\alpha U)) \tag{6}$$

$$\ln(I/U^2) = \ln(R^{Fit}) + S^{Fit}(1/U) \tag{7}$$
$$S^{Fit} = -0.95 b_{FN} \varphi^{3/2}/\alpha \tag{8}$$
$$R^{Fit} = A_n \tau^{-2} a_{FN} \varphi^{-1} \exp(1.03\eta) \alpha^2 \tag{9}$$

Coordinates $g_n = A_n/(2\pi r_a^2)$ vs $f = F/F_R$, which can characterize the functional dependence of the notional EA on the field, were proposed in [6] for a hemisphere of radius r_a on a plane.

For emitters of real shape, a reliable way to experimentally find this dependence is to use exclusively homogeneous emitters or single emitter.

The investigated tungsten tips were made using of electrochemical sharpening method with sodium alkali. The initial radius of curvature of the emitter r_a is 50 nm.

III. Results and Discussion

The details of the experimental setup and the real-time processing technique for emission data are described in [7]. The vacuum level was no worse than $4 \cdot 10^{-8}$ Torr.

A single-tip emitter was placed in the experimental setup at a distance d_{a-k} between the tip of the tip and the anode. This value was used as the d_{sep} parameter for the FEF calculations. It is clear that the γ obtained in this way in the case of a one-point system does not have a strict physical meaning. However, it can be easily converted to the characteristic length ξ using the d_{sep} parameter (see Table 1).

In contrast to work [8], we have determined the effective values of single-wire tungsten in modes with sufficiently high emission currents. Fig. 1a shows a characteristic graph in MG coordinates and achieved current levels up to 45 μA for sample C7. The approximation of the graph and the trend line are plotted according to the formulas for the MG eq. (1-5).

Table 1 shows some calculated effective parameters for three emitters for a given work funtion 4.6 and 5.6 eV, for convenience of comparison with work [8]. All measurements were made taking into account the satisfaction of the emission range for the field f.

Table 1. Comparison of the main emission parameters of single-tip tungsten emitters using the FN and MG-plot.

Sample;;	φ, eV;	FN plot		MG plot	
	$f=F/F_R$	γ	A_n, nm^2	γ	A_n, nm^2
C4	4.6	5002	335	4937	394
	f	0.315	0.450	0.291	0.416
d_{a-k}= 1360 μm	5.6	6718	352	6670	379
I_{max}= 24 μA	f	0.284	0.406	0.266	0.380
C7	4.6	1907	4396	1895	4701
	f	0.290	0.384	0.272	0.360
1350	5.6	2562	4612	2558	4652
45	f	0.261	0.346	0.247	0.328
B3	4.6	1067	213	1053	248
	f	0.335	0.406	0.311	0.376
600	5.6	1433	223	1423	238
6	f	0.304	0.368	0.285	0.345

Comparison of the methods of processing the IVC using the FN and MG-plot showed the influence of the shape of the tip on the bending of the characteristic. As a result, it was found that the effective area values are slightly higher when processed in MG-plot than FN. In the approximation of a flat (one-dimensional) case, on the contrary, the MG coordinate method gives FEF values greater and EA less than for FN-plot.

When single-tip system operate, strong fluctuations of characteristics are observed. To increase the reliability of the obtained characteristics, a statistical analysis of the parameters was applied in real-time (Fig. 1b).

Plotting the dependence in logarithmic coordinates (Fig. 2a) showed a functional dependence of the emission area on the field. It is clear that the true geometric FEF is unchanged for a one-sided system. Therefore, to find f_a at the apex of the tip, we used the FEF value determined at the initial IVC segment IVC $f = U\gamma^{eff}/(d_{a-k}F_R)$.

Fig. 2b showed a large scatter of values in the determination of the trend line. Therefore, real-time statistical analysis was applied, which showed stable averages of the slope $\langle k_A\rangle$. It can be shown that this value characterizes the deviation of the pre-

exponential voltage-exponent from the flat case $k_p = 2-\eta/6$ used in the MG equation [2]. The values of k_A were obtained on the order of or greater than 1, which is in good agreement with theoretical predictions for an emitter with an ellipsoidal shape.

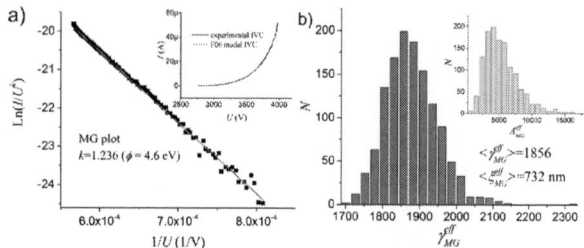

Fig.1. (a) Current-voltage characteristics of a C7 emitter and MG-plot. (b) Statistical data on effective parameters byMG-plot in real-time.

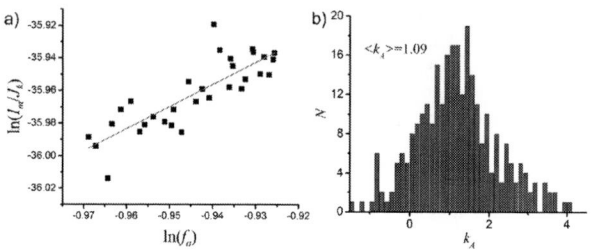

Fig.2. (a) Logarithmic dependence of the notional emission area $A_n = I_m/J_k$ on the logarithm of the dimensionless field. (b) The mathematical expectation of of the slope of the graph in coordinates $\ln(I_m/J_k)$ vs $\ln(f)$.

IV. CONCLUSION

We obtained the statistical distributions of the effective values of the FEF parameters and the emission area obtained by the method of plotting in MG plot coordinates. An experimental technique for determining the shape of a single-tip emitter is implemented based on plotting the dependence of the notional area on the emission field.

REFERENCES

[1] R.G. Forbes, A.G. Kolosko, S.V. Filippov, E.O. Popov, "Reinvigorating our approach to field emission area extraction (because Murphy-Good plots are better than Fowler-Nordheim plots)", 32nd International Vacuum Nanoelectronics Conference (IVNC) & 12th International Vacuum Electron Sources Conference (IVESC), 2019, p. 23.

[2] R.G. Forbes, E.O. Popov, A.G. Kolosko and S.V.Filippov, R. Soc. Open Sci. 8, 201986 (2021).

[3] R.G. Forbes and J.H. Deane, Proc. of R. Soc. A, 463, 2907-2927 (2007).

[4] C.A. Spindt, I. Brodie, L. Humphrey and E.R. Westerberg, J. of Appl. Phys., 47, 5248-5263 (1976).

[5] E.O. Popov, A.G. Kolosko and S.V. Filippov, Tech. Phys. Lett., 46, 838-842 (2020).

[6] K.L. Jensen, Introduction to the Physics of Electron Emission (Chichester, UK: Willey, 2018).

[7] E.O. Popov, A.G. Kolosko, S.V. Filippov, E.I. Terukov, R.M. Ryazanov and E.P. Kitsyuk, J. Vac. Sci. Technol. B 38, 4, 043203-1-10 2020.

[8] M. Madanat, M. Al Share, M.M. Allaham, M.S. Mousa, J. Vac. Sci. Technol. B, 39, 024001 (2021).

HIGH CURRENT FIELD EMISSION ARRAYS FOR CROSSED-FIELD DEVICE EXPERIMENTS

Ranajoy Bhattacharya[1], Mason Cannon[1], Rushmita Bhattacharjee[1], Winston Chern[2],
Nedeljko Karaulac[2], Girish Rughoobur[2], Akintunde I. Akinwande[2] and Jim Browning[1]

[1]*Boise State University, Boise, ID, 83725 USA*
[2]*Massachusetts Institute ofTechnology, Cambridge, MA, 02139 USA*
*Contact: jimbrowning@boisestate.edu, phone +1-208-426-2347

Abstract— **Specially designed silicon gated field emitter array (Si-GFEA) tips are used as large current electron sources for crossed-field device experiments. Thirty-six discrete field emitter arrays, each made of 100×100 tips, are integrated in to a single die using a mesh configuration for uniformity and reliability. The I-V characterization shows the devices are capable of producing current up to 5 mA at a gate voltage of 50 V and anode voltage of 200 V. However, after ultra-violet light exposure of 100 minutes, the anode current increases to > 50 mA. This enhancement can be attributed to the residual gas desorption stimulated by UV exposure. Eight such die are being integrated into a planar crossed-field device configuration with plans to use 32 die in a magnetron experiment.**

I. INTRODUCTION

Crossed-field vacuum electron devices such as crossed-field amplifiers (CFAs) and magnetrons provide superior performance in terms of power density and efficiency. Improvements of these devices in terms of power density, phase-locking or control, noise reduction, and faster start-up times are of particular interest. Magnetrons have very high efficiency and are used in radar, but phase control would provide greater opportunities for application. Improving performance by understanding electron stability in crossed-field geometries and by modulating electron injection in those devices is the goal of our research. As the first step, a simple planar crossed-field geometry optimized by previous simulation work[1] will be studied experimentally. However, large current (>150 mA) injected beam sources are required for the experiment. We propose using gated field emission arrays (GFEAs) as electron sources in simple planar crossed-field configurations[1] and in industrial magnetron experiments[2]. In this work, we present results from GFEAs characterization on die specifically designed for use in our experiments. The GFEA die were fabricated using a novel method[3] and are capable of producing a field emission current up to > 50 mA per die. Each die is 7.5 mm long, 2.5 mm wide and 0.7 mm thick as seen in Figure 1 (a). A single die is segmented into thirty-six active areas in a mesh configuration for the gate metal where each active area has 100×100 silicon nano wires with tips of radius ≈ 6 to 10 nm. At the top and bottom of each die (as shown) are gate contact pads.

II. EXPERIMENTAL SETUP

Experiments were carried out in in stainless steel high vacuum chamber equipped with a 3-axis manipulator probe arm, electrical and thermocouple feedthroughs, a residual gas analyser and a UV light source. The chamber is also equipped with turbo pump backed by a rotary pump to main a high vacuum of $\approx 5 \times 10^{-8}$ torr. Measurements were carried out using a Keysight B2902A source measurement unit (SMU). An in-house developed test jig fabricated from a Low Temperature Co-Fired (LTCC) was used to mount the test die. The LTCC test jig can be seen in Figure 1(b) showing 8 die with the silver paste traces and an interconnect. The test jig has locations for each die in a slot. The gate is connected on the top side of the structure using silver and copper tape. Silver paint is coated on the backside of the silicon die to improve electrical connection, and copper tape is again used to connect to substrate. A fixed 200 V DC and 0-50 V pulsed sweep voltage were applied to collector and gate, respectively. After the die are placed in the jig and electrically tested for connection, the jig is placed in the vacuum test chamber. In some cases a large ZnO:Zn phosphor screen is used as the collector and is placed

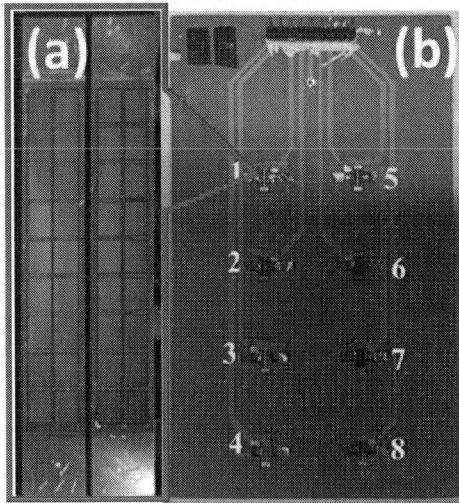

Fig. 1 (a). Magnified image of the GFEA die showing the mesh configuration and (b) photograph of the LTCC test jig with 8 die and the electrical connections.

over all 8 die. In other cases, a copper plate anode mounted on a stainless-steel probe arm is placed over the die.

III. EXPERIMENTAL RESULTS

These GFEAs were characterized by measuring the I-V curves. The collector (3-5 mm above the emitters) was kept at 200 V DC while the gate was swept to 50 V pulsed (1% duty cycle, 15 ms pulse width). Observed average emission current for an applied gate voltage of 50 V average $\approx$ 5 mA per die for more than 20 die. However, after a 100-minute exposure to ultra-violet (UV) light[4], observed emission current increased by >10× to >50 mA per die due to water desorption[4] with a die average current density of $\approx$ 260 mA/cm^2. A before and after UV comparison of I-V characterization can be seen in Figure 2. The UV light source has a 185 nm wavelength and power density of 350 μW/cm^2.

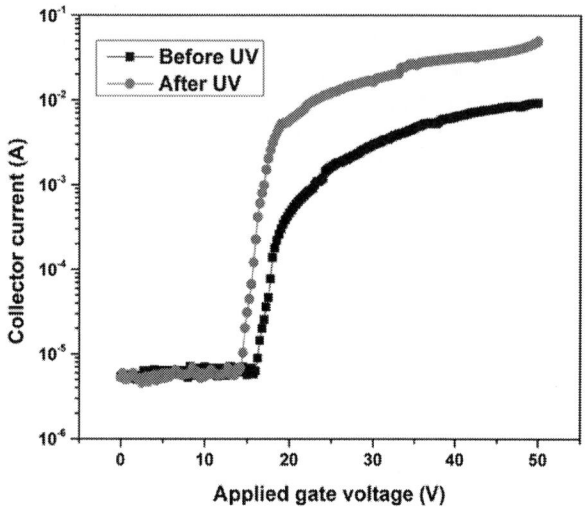

Fig.2. I-V characterization graph for before and after UV exposure.

For the planar crossed-field device (CF) experiment, a simple, planar electrode system 10 cm wide and 20 cm long with a variable gap of 0.5 cm to 2 cm has been developed. The schematic of the complete CF structure is shown in Figure 4. Here a separate emitter holder is used as an injected beam source to launch electrons into the crossed-field gap between a sole electrode and an anode. This region of interest will be used for our study. Mainly, three parameters, anode-sole gap, injected beam current density, and magnetic field tilt will be studied in the experiment and compare with simulation and theory. For the experiment, an existing vacuum test chamber has been modified. A 3 kV, high voltage DC power supply will be used as the voltage source. Two existing Helmholtz coils will be used as the magnetic field source. An in-house developed pulsed, DC driver circuit will be used to drive the emitters. A LabView code will be used to drive the electronics as well as to record the data. A segmented anode and a segmented end collector are used to measure the electron current at various locations in the device. In addition, openings in the anode (Figure 3) allow the use of current probes and RF probes to measure the beam and noise in the gap.

Figure 4 shows the fabricated test jig. This test jig will be used for the experiment.

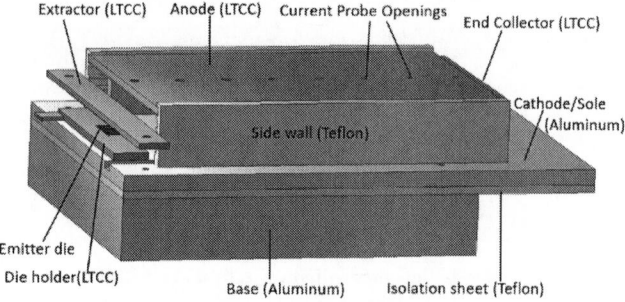

Fig.3. Schematic of the planar crossed-field device experimental test jig.

Fig.4. Image of the test fixture fabricated from LTCC with interconnects for the GFEA die, segmented anode and end collectors as indicated.

IV. CONCLUSIONS

In this work, we have carried out I-V characterization experiments on specially designed Si-GFEAs which will be used as an injected beam electron source in a simple planar crossed-field device experiment and in a magnetron experiment. From the experiments, it was observed that after UV exposure, the field emission current from the die was > 50 mA at 50 V, having high potential for CFAs. Higher operating voltages are planned along with demonstration of the CF injected beam configuration.

ACKNOWLEDGMENT

Material support for his work is provided by the Air Force Office of Scientific Research under grant #FA9550-19-1-0101.

REFERENCES

[1] R. Bhattacharya et. al., "Analysis of Injected Electron Beam Propagation in a Planar Crossed-Field Gap", Appl. Sci. 2021, 11(6), 2540

[2] Daylon Black et. al., "A cathode support structure for use in a magnetron oscillator experiment" Int J Appl Ceram Technol. 2020;17:2393–2406.

[3] Stephen A. Guerrera, and Akintunde I. Akinwande. "Silicon Field Emitter Arrays With Current Densities Exceeding 100 A/cm2 at Gate Voltages Below 75 V," IEEE Electron Device Letters, Vol. 37, no. 1, January 2016.

[4] R. Bhattacharya et. al., "Ultra-Violet Light Stimulated Water Desorption Effect on Emission Performance of Gated Field Emitter Array", Journal of Vacuum Science & Technology B **39**, 033201 (2021).

Lifetime and Breakdown Mechanisms in Double-Gated Si FEAs

Girish Rughoobur and Akintunde I. Akinwande
Microsystems Technology Laboratories
Massachusetts Institute of Technology
Cambridge, MA 02139, USA
Email: grughoob@mit.edu

Abstract—**We demonstrate the lifetime and breakdown mechanisms in dense (10^{12} m^{-2}) self-aligned double-gated Si field emitter arrays. We perform measurements at relatively low currents of 10 nA for over 300 hrs but found that increasing the current to 50 nA causes catastrophic breakdown in less than 100 hrs. Over the lifetime measurements, we observe charge accumulation that could be between the two gates, causing noisy current-voltage characteristics. In addition we assess additional mechanisms that could lead to early breakdown in such devices.**

Keywords—**double-gate, lifetime, breakdown**

I. Introduction

Nano-fabricated cold cathodes have emerged as ideal candidates to replace thermionic electron sources due to their low voltage operation, faster response and scalability. Extremely dense (10^{12} m^{-2}) Si field emitter arrays (FEAs) with self-aligned extractor and focus gates reported in [1] have been recently demonstrated to operate at even lower voltage ($\sim$10 V). These devices had two self-aligned gated apertures with approximate diameters 350 nm and 550 nm, separated by an insulator with thickness of 350 nm. High aspect-ratio nanowire current limiters were also integrated with the emitter to reduce the tip burnout due to spatial variation of the emitter radius. [2] Despite the successful demonstration of the focusing capabilities using the second gate, the concern is the reduced lifetime at increased emission currents, compared to similar single-gated FEAs that could operate at high current densities exceeding 10^{6} A·m^{-2} for over 100 hrs [3]. In the case of double-gated Si FEAs there gate leakages are higher even when the beam is focused optimally, as stray electrons that have low vertical velocity will be captured by the extractor gate. When out of focus, stray electrons emitted at a wide-angle will be captured by the focus gate. These additional leakage mechanisms will therefore dramatically reduce the lifetime of the double-gated Si FEAs by either charge accumulation or Joule heating of the gates. In this work we perform the lifetime characterization of these extremely dense self-aligned double-gated Si FEAs to identify the potential mechanisms of failure in such devices.

Sponsors: IARPA/AFRL under the contract number: FA8650-17-C-9113.

II. Experimental

A. Device $I-V$ Initial Performance

The double-gated Si FEAs (500×500) were characterized as a four terminal device using four source measurement units (SMUs, Keithley 2657A), with the emitter voltage, V_E, swept between 0-35 V, and the extractor gate voltage, V_G, biased at 0 V. The focus voltage, V_F was maintained at a ratio of V_E, whereas the anode voltage, V_A was at 3 kV. The current-voltage ($I-V$) characteristic curves are shown in Fig. 1.

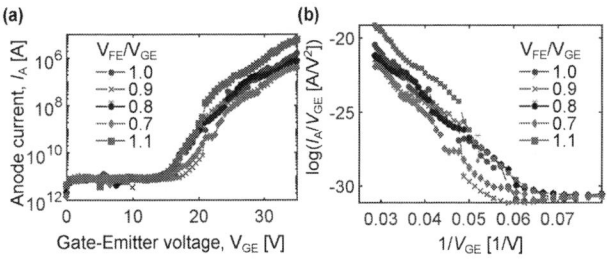

Fig. 1. (a) $I-V$ characteristics measured with different ratios of V_{FE} to V_{GE}; and (b) FN plots for the data for different V_{FE}/V_{GE} ratios.

The measured absolute value of slope from the Fowler-Nordheim (FN) plot, b_{FN}, was 310 V for $V_{GE}/V_{FE} = 1$, with maximum current of 1.6 μA measured at $V_{GE} = 35$ V. The extracted field-factor, β, was $\sim$0.18 nm^{-1}. This was consistent with data we have reported previously [1].

B. Lifetime Measurements

We operated the devices for approximately 100 hrs as shown in Fig. 2(a) and record the $I-V$ characteristics at every 10-hrs intervals, during which the measurement is stopped for approximately 1 hr. In these measurements, I_A was maintained at the current level by varying V_{GE}. For 10 nA, V_{GE} had a range from 16 V to 30 V, whereas for 50 nA [Fig. 2(b)], the corresponding range was from 25 V to 35 V. However, the current variations were significant at lower current possibly due to noise in the measurement system. By contrast for 50 nA anode current, the variations reduced as expected. The device failed after 95 hrs at 50 nA [Fig. 2(b)], at which point the gate leakage increased dramatically. By inspecting the $I-V$ characteristics taken in between the measurements, we gradually observe that after 50 hrs, the $I-V$ curves were

978-1-6654-2590-2/21 $31.00 © 2021 IEEE

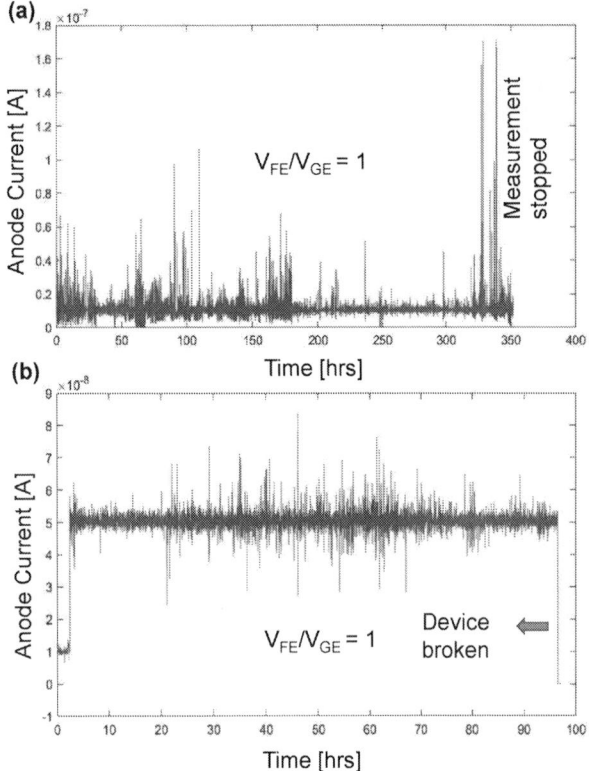

Fig. 2. (a) Lifetime data for 10 nA; and (b) Lifetime data for 50 nA.

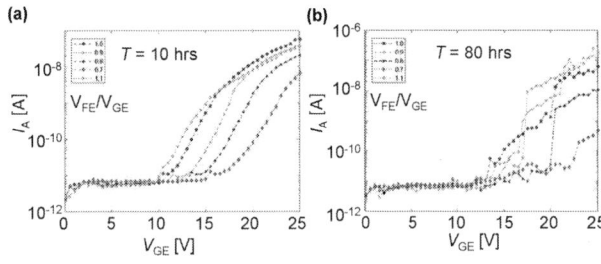

Fig. 3. Comparison of $I - V$ data at 10 hrs in (a); and at 80 hrs for the 50 nA emission current in (b).

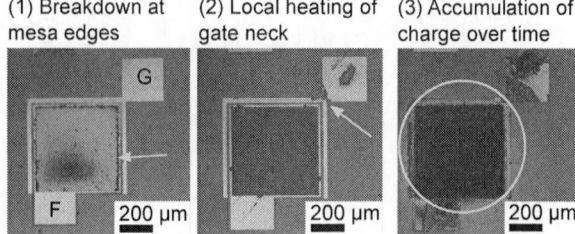

Fig. 4. Breakdown mechanisms illustrated with optical images (1) emitters at mesa edge burning out; (2) heating at narrow neck of extractor gate due to gate leakage and (3) full device breakdown due to charge accumulation over long measurement times.

Case (1) could be due to the non-uniform fabrication process that results in the edge emitters being sharper and these eventually burn-out earlier when operating at high currents. The melted tips at the edges could cause an electrical short between the gate to the emitter or between the two gates. Previous issues with a non-vertical mesa were resolved in our earlier work [4]. Case (2) could be as a result of large gate leakage, especially when electrons emitted at a wide angle are captured by either the focus or the extractor gates. The large leakage causes localized "hot-spots" on the electrode especially in the narrow regions which ultimately lead to the loss of electrical contact. Case (3) is the complete failure of the array that occurs possibly due to the charge injection in the dielectric separating the two gates. Over time, this charge causes the time dependent breakdown of the insulator separating the two gates. Possible solutions to these breakdown mechanisms could include the use of a metal gate, lithography/etch uniformity improvement, oxide with higher dielectric constant such as hafnia, thicker electrodes, and optimizing the shape of the electrode to minimize resistive junctions.

III. CONCLUSION

We have measured the lifetimes of self-aligned double-gated Si FEAs with emitter densities of 10^{12} m^{-2}. The experiments were conducted at relatively low currents of 10 nA and 50 nA. We demonstrated that at 10 nA the device can be operated for over 300 hrs, whereas at 50 nA, the device failed after 95 hrs, probably due to charge accumulation between the two gates. We also summarized the possible failure modes during $I - V$ characterization of these dense double-gated Si FEAs.

REFERENCES

[1] G. Rughoobur, N. Karaulac, L. Jain, O. O. Omotunde, and A. I. Akinwande, "Nanoscale silicon field emitter arrays with self-aligned extractor and focus gates," *Nanotechnology*, vol. 31, no. 33, p. 335203, Jun. 2020.

[2] N. Karaulac, G. Rughoobur, and A. I. Akinwande, "Highly uniform silicon field emitter arrays fabricated using a trilevel resist process," *Journal of Vacuum Science & Technology B*, vol. 38, no. 2, p. 023201, Mar. 2020.

[3] S. A. Guerrera and A. I. Akinwande, "Nanofabrication of arrays of silicon field emitters with vertical silicon nanowire current limiters and self-aligned gates," *Nanotechnology*, vol. 27, no. 29, p. 295302, Jun. 2016.

[4] G. Rughoobur and A. I. Akinwande, "Arrays of Si field emitter individually regulated by Si nanowires - High breakdown voltages and enhanced performance," in *2018 31st International Vacuum Nanoelectronics Conference (IVNC)*. IEEE, Jul. 2018.

no longer smooth as shown in Fig. 3. At 10 hrs, the b_{FN} values varied from 113 V to 340 V when V_{FE}/V_{GE} changed from 0.7 to 1.1. At $T = 80$ hrs, sudden jumps in the $I - V$ curves were evident. Together with larger influence of V_{FE}/V_{GE}, this could indicate charge accumulation at the dielectric between the two gates, possibly due to electrons emitted at a wide-angle, which eventually leads to the catastrophic breakdown.

C. Breakdown Mechanisms

During the device electrical characterizations, we identified three principal breakdown mechanisms in the double-gated Si FEAs as illustrated in Fig. 4:

1) Emitters at the edge of the array;
2) Local heating causing loss of electrical contact; and
3) Charge accumulation over time with entire array failing.

Influence of Geometrical Arrangements of Si Tip Arrays Fabricated by Laser Micromachining on their Emission Behaviour

Matthias Hausladen[1,*], Vitali Bomke[1], Philipp Buchner[1], Michael Bachmann[2], Alexandr Knápek[3], and Rupert Schreiner[1]

1 Faculty of Applied Natural Sciences and Cultural Studies, OTH Regensburg, D-93053 Regensburg, Germany
2 Ketek GmbH, 81737 Munich, Germany
3 Institute of Scientific Instruments of the Czech Academy of Sciences, Královopolská 147, 612 64 Brno, Czech Republic
Contact: matthias.hausladen@oth-regensburg.de, phone +49-941-943-1454

Abstract— **Densely packed emitters on a field emission array lead typically to mutual shielding. Taking biology as a role model for geometric arrangements could be a way to reduce this effect. For comparison, two electron sources, one with a spiral and a second with conventional rectangular (orthogonal) arranged emitters, were fabricated and investigated. Emission currents of 6 µA in the spiral ordered array and 120 µA in the rectangular array were reached with an extraction voltage of 400 V. From a mid-term measurement over 1 h a current stability of ±8.8 % (spiral) respectively ±5.7 % (rectangular) with a mean degradation of -3.0 µA/h (spiral) and -0.12 µA/h (rectangular) could be observed.**

Keywords — silicon field emission, field emitter array, parastichy tip arrangements, aperture field emission grid, laser micromachining

I. INTRODUCTION

To reduce the shielding effect of densely packed emitters in a field emission array (FEA), one approach could be to rearrange the tips. Harris et al. who studied the β-factor for rectangular and triangular arrays concluded that a regularly triangular tip distribution is preferable to a rectangular one [1]. For this research, we use a derived phyllotaxis model from botany to determine the tip positions, called parastichy [2]. The basic arrangement takes place using a golden spiral, which is based on the Fibonacci sequence and where adjacent elements (emitters) of the spiral forms a golden triangle that changes as the spiral progresses. Because of the large base diameter of the emitters, caused by the laser beam shape of the manufacturing process, we decided to modify the geometric arrangement in such a way, that the tip positions are determined by two interacting spirals. The aim of this study is to investigate whether this arrangement improves the emission properties of an array compared to the rectangular arrangement, while the number of tips remains the same.

II. FABRICATION PROCESS & ELECTRON SOURCE ASSEMBLY

A cathode layout based on the intersections of the primary spiral and a contrary directed secondary spiral (parastichy model) was created (Fig. 1a). The resulting positions should lead to a reduction of shielding or can increase the density of emitter tips in a certain area. For this, we improved our laser micromachining process [3], so that we can fabricate complex tip arrangement layouts by using DXF files from a CAD software. For comparison purposes, a second cathode with a conventional rectangular 8 x 8 tip arrangement was

The research work was funded by the Bavarian Research Foundation under project-number AZ-139619.

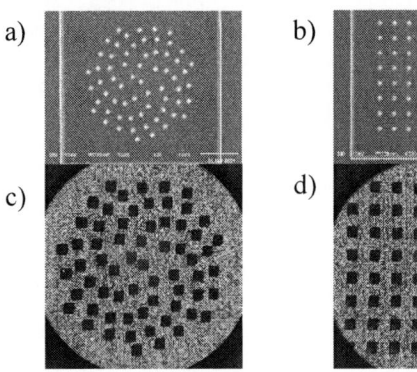

Fig. 1: (a, b) Top view SEM images of the cathode tip arrangements. (c, d) Top view microscope images of the grids with the aligned cathode underneath. The grid apertures appear black due to the lateral light deflection at the tips.

TABLE I
CATHODE AND GRID DIMENSIONS

Dimension	FEA arrangement layout	
	Spiral	Rectangular
Cavity outline (at surface, at ground)	4.2 x 4.2 mm² 3.95 x 3.95 mm²	4.2 x 4.2 mm² 3.9 x 3.9 mm²
Cavity depth (surface-to-ground)	291 µm	305 µm
Tip height	212 µm	215 µm
Tip to surface distance	79 µm	90 µm
Tip radii (array-center array-edge)	310 nm 210 nm	105 nm 145 nm
Dimension	**Grid arrangement layout**	
	Spiral	Rectangular
Cavity outline (at surface, at ground)	4.7 x 4.7 mm² 4.5 x 4.5 mm²	4.7 x 4.7 mm² 4.5 x 4.5 mm²
Cavity depth (surface–to-ground)	427 µm	412 µm
Residual grid thickness	98 µm	113 µm
Aperture outline (upper lower)	190 x 195 µm² 137 x 150 µm²	186 x 193 µm² 135 x 145 µm²

designed (Fig. 1b). Both cathode designs comprise 64 emitter tips within an outline area of 3 x 3 mm² and were manufactured on a silicon wafer (<100>, n-doped, 5-10Ωcm, 525 µm). After loading the DXF into the laser ablation system, the basic contour of the layout is known to the system. For two-dimensional laser ablation the contour is hatched by parallel straight lines in a distance of 5 µm, which are rotated 3 times by 120 ° and finally transferred into the substrate by scanning the line pattern with the beam deflection system multiple times. The resulting cavities of the cathodes have a depth of ≈ 295 µm and the emitters reach heights of ≈ 215 µm (Table I). Consequently, the tip-to-surface distance is in the range of 85 µm. To remove ablation dust and for tip sharpening, the cathodes were wet etched with hydrofluoric acid (HF) and tetramethylammonium hydroxide (TMAH) subsequently. For equal measurement conditions, special grids, with individual apertures per emitter, were designed to avoid emission dependencies, caused by the tip position underneath the grid (Fig. 1c, d). The fabrication of the grids was done in the same manner as for the cathodes and consists of the same material (Table I). A pattern of parallel straight lines in an area of 4.6 x 4.6 mm² was used to thin the substrate down to 105 µm, followed by a 150 x 150 µm² square pattern, which opened the substrate at the emitter positions. Afterwards the grids were wet etched with HF and TMAH. Each of the resulting FE electron sources consists of a cathode, a mica spacer (insulator), and an extraction grid and is clamped by a ceramic sample holder, composed by two ceramic plates and a copper shield. In order to align the parts, 2 fitting holes were created in each part during the laser process, which ensures that each emitter tip is centered underneath its aperture (Fig. 1c, d). The electron source is assembled by stacking the components onto an alignment fixture, providing 2 dowel pins, in sequential order: cathode ceramic plate, cathode, mica spacer (50 µm), extraction grid, grid ceramic plate and finally a copper shield. Finally, the stack is fixed with 3 screws and removed from the alignment fixture.

III. EMISSION CHARACTERIZATION

The measurements were performed in triode configuration. The currents were measured with a Keithley 6485 (I_C) and two Keithley 6487 (I_G, I_T) where the transmission current was measured field-free (no voltage between grid/shield and copper anode). The negative extraction voltage was applied to the cathode by a Keithley 6517B (U_C). The measurement points were recorded every 2 seconds with an averaging time of 100 ms. For device protection, a 150 kΩ resistor was added to the grid path. The IV-Plot (Fig. 2) show maximum currents of 6 µA for the parastichy FEA respectively 120 µA for the rectangular array at an extraction voltage of 400 V (Table II). The onset voltage of the parastichy FEA is with ≈ 200 V also higher compared to the rectangular one (≈ 136 V). For a mid-term measurement over 1 h and a mean current of ≈ 17 µA, a slightly higher standard deviation (±8.8 %) for the parastichy arrangement compared to the rectangular FEA (±5.7 %) was observed. Furthermore, in case of the degradation, a significant higher value was found for the parastichy sample. With -3.0 µA/h it is 25 times higher than for the rectangular layout (-0.12 µA).

IV. DISCUSSION

The influence of the geometrical arrangement of the tips on the emission behaviour was subject of this investigation. However, it seems that this plays a subordinate role, since the emission is dominated by the impact of the tip radii. From the

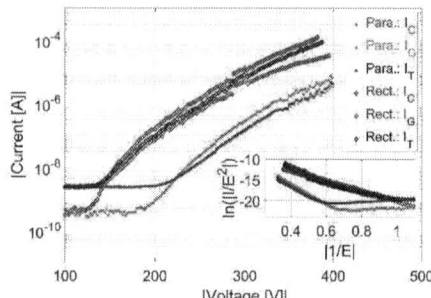

Fig. 2: Triode emission characteristics of both electron sources with integrated FN plot inset.

TABLE II
ELECTRICAL CHARACTERISTICS

Parameter	Electron source	
	Spiral	Rectangular
Tip-grid distance (50 µm spacer)	≈ 129 µm	≈ 140 µm
Onset voltage (@1 nA)	200 V	136 V
Maximum current I_C (@-400V)	6.4 µA	120 µA
Voltage conversion factor γ	$2.3 \cdot 10^9$ m⁻¹	$3.5 \cdot 10^9$ m⁻¹
Mean current (mid-term 1h)	16.9 µA	16.2 µA
Standard deviation (mid-term 1h)	±1.5 µA (±8.8 %)	±932 nA (±5.7 %)
Degradation (mid-term 1h)	-3.0 µA/h	-120 nA/h

geometrical values of the tip radii (Table I) can be concluded, that the rectangular arranged cathode must have a higher emission performance due to its sharper tips. Since the laser beam is deflected along straight lines, it is interrupted within the tip areas (DXF contour). Consequently, the thermal stress on the substrates surface is different, because of the irregular tip-to-tip distances. This leads to slightly different melting processes on each emitter tip and results in bigger deviations in tip radii during the subsequent wet etch process for tip sharpening.

V. CONCLUSIONS

It has turned out, that the current laser micromachining process is not a suitable fabrication approach for the investigation of parastichy arranged FEAs, caused by the thermal induced stress given by the geometrical layout (tip-to-tip distance). Therefore, either the given process has to be optimized to create more comparable emitter tips in both designs, or a different fabrication method has to be utilized to investigate the influence of the geometrical arrangement of the FEAs. Nevertheless, parastichy arrangements could be a promising approach to place field emission tips even denser.

REFERENCES

[1] J. R. Harris, K. L. Jensen, D. A. Shiffler, and J. J. Petillo, "Shielding in ungated field emitter arrays," Appl. Phys. Lett., vol. 106, no. 20, p. 201603, May 2015, doi: 10.1063/1.4921709.

[2] V. Kolařík, M. Horáček, A. Knápek, S. Krátký, M. Matějka, and P. Meluzín, "Spiral arrangement: From nanostructures to packaging," Journal of Electrical Engineering, vol. 70, no. 1, pp. 74–77, Mar. 2019, doi: 10.2478/jee-2019-0011.

[3] C. Langer et al., "Silicon chip field emission electron source fabricated by laser micromachining," Journal of Vacuum Science & Technology B, vol. 38, no. 1, p. 013202, Jan. 2020, doi: 10.1116/1.5134872

Silicon Field Emitter Arrays Fabricated Using a Layout-Independent Process

Nedeljko Karaulac, Winston Chern, Girish Rughoobur, and Akintunde I. Akinwande
Microsystems Technology Laboratories
Massachusetts Institute of Technology
Cambridge, MA 02139, USA
Email: karaulac@mit.edu

Abstract—**We present a layout-independent fabrication process for silicon field emitter arrays (FEAs) that improves the scalability of emission current with array size. Using this process, we were able to fabricate and measure FEAs with different array sizes ranging 1 μm^2 to 1 mm^2, which represents a range of six orders of magnitude in area.**

Keywords—**silicon, field emitter arrays, tungsten, CMP**

I. INTRODUCTION

Fabricating silicon FEAs is a significant challenge, often resulting in FEAs with unpredictable current-voltage characteristics that do not scale linearly with array size. This result is due to the non-uniformity of some critical process steps in the fabrication process, such as photolithography, reactive ion etching (RIE), and chemical-mechanical polishing (CMP). Consequently, the uniformity of the fabrication process must be improved before silicon FEAs can be batch manufactured reliably and used in commercial systems. In previous work, we improved the uniformity of the FEA emission current by developing a more uniform photolithography process based on trilevel resist that reduced the distribution of emitter tip radii [1]. However, we were not able to measure any emission current from our small FEAs, indicating a problem with the fabrication process. We hypothesized that the problem was due to the non-uniformity of two other process steps: CMP and RIE. Unfortunately, the uniformity of these steps is sensitive to the pattern density and therefore depends on the layout design of the FEAs. As a result, we fabricated silicon FEAs using a new layout-independent fabrication process presented in this work in an attempt to address the CMP and RIE non-uniformity.

II. EXPERIMENTAL

A. Device Fabrication

Fig. 1 highlights the key steps of the layout-independent fabrication process. This process closely follows the process reported in [2], except for one major difference: instead of etching the silicon mesa at the beginning of the process and thereby establishing the FEA layout, we first fabricate silicon field emitters with 1 μm pitch everywhere across the wafer (Fig. 1(A)). Then, at the end of the process (Fig. 1(B-D)), we selectively etch emitters where we do not need them, and we replace them with dielectric in order to realize any arbitrary layout desired. Because the layout design is determined at the

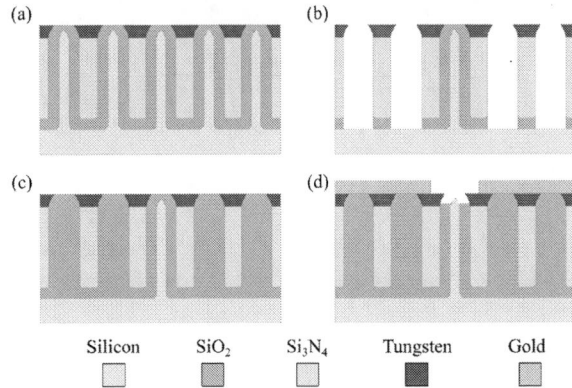

Fig. 1. Process flow showing the main steps of the layout-independent fabrication process: (a) Fabrication of silicon field emitters; (b) Wet etch of oxide and silicon field emitters that are not included in the layout design; (c) Fill-in of voids with dielectric; (d) Deposition of metal to create contact pad and tip release.

end of the process, any CMP and RIE non-uniformity that depends on pattern density and layout design is eliminated during the fabrication of the field emitters.

Another novel feature of this process is the fabrication of a self-aligned tungsten metal gate. The gate is formed by depositing 800 nm of tungsten using sputter deposition (Fig. 2(A)), and then the self-aligned apertures are formed using CMP (Fig. 2(B)). The metal gate is necessary in order to create high etch selectivity between the gate and emitters when selectively etching the silicon emitters and nanowires in regions where they are not desired. If a poly-Si gate was used as in [2], then the gate in Fig. 1(B) would be destroyed during the silicon nanowire etch, and a good metal/poly-Si contact would not be guaranteed in Fig. 1(D). To define the field emitter arrays, we first mask the FEA regions using photolithography. Then, we use HF to remove the SiO_2 surrounding the field emitters we wish to selectively etch, and next we etch the silicon field emitters using TMAH. The TMAH does not etch SiO_2 or tungsten. Fig. 2(C) shows the bottom right corner of an FEA after the TMAH wet etch; the black holes are the remaining voids, and the field emitters in the array are still embedded in SiO_2. The voids are filled with

978-1-6654-2590-2/21 $31.00 © 2021 IEEE

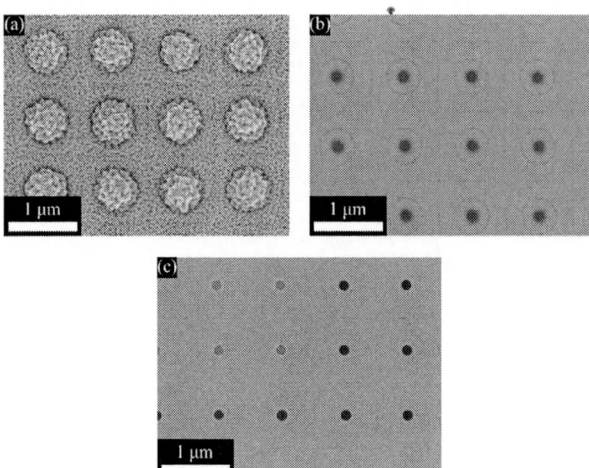

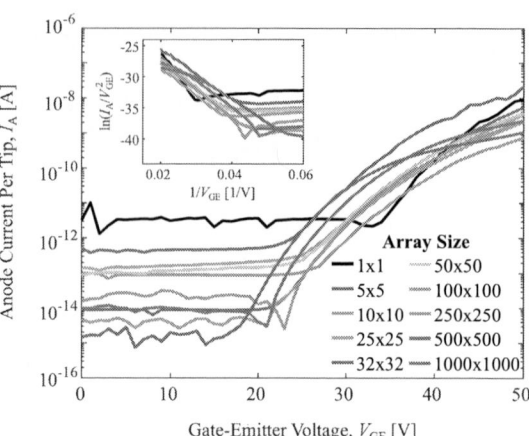

Fig. 2. SEM images showing (a) gate formation after sputter deposition of tungsten (b) self-aligned apertures after tungsten CMP (c) voids remaining after HF and TMAH wet etch.

Fig. 3. Transfer characteristics and associated FN plots of FEAs with different array sizes. All plots and extracted parameters are normalized by the number of tips in each array.

TABLE I
EXTRACTED PARAMETERS FOR DIFFERENT ARRAY SIZES

Array Size	I_A (nA)*	b_{FN}	$\log(a_{FN})$	β (10^6/cm)	r (nm)
1×1	9.25	783	-10.38	0.68	11.3
5×5	21.46	432	-16.66	1.22	5.06
10×10	0.74	437	-19.94	1.21	5.15
25×25	3.37	445	-18.56	1.19	5.26
32×32	5.49	498	-16.93	1.06	6.13
50×50	3.82	457	-18.06	1.16	5.45
100×100	2.08	436	-19.11	1.21	5.13
250×250	2.16	458	-18.32	1.15	5.48
500×500	2.27	407	-18.55	1.30	4.67
1000×1000	1.07	375	-18.64	1.41	4.18

*Extracted at V_{GE} = 50 V. All extracted parameters are normalized by the number of tips in each array.

spin-on glass before metalization in order to prevent a short between the gate and substrate and also to structurally support the metal contact pad.

B. Electrical Characterization

Following the layout-independent process, we fabricated and measured 10 different array sizes ranging from 1×1 (i.e., a single emitter) to 1000×1000. Fig. 3 shows the transfer characteristics of the FEAs. For these measurements, the anode-emitter voltage, V_{AE}, was fixed at 1000 V while the gate-emitter voltage, V_{GE}, was swept from 0 V to 50 V. The Fowler-Nordheim (FN) plot was used to extract the FN coefficients, b_{FN} and a_{FN}. Table I lists a summary of important parameters extracted from the plots in Fig. 3.

With the exception of a couple outliers, the anode current per tip (@V_{GE} = 50 V) for most FEAs was on the order of 1–10 nA, which is equivalent to 0.1–1 A/cm^2. In addition, the slope of the FN plots, b_{FN}, was in the range of 400–500 V. From b_{FN}, the field factor, β, can be extracted by assuming the electron affinity of silicon (4.05 eV) is the work function, and the tip radius can be extracted using $\beta = k/r^n$ [2]. The extracted tip radius for these field emitters is approximately 4.0–6.0 nm, which is consistent with prior works [1–2]. These measurement results show that we were able to successfully fabricate FEAs spanning six orders of magnitude in area from 1 μm^2 to 1 mm^2. This represents a significant step forward in terms of scalable and predictable FEA current performance. Although we were able to achieve current performance similar to previous results, additional improvements will require further process optimization and development.

III. CONCLUSION

We demonstrated the successful fabrication of silicon FEAs using a layout-independent process. The key feature of this

process is that we first fabricate silicon field emitters everywhere across the wafer, and then at the end of the process, we selectively etch the field emitters that are not needed in order to realize any arbitrary layout. Another novel feature of this process is the fabrication of a self-aligned tungsten metal gate. Using this process, we were able to fabricate and measure FEAs with different array sizes ranging 1 μm^2 to 1 mm^2, which represents a range of six orders of magnitude in area.

ACKNOWLEDGMENT

This material is based upon work supported by AFOSR Grant No. FA9550-18-1-0436. This work was carried out in part through the use of MIT's Microsystems Technology Laboratories and MIT.nano faciltes.

REFERENCES

[1] N. Karaulac, G. Rughoobur, and A. I. Akinwande, "Highly uniform silicon field emitter arrays fabricated using a trilevel resist process," *Journal of Vacuum Science & Technology B*, vol. 38, no. 2, p. 023201, Mar. 2020.

[2] S. A. Guerrera and A. I. Akinwande, "Nanofabrication of arrays of silicon field emitters with vertical silicon nanowire current limiters and self-aligned gates," *Nanotechnology*, vol. 27, no. 29, p. 295302, Jul. 2016.

978-1-6654-2590-2/21 $31.00 © 2021 IEEE

Current dependent performance test used on different types of silicon field emitter arrays

A. Schels[*], S. Edler
and W. Hansch
Institute of Physics
Faculty of Electrical Engineering
and Information Technology
Universität der Bundeswehr München
85577 Neubiberg, Germany

*Contact: Andreas.Schels@ketek.net

M. Bachmann, F. Herdl,
F. Düsberg, M. Eder, M. Meyer,
M. Dudek, A. Pahlke
KETEK GmbH
81737 Munich, Germany

R. Schreiner
Faculty of Applied Natural Sciences
and Cultural Studies
OTH Regensburg
93053 Regensburg, Germany

Abstract— **A current dependent performance test is used to investigate the influence of doping and emitter geometry on the lifetime of silicon field emitter arrays. The measurements reveal an improved performance for lower n-type dopant concentrations. Furthermore, two new types of field emitters are introduced by slightly varying the original fabrication process [1]. The comparison shows superiority of tip like emitters over blade like structures.**

I. INTRODUCTION

Due to several advantages compared to thermionic electron sources, silicon field emitter arrays (FEAs) are promising candidates for applications in X-ray sources [2], [3] or electron sources [4]. One important but still insufficient investigated parameter for the usability of FEAs in those applications is the durability of the emitter over the operating time. Due to phenomena like adsorbates at the tips surface [5], ion bombardment [6] and thermally induced stress at high current densities [6], the lifetime of FEAs is strongly dependent on the emitted current. Herdl et al. recently developed a novel current dependent performance test (CDPT), which allows a systematic comparison between various FEAs [7]. In this contribution the method is used to investigate the influence of different doping levels and geometries on the current dependent degradation of field emitter arrays.

II. SAMPLE FABRICATION AND MEASUREMENT

The measured samples are fabricated mainly by saw dicing and anisotropic wet chemical etching of silicon with tetramethylammonium hydroxide [1]. In order to evaluate the presented FEAs type S samples with 1296 tips (see Fig. 1a and 1b) were fabricated in two different doping levels. The resistivity of the phosphorus (n++ or n) doped Si 100-wafers used for this was 10 - 20 Ωcm and 0.3 - 0.5 Ωcm, respectively. Additionally, the fabrication process was slightly changed by altering the pitches of the saw dicing and adjustment of the etching time, resulting in the herein called sheet FEAs (see Fig. 1c) with a total number of 2045 sheets and line FEAs (see

Fig. 1d) with 79 lines. Both types were fabricated with phosphorus (n+) doped wafers with a resistance of $3 - 6$ Ωcm.

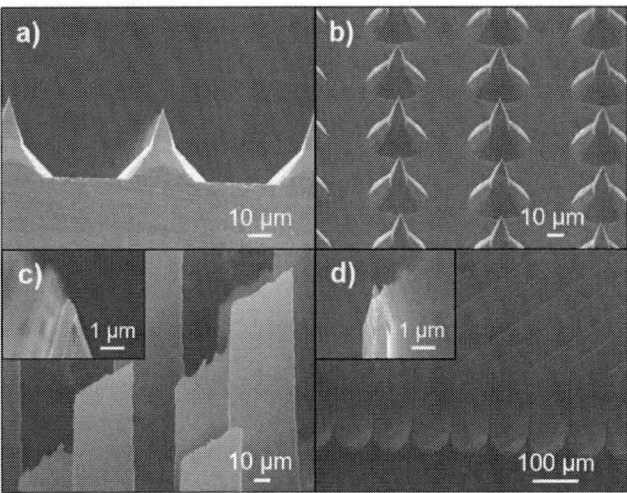

Fig. 1 SEM images of the investigated structures: a)-b) standard S-FEA [1], c) sheet FEA with the enlarged edge of the sheet as inset and d) line FEA with the enlarged edge of the line as an inset.

The CDPT consists of several constant current measurements for one hour each (CCM) at gradually increasing regulated current levels combined with IV-characteristics prior and after each CCM. All measurements are realized in a vacuum setup were the pressure was regulated to 10^{-5} mbar, to roughly meet the conditions present in a hermetically sealed housing. [4] If the set current is not reached anymore with the maximum voltage of 1400 V the measurement is aborted, defining the last current level as the maximum current. The current regulation circuit presented earlier [1], [8] was expanded by a voltage divider enabling the measurement of the applied extraction voltage at the FEA. Since the impact of the IV-sweeps on the degradation is mostly negligible [7] the temporal course of the voltage during the CCM quantified by a

978-1-6654-2590-2/21 $31.00 © 2021 IEEE

linear regression is used as a quantification of the FEA degradation. In order to take a variable extractor tip distances into account the electric field shift (EFS) is used instead of a extraction voltage shift for comparing different FEA types. [7]

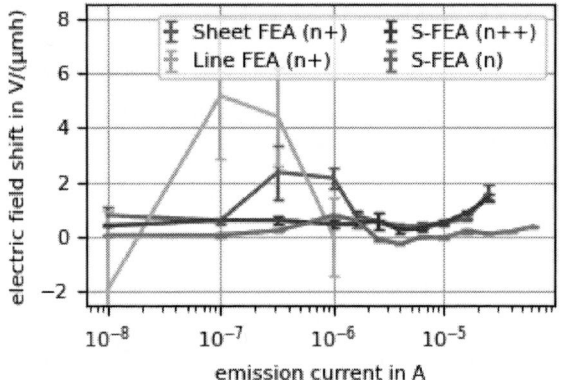

Fig. 2 Extracted electric field shifts of the CCM measurements of the discussed types of FEAs. The data of the S-FEAs is based on three samples each, whereas the data for the sheet and sheet line FEAs is based on two samples each.

An additional parameter indicating the performance of the FEAs measured with this method is the onset electric field (OEF). It combines the accumulated shift of the electric field for different currents with the characteristics of the FEA and can be calculated by dividing the y-intercept voltages from the linear regressions by the extractor-tip distance. Therefore, higher maximum OEFs can be achieved by lower extractor-tip distances.

III. Results and Discussion

For statistical reasons, three FEAs of each type are measured with the CDPT. The estimated mean and error values of the EFSs for the presented FEA types are shown in Fig. 2. Comparing the S-FEAs with different doping levels (n++ and n), higher dopant concentration leads to slightly higher degradation rates already at lower currents. At currents above 10 µA this deviation enhances by a strong degradation of the n++ doped FEAs, resulting in a lower maximum current as for the lower doped ones. This is probably caused by a higher expected adsorbate coverage for high dopant concentrations [9] leading to more degradation [5]. Furthermore, joule heating is reduced leading to a lower apex temperature. This can enhance the adsorbates coverage. The higher doping level also results in less resistive self-limitation, which might explain this behaviour.

For the n+ doped sheet FEAs (only two measured) nearly congruent values to the n++ doped S-FEAs are observed. With the given medium dopant concentration EFS values in between the two different S-FEA types are expected for the same geometry. Therefore, it can be concluded that the geometry of the sheet FEA might not be advantageous for the durability of the field emitters. Expanding the sheets to lines for the line FEAs further confirms this conclusion. The observed EFSs for these FEAs (only two measured) are quite high overall, leading to a fast degradation and a maximum current of just 1 µA. For both types this might correlate with the lower field enhancement factor.

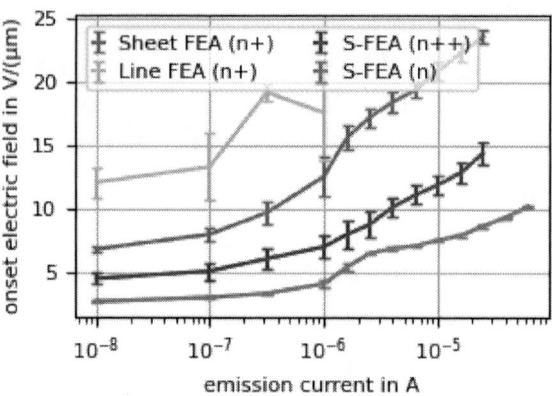

Fig. 3 Onset electric field extracted from the CCM measurements for the different types of FEAs. The data of the S-FEAs is based on three samples each, whereas the data for the sheet and sheet line FEAs is based on two samples each.

In Fig. 3 the OEFs of the four different FEA types are presented. As before, the observed OEFs of the n++ doped S-FEA are higher compared to the n doped S-FEAs, indicating an even better performance during characteristics measurement. The sheet and line FEAs show even higher OES than the n++ doped S-FEAs despite their lower doping concentration. Those high homogeneous electric fields are due to their low field enhancement factors, as well as their strong degradation (Fig. 2). Together with the results of the EFS this shows an overall better performance for S-FEAs with low dopant concentrations compared to the other geometries and higher doping levels.

IV. Conclusion

The current dependent performance test enables a meaningful comparison of the presented FEA types. The EFS and OEF investigation reveals a better performance of emitter types with lower conductivity. For different geometries, the tip like geometry seems to be superior to the sheet or line like, which might be attributable to a higher field enhancement.

Acknowledgment

The research work was funded by the Bavarian Research Foundation under project-number AZ-139619.

References

[1] S. Edler *et al.*, *J. Vac. Sci. Technol. B*, vol. 39, no. 1, p. 027001, 2021

[2] S. Cheng, F. A. Hill, E. V. Heubel, and L. F. Velasquez-Garcia, *J. Microelectromechanical Syst.*, vol. 24, no. 2, pp. 373–383, 2015

[3] A. Basu, M. E. Swanwick, A. A. Fomani, and L. F. Velásquez-García, *J. Phys. D. Appl. Phys.*, vol. 48, no. 22, p. 225501, 2015

[4] M. Bachmann *et al.*, *J. Vac. Sci. Technol. B*, vol. 38, no. 2, p. 023203, 2020

[5] S. Edler *et al.*, *J. Appl. Phys.*, vol. 122, no. 12, p. 124503, 2017

[6] W. I. Karain, L. V. Knight, D. D. Allred, and A. Reyes-Mena, *J. Vac. Sci. Technol. A Vacuum, Surfaces, Film.*, vol. 12, no. 4, pp. 2581–2585, 1994

[7] F. Herdl *et al.*, *unpublished*, p. submission to this conference, 2021

[8] C. Prommesberger *et al.*, *IEEE Trans. Electron Devices*, vol. 64, no. 12, pp. 5128–5133, 2017

[9] A. Rothschild, Y. Komem, and N. Ashkenasy, *J. Appl. Phys.*, vol. 92, no. 12, pp. 7090–7097, 2002

Gap in pagination due to unavailable paper.

Page 175

Carbon Nanotube Fiber Cathodes and Saturation of their Field Emission Current:

Evgenii P. Sheshin, Ilya N. Kosarev[*], Bulat I. Masnaviev and Dmitry I. Ozol

Moscow Institute of Physics and Technology, Dolgoprudny, Moscow Region, Russian Federation

[*]Contact: ilyakosarev@gmail.com

Abstract— **Several different types of carbon nanotube fiber cathodes of different diameters (20, 40, 400 μm) were measured for emissive properties. Currents up to 1.5 mA are obtained. Durability properties were qualitatively estimated. Current "saturation" was observed for 20-μm fiber, probably due to resistivity.**

I. INTRODUCTION

It is well known that carbon materials can have good auto-emission properties [1]. This makes them promising for the manufacture of cathodes of various devices - in particular, cathodoluminescent sources of visible light [2] and ultraviolet radiation [3],[4], the latter are especially important for disinfection.

Carbon nanotube (CNT) fibers are an interesting and promising novel carbon material. Several different types of carbon nanotube fiber cathodes of different diameters (20, 40, 400 mkm) were measured for emissive properties and durability under high vacuum and strong electrical field. .

II. EXPERIMENT AND RESULTS

Carbon nanotube fibers of different diameters (20, 40, 400 nm) were used in the experiment. Fibers were provided by Technological Institute for Superhard and Novel Carbon Materials (TISNCM), located in Troitsk, Russia. Mixture of sulfur, ferrocene and hydrocarbon, mixed at high temperatures, react in a way, causing CNTs to appear. CNTs then agglomerate and coagulate into a bundle, called a "sock". Sock is wound on a mechanism, which rotates with a constant speed. That speed can be varied, but should not exceed the speed of maximum growth, which equals around 10-20 m/min. The thread then is purified of residue reactants and iron by annealing at the temperature of 420°C and consequent wash in hydrochloric acid [5],[6].

Cathodes were prepared using CNT fibers mentioned above according to the following procedure: fibers were placed on a clean glass surface and then cut a length of 5-10 mm with a sharp blade widthwise; after that CNFs were placed in a nickel tube with a diameter of 300-400 micrometers and glued to the tube using aquadag. To ensure solidity the tube was additionally flattened. Afterwards, nickel tubes were welded into a cathode – anode construction. Looks of the cathodes are shown on the Fig. 1.

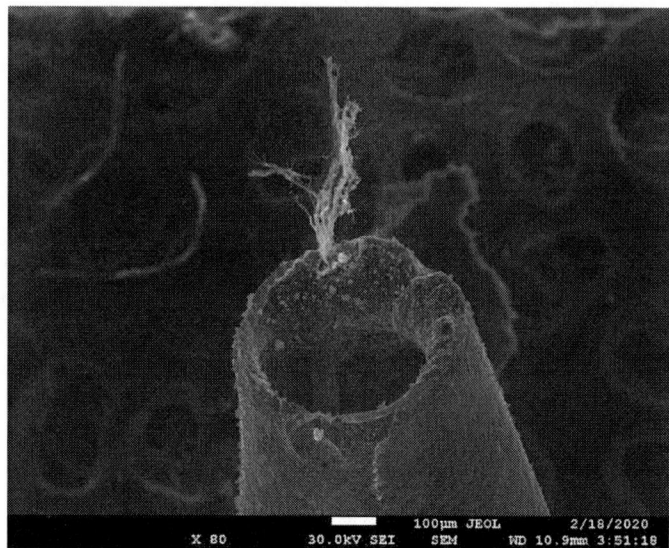

Fig. 1 SEM image of a nickel tube with a CNT-fiber cathode inside

The construction was put on high-voltage electrical leads, and the whole construction was put into a vacuum camera. The chamber was pumped out using a low-vacuum and a turbomolecular pumps, connected consequently to the chamber. Vacuum of 10^{-5} Torr was achieved.

SEM-images of various fibers before the experiment are shown in Fig. 2.

Nanotube fibers of 20 and 40 μm diameter showed the best emission characteristics with the emission starting at 450 V/cm. Currents up to 1.5 mA are obtained. Durability properties were qualitatively estimated. The fibers with the diameter of 20 μm show the best emissive properties. Fiber with the diameter of 40 μm showed better durability and stability properties. The one with the diameter of 400 μm has shown the best durability and stability properties, but the emission current is an order of magnitude lower.

Fibers with the diameter of 20 μm showed the best emission current but low stability during the experiment. It could happen due to fiber integrity deterioration. High current and presence of residual gases cause damage to fibers, which results in fiber's structural and material loss.

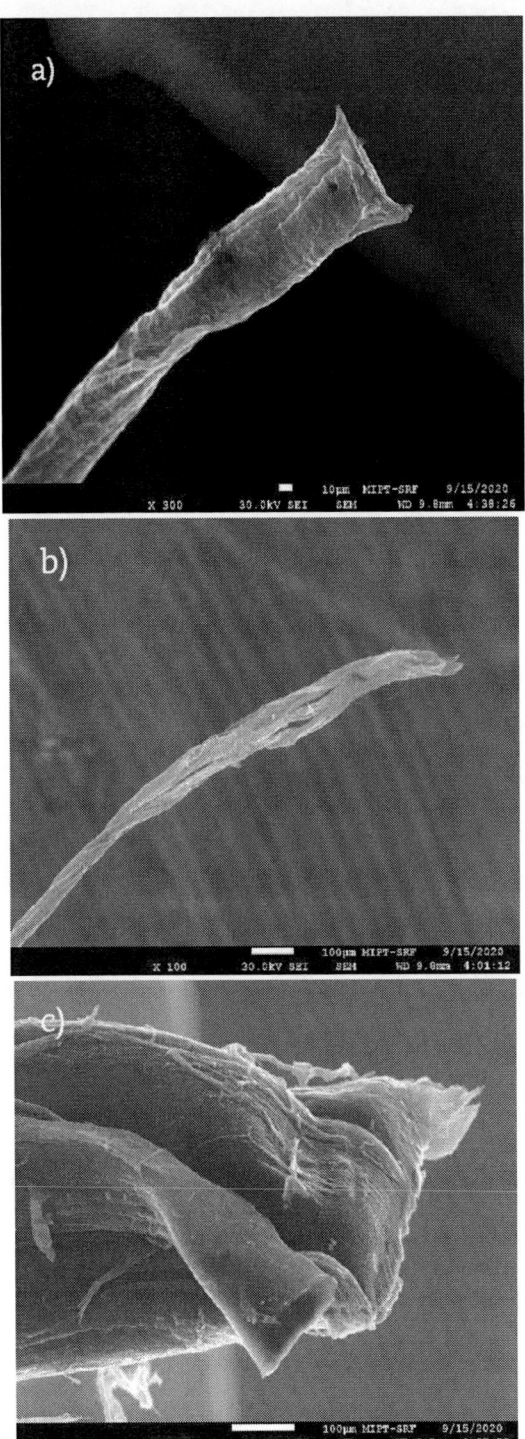

Fig. 2 a),b),c) – SEM images of cathodes (20, 40, 400 μm respectively) before the experiment.

The obtained current-voltage characteristics are shown in Fig. 3.

While 400 μm CNF cathode faced no significant damage during the experiment, 20 and 40 μm CNF cathodes did not cope as good. During the experiment the current value increased from 1 mA to 1.3 mA, 1.44 mA, 1.2 mA for 20μm, 40 μm, 400

μm CNF cathodes respectively with voltage staying constant. This phenomenon requires further

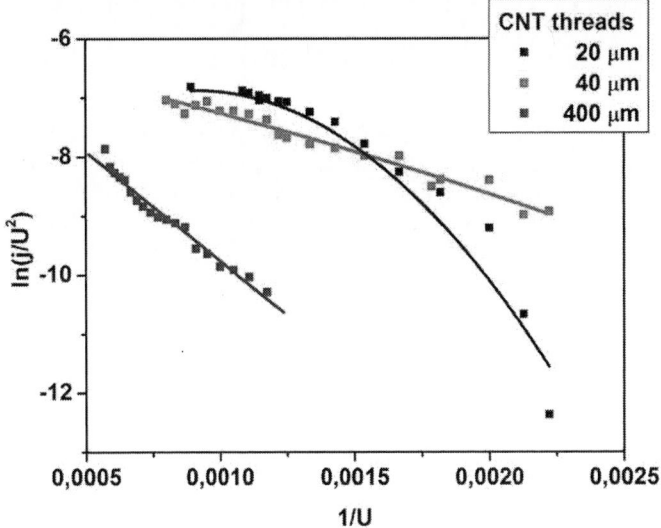

Fig. 3 Current-voltage characteristics of the three CNT cathodes in Fowler-Nordheim coordinates.

investigation. Fibers with the diameter of 20 μm showed low stability during the experiment.

Current "saturation" of 20 μm CNT-fiber (see Fig.3) is probably due to ohmic resistance by the mechanism described by Forbes [7]: voltage drop along the filament due to the resistivity of the fiber's material, increasing with increasing current, effectively reduces the field enhancement factor.

III. CONCLUSIONS

Cathodes made of CNT-fibers showed incredible field emissive properties. It makes them a promising candidate for use in cathodoluminescent lamps of visible and UV-range and allow for lower voltages and higher currents, despite small dimensions.

REFERENCES

[1] N. Egorov, and E. Sheshin, Field Emission Electronics (Springer, Cham, 2017).

[2] E. Sheshin, A. Kolodyazhnyj, N. Chadaev, A. Getman, M. Danilkin, and D. Ozol "Prototype of cathodoluminescent lamp for general lighting using carbon fiber field emission cathode" Journal of Vacuum Science & Technology B 37, 031213 (2019)

[3] D. Ozol, E. Sheshin, M. Danilkin and N. Vereschagina "Cathodoluminescent UV Sources for Biomedical Applications" In: Tiginyanu I., Sontea V., Railean S. (eds) 4th International Conference on Nanotechnologies and Biomedical Engineering. ICNBME 2019. IFMBE Proceedings, vol 77. Springer, Cham.

[4] N. Vereschagina, M. Danilkin, M. Kazaryan, D. Ozol, E. Sheshin and D. Spassky "Cathodoluminescent UV-radiation sources", Proc. SPIE 10614, International Conference on Atomic and Molecular Pulsed Lasers XIII, 106141F (16 April 2018)

[5] D. Conroy, A. Moisala, S. Cardoso, A. Windle and J. Davidson "Carbon nanotube reactor: Ferrocene decomposition, iron particle growth, nanotube aggregation and scale-up" Chemical Engineering Science, Volume 65, Issue 10, 2010, pages 2965-2977

[6] V. Mordkovich, N. Kazennov, V. Ermolaev, E. Zhukova, A. Karaeva, "Scaled-up process for producing longer carbon nanotubes and carbon cotton by macro-spools", Diamond and Related Materials, Volume 83, 2018, Pages 15-20,

[7] Forbes, R.G. "The theoretical link between voltage loss, reduction in field enhancement factor, and Fowler-Nordheim-plot saturation." Applied Physics Letters, 110(13), 2017.

Field emission properties of sharp tungsten cathodes coated with a thin resilient oxide barrier

Daniel Burda[1,2], Mohammad M. Allaham[1], Alexandr Knápek[1], Dinara Sobola[1,2], Marwan Suleiman Mousa[3]

[1]Institute of Scientific Instruments of the CAS, Královopolská 147, 612 64 Brno, Czech Republic
[2]Faculty of Electrical Engineering and Communication, BUT Brno, Technická 2848/8, 616 00 Brno, Czech Republic
[3]Department of Physics, Mutah University, Al-Karak 61710, Jordan
*Contact: burda@isibrno.cz

Abstract— **This research is aimed towards more in-depth understanding of field emission properties of tungsten single tip field emitters (STFEs) coated with tens of nanometer thin barrier of selected refractory oxides such as Al₂O₃. Introducing the additional barrier into metal-vacuum interface system of the emitter can be beneficial for an improvement of its performance. The pristine tungsten emitters were prepared using a two-step electrochemical drop-off etching technique. Thin oxide barriers were prepared by using low-temperature atomic layer deposition (ALD). Field emission was studied in field emission microscope (FEM) working in UHV vacuum (< 1×10⁻⁷ Pa), experimental field emission data were analyzed by the so-called *Murphy-Good plots*, revealing the non-orthodox behavior of the prepared emitters.**

I. INTRODUCTION

In cold field electron (CFE) emission, most of the electrons tunnel through *Schottky-Nordheim* (SN) barrier from electron states below the Fermi level. The theoretical assumptions that form the foundation of field emission theory are called *Fowler-Nordheim equations*. In this study we employ the so-called *extended Murphy-Good equation* (EMG), details regarding the equation can be found in [2,3]:

$$I(V) = \{A_f^{SN}(\theta \exp \eta)V_R^{-\kappa}\}V^{\kappa} \exp(-\eta V_R/V) \quad (1)$$

where A_f^{SN} is the formal emission area assuming SN barrier tunneling, V_R is the reference measured voltage [2] needed to pull the top of the SN barrier down to Fermi level [3], $\theta(\varphi)$ and $\eta(\varphi)$ are work function φ dependent scaling parameters and $\kappa = 2 - \eta/6$ [2]. By dividing both sides of (1) by V^{κ} and applying natural logarithm, equation becomes:

$$\ln\{I/V^{\kappa}\} = \ln\{A_f^{SN}(\theta \exp \eta)V_R^{-\kappa}\} - \eta V_R/V \quad (2)$$

which is the theoretical form of the so-called *Murphy-Good (MG) plot* [2]. Assuming the local work function of the emitter interface is known, the so-called *orthodoxy test* [3] based on (2) can be performed on measured current-voltage (*I-V*) characteristics. When an emitter is declared orthodox, the emitter related parameters can be extracted, mainly: the formal emission area A_f^{SN} and characteristic voltage conversion length ζ_C [2]. The exact parameters and criteria of the orthodoxy test are further described elsewhere [4]. Through this abstract *electron emission convention* is used.

II. METHODOLOGY

A. Electrochemical etching

The base tungsten emitter was prepared by the means of two-step electrochemical etching of polycrystalline tungsten wire (purity 99,9+%, diameter of 0.3 mm) in 3M aqueous solution of NaOH. Using this technique, sharp symmetrical tungsten tips with tip radius of 80 nm or less can be obtained. Details on the in-house setup can be found elsewhere [5].

To ensure the tip surface is clean before the next step, it is carefully submerged in 38% hydrofluoric acid (HF) for 20 minutes, the acid readily removes oxides, hydroxides and organic contaminants.

B. Atomic layer deposition

Atomic layer deposition (ALD) is an ultra-high vacuum deposition technique, the growth of layers is based on sequential, self-limiting surface reactions of two vapour phase precursors. ALD allows to grow uniformly thin layer on very complex or sharp structures [6]. The nature of the layer formation allows for precise thickness control at the Angstrom level [6], as the overall thickness is controlled by number of cycles of sequential reactions of the two precursors.

Layers of Al₂O₃ were prepared at CEITEC Nano facility in Ultratech-Cambridge Nanotech Fiji 200 ALD tool. The used precursors were trimethylaluminum (TMA) and water vapor. The temperature was set to 150 °C during the process and the number of cycles was set to 150. Each one ALD cycle comprised of the following steps: 1) injection of TMA, duration 0.06 s, 2) purge, duration 20 s, 3) injection of H₂O, duration 0.06 s and 4) purge, duration 20 s. The process resulted in 15 nm thin layer of Al₂O₃.

C. Field emission microscope measurement

The emitter had been connected as the cathode in triode configuration of FEM. The extractor electrode with hole of 1 mm of diameter was precisely mounted just tenths of mm above the emitter tip. Al-coated Ce:YAG scintillator was used as the anode. The emission current was measured between the scintillator and the ground by Keithley 485 Pico ammeter. The macroscopic distance between the tip of cathode and the scintillator was set to 40 mm.

978-1-6654-2590-2/21 $31.00 © 2021 IEEE

Clean tungsten emitter was loaded into FEM and its *I-V* characteristics were recorded. Then, the emitter was removed from FEM and subsequently coated using ALD. After, the *I-V* characteristics of coated emitter were obtained.

III. RESULTS AND DISCUSSION

In order to protect the emitter from exploding during the measurement, the *I-V* characteristics shown in the Fig. 2 were captured in this manner: After the first emission, the extraction voltage had been increased, so that the ammeter would collect the current of no more than 10 μA. then the actual *I-V* measurement started, recording the decrease of current while decreasing the extraction voltage at the constant cathode voltage. The coated emitter had showed very strong *switch-on effect* [6], its first emission current was 5 μA at V_{ext} = 977 V.

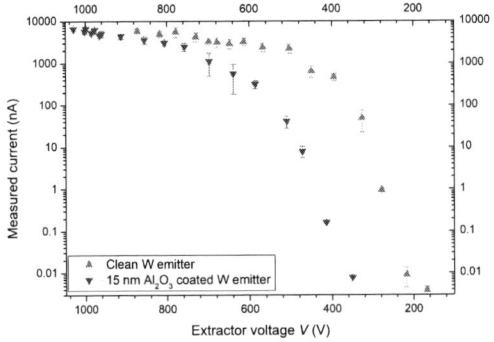

Fig. 1 *I-V* characteristics of clean and coated emitter, cathode voltage 5030 V.

The local work function values are assumed to be 4.5 eV and 4.0 eV for uncoated tungsten and for ultrathin Al_2O_3 layers [7], respectively. Murphy-Good plots of experimental data are shown in the Fig. 3. Both of the emitters showed nontrivial behavior, passing the orthodoxy test only when extraction voltage was set relatively low. But this behavior is not uncommon in very sharp tungsten emitters, at high extraction voltage, slight macroscopic changes in electrostatic field at the tip can occur and the emitter can no longer be properly described by the theory. Extracted values from linear parts of Murphy-Good plots are shown in the Table 1.

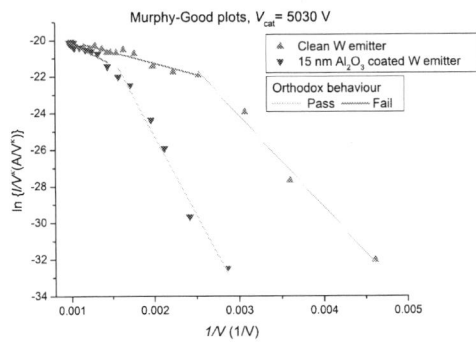

Fig. 3 Murphy-Good plots of experimental *I-V* characteristics of measured emitter(s).

The projection images of the Al_2O_3 coated emitter are illustrated in the Fig. 4. The images resembled of the multi-spot emission patterns, fluctuations in the current were also visible as occasional bright spots.

TABLE I
EXTRACTED VALUES OF EMITTER PARAMETERS FROM LINEAR PARTS OF
MURPHY-GOOD PLOTS

	Result	A_f^{SN} (nm^2)	ζ_C (nm)
Clean W	Passed	63.7	75.9
Al_2O_3	Passed	369.3	158.4

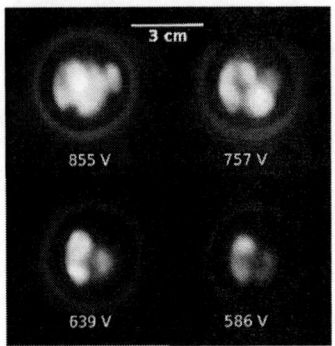

Fig. 4 Evolution of emission patterns in projection images of Al_2O_3 coated emitter. Four values of extractor voltage V_{ext}, constant cathode voltage V_{cat} = 5030 V, exposure: f/2, 1/250 s, ISO 1600.

IV. CONCLUSIONS

Presented research shows the possibilities of fabrication of ultrathin Al_2O_3 oxide coated emitter. The coated emitter bears a set of interesting properties, it exhibits strong *switch-on effect*, the orthodox behavior can be observed when operated in low extractor voltage. Finally, atomic layer deposition enables for fine tuning of the thickness of the coating and this feature could prove very useful in fabrication of more advanced emitters.

ACKNOWLEDGMENT

The infrastructure was supported by RVO:68081731. The research described in this paper was supported by the Ministry of the Interior of the Czech Republic (project. No. VI20192022147) and partially supported by the Internal Grant Agency of Brno University of Technology, grant No. FEKT-S-20-6352. We acknowledge CzechNanoLab Research Infrastructure supported by MEYS CR (LM2018110).

REFERENCES

[1] M. Drechsler, "Erwin Müller and the early development of field emission microscopy", *Surface Science*, vol. 70, no. 1, pp. 1-18, 1978.

[2] R. G. Forbes, "The Murphy–Good plot: a better method of analysing field emission data", *Royal Society Open Science*, vol. 6, no. 12, Dec. 2019.

[3] M. M. Allaham, R. G. Forbes, A. Knápek, and M. S. Mousa, "Implementation of the orthodoxy test as a validity check on experimental field emission data", *Journal of Electrical Engineering*, vol. 71, no. 1, pp. 37-42, Feb. 2020.

[4] M. M. Allaham, R. G. Forbes and M. S. Mousa, "Applying the Field Emission Orthodoxy Test to Murphy-Good Plots", *Jordan Journal of Physics*, vol. 13, no. 2, pp. 101-111, Aug. 2020.

[5] A. Knápek, J. Sýkora, J. Chlumská, and D. Sobola, "Programmable set-up for electrochemical preparation of STM tips and ultra-sharp field emission cathodes", *Microelectronic Engineering*, vol. 173, pp. 42-47, 2017.

[6] C. S. Athwal and R. V. Latham, "Switching and other nonlinear phenomena associated with prebreakdown electron emission currents", *Journal of Physics D: Applied Physics*, vol. 17, no. 5, pp. 1029-1043, May 1984.

[7] W. Song and M. Yoshitake, "A work function study of ultra-thin alumina formation on NiAl(110) surface", *Applied Surface Science*, vol. 251, no. 1-4, pp. 14-18, 2005

Using High Aspect Ratio AFM Probe for Digital Twin Development of SiC FEA

Konstantin Nikiforov[1*], Nikolay Egorov[1], Ivan Sokolov[1], Valery Strebko[1], Vladimir Mikhailovskiy[1], Denis Danilov[1], Vladimir Golubkov[2], Vladimir Ilyin[2], and Alexey Ivanov[2]

[1]*Saint Petersburg State University, 7-9 Universitetskaya nab., St. Petersburg, 199034 Russia*
[2]*Saint Petersburg Electrotechnical University "LETI", 5 ul. Professora Popova, St. Petersburg, 197376 Russia*
*Contact: k.nikiforov@spbu.ru, phone +7-812-428-4235

Abstract— **The nanopillar growth method of deposition of metal-organic molecules by a sharply focused electron beam is used to form high aspect ratio AFM probe. It is applied to AFM-based digital twin development of two-tier silicon carbide field emission array (FEA), made in SPETU LETI by a two stage reactive ion etching technique in a fluoride atmosphere with high-density emission tips (of the second stage) and vertically oriented surface morphology. In digital twin approach the experimental information describing the detailed surface morphology are represented by a numerical multiscale simulation.**

I. INTRODUCTION

Numerical and natural experiments, mathematical and physical modelling, computer simulation and digital twin development are relevant for silicon carbide field emitters both uncoated and coated with various carbon-based thin films [1]–[4]. In the context of this work, it is important that the bulk of the AFM data in silicon carbide surface science is obtained for plane surfaces with lateral morphology: for example, studying crystal and electronic structure of graphene films grown on 6H-SiC (0001) [5], the structure and transport properties of graphene, nano-carbon and multi-graphen films prepared by sublimation on the surface of 6H-SiC [6]–[8], formation of periodic steps on 6H-SiC (0001) surface by annealing in a high vacuum [9].

Another part of the research concerning the emission properties of SiC with non-lateral, but vertically-oriented surface nanostructures (for which obtaining AFM data is rather difficult), contains characteristics of field electron emission, which indicate the observance of the Fowler-Nordheim law (for example, [3, 4]), and results were published on the calculation of the field characteristics of SiC surfaces with model inhomogeneities [10, 11], or with directly imported AFM-data in the frame of numerical simulation of SiC FEA surface morphology [12].

In this work, the 6H-SiC two-tier microsized matrix structure [12] is considered, TEM image and current-voltage characteristic of which are shown on fig. 1, measured in diode configuration. Total energy distribution (TED) of field emission electrons from cathode material is compared with various carbon-based emission nanostructures: highly oriented pyrolytic graphite (HOPG), multi-layered graphene-like structures (GLS), multi-wall CNT (mwCNT) [13] (fig. 1, inset).

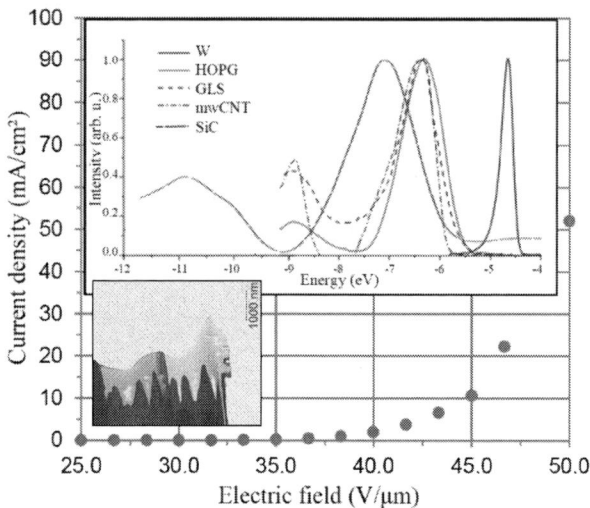

Fig. 1 Current-voltage characteristic of 6H-SiC FEA (STEM image shown) and bimodal total energy distributions of different materials (inset).

All considered TEDs demonstrate a similar bimodal form (excluding tungsten TED presented on the graph as reference data).

In order to develop a digital twin of field emission electron source based on silicon carbide two-tier microsized matrix structure we use:

1) High aspect ratio AFM probe to obtain surface morphology with vertically oriented structural elements with the aim of increasing accuracy of developed surface model [12];

2) TEM studying the surface structure at atomic level, including checking the presence of carbon-based coatings on the SiC surface, since the above mentioned similarity of bimodal TEDs is very important taking into account several ways for the formation of carbon nanostructures on the SiC surface, possibly realizable both during the manufacturing and during the operation of SiC microsized matrix cathodes.

II. METHODS AND RESULTS

The high aspect ratio AFM probe is fabricated using an electron beam induced deposition (EBID) bottom-up assembly method. The method enables precise control of the inclination angle of the nanopillar (fig. 2a), explained through the dissociation of adsorbed molecules by scattered and secondary electrons far from the point of beam incidence [14], thereby reducing the flow of diffusion to a probe's peak. As result, it allows the probe to be made perpendicular to the probed surface for AFM-scanning. Then the 3D shape of the second-level tips on the emission surface is determined from AFM data (fig. 2b) and TEM Lamella (fig. 2c) is analysed for emission surface structure modelling at atomic scale [15]–[16]. Electric field calculation is performed (fig. 2d) using TetGen and Tapsim [17].

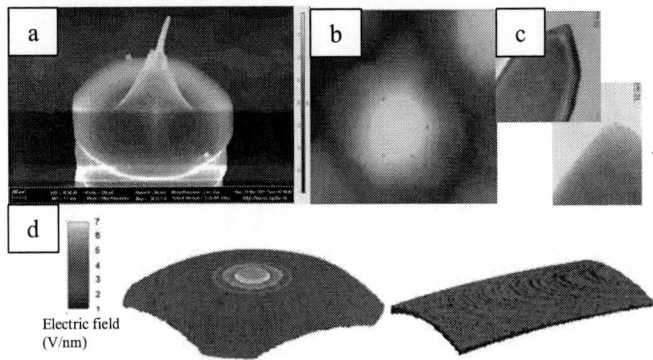

Fig. 2 a) High aspect ratio AFM probe; b) AFM data of single scanned tip; c) TEM images of single tip; d) Electric field calculated distribution over the apex surface of single tip on the SiC FEA surface.

III. Conclusions

Digital twin development of SiC FEA is based on directly imported AFM-data of surface morphology and TEM information of surface structure at atomic scale. For more accurate scanning of vertically oriented second-level tips on the emission surface, a high aspect ratio elongated probe on standart AFM-cantilever was grown by the EBID method. Mathematical multiscale model of emission surface includes at atomic scale crystallographic structure of single tip apex, and will further make it possible to apply ab initio methods of computational quantum chemistry to determine the emission characteristics (ionization potential, work function distribution and energy spectra). The regularities of TED behavior are common for SiC and carbon nanoclusters obtained in earlier studies [13] (multi-wall CNT, HOPG, multi-layered GLS), but TEM and STEM studies didn't confirm the presence of carbon nanostructures on the SiC FEA surface.

Acknowledgment

The reported study was funded by RFBR, project number 20-07-01086. Scientific research were performed at the Research park of St. Petersburg State University «Centre for Nanofabrication of Photoactive Materials (Nanophotonics)», «Interdisciplinary Center for Nanotechnology» and «Computing Center».

References

[1] N. Egorov and E. Sheshin, "Carbon-based field emitters: properties and applications," in Vacuum Electron Sources, ch. 10, G. Gärtner, W. Knapp and R. G. Forbes, Eds. Springer, 2020. (references)

[2] K. Nikiforov, V. Trofimov, N. Egorov, V. Golubkov, V. Ilyin, and A. Ivanov, "The energy spectrum of field emission electrons from 4H silicon carbide," in 2020 33rd International Vacuum Nanoelectronics Conference (IVNC), Lion, art. number. 9203525, 2020. (references)

[3] V. A. Golubkov, A. S. Ivanov, V. A. Ilyin, V. V. Luchinin, S. A. Bogdanov, V. V. Chernov, and A. L. Vikharev, "Stabilizing effect of diamond thin film on nanostructured sislicon carbide field emission array", J. Vac. Sci. Technol. B, vol. 34, p. 062202, 2016.

[4] O. A. Ivanov, S. A. Bogdanov, A. L. Vikharev, V. V. Luchinin, V. A. Golubkov, A. S. Ivanov, and V. A. Ilyin, "Emission properties of undoped and boron-doped nanocrystalline diamond films coated silicon carbide field emitter arrays," J. Vac. Sci. Technol. B, vol. 36, p. 021204, 2018.

[5] V. Yu. Davydov, D. Yu. Usachov, S. P. Lebedev, A. N. Smirnov, V. S. Levitskii, I. A. Eliseyev, P. A. Alekseev, M. S. Dunaevskiy, O. Yu. Vilkov, A. G. Rybkin, and A. A. Lebedev, "Study of crystal and electronic structure of graphene films grown on 6H-SiC (0001)," Semiconductors vol. 51(8), pp. 1072–1080, August, 2017.

[6] A. A. Lebedev, "Growth, study, and device application prospects of graphene on SiC substrates," Nanosystems: physics, chemistry, mathematics, vol. 7(1), pp. 30–36, January, 2016.

[7] N. V. Agrinskaya, V. A. Berezovets, V. I. Kozub, I. S. Kotousova, A. A. Lebedev, S. P. Lebedev, and A. A. Sitnikova, "The structure and transport properties of nano-carbon films prepared by sublimation on the surface on 6H-SiC," Semiconductors vol. 47, pp. 301–306, February, 2013

[8] A. A. Lebedev, N. V. Agrinskaya, V. A. Beresovets, I. V. Kozub, S. P. Lebedev, and A. A. Sitnikova, "Low temperature transport properties of multigraphene structures on 6H-SiC obtained by thermal graphitization: evidences of a presence of nearly perfect graphene layer," J. Mater. Sci. Eng. A, vol. 3, 11A, pp. 757–762, November, 2013.

[9] S. P. Lebedev, V. N. Petrov, I. S. Kotousova, A. A. Lavrent'ev, P. A. Dement'ev, A. A. Lebedev, and A. N. Titkov, "Formation of periodic steps on 6H-SiC (0001) surface by annealing in a high vacuum", Mater. Sci. Forum, vol. 679–680, pp. 437–440, March, 2011.

[10] I. L. Jityaev and A. M. Svetlichnyi "Study of the electric field strength in planar multigraphene/SiC field emission nanostructures with different arrangement of the electrode planes," J. of Physics: Conf. Series, vol. 1038, p. 012055, October, 2018.

[11] E. Volkov, I. Jityaev, and A Kolomiitsev "Design features of matrix nanoscale pointed graphene/SiC field emission cathodes," IOP Conf. Series: Materials Science and Engineering, vol. 93 p. 012031, 2015.

[12] M. Chumak, M. Sayfullin, and K. Nikiforov, "Numerical simulation of surface morphology of two-tier microsized matrix structure of SiC FEA," in 2020 33rd International Vacuum Nanoelectronics Conference (IVNC), Lion, art. number. 9203144, 2020.

[13] G. Fursey, I. Zakirov, N. Bagraev, N. Egorov, V. Trofimov, V. Bocharov, A. Nashchekin, and E. Popov, "The energy spectrum of field emission electrons from HOPG, mwCNT and graphene-like structures," in 2018 31st International Vacuum Nanoelectronics Conference (IVNC), Osaka, art. number. 8520131, 2018.

[14] G. Zhdanov, A. Manukhova, and M. Lozhkin, "Controlling the growth dynamics of carbon nanotips on substrates irradiated by a focused electron beam," Bulletin of the Russian Academy of Sciences. Physics, vol. 78, No. 9, pp. 881–885, September, 2014.

[15] K. Nikiforov, N. Egorov, and C.-C. Shen, "Surface Reconstruction of a Field Electron Emitter," J. Surf. Investig. X-ray Synchrotron and Neutron Techniques, vol. 3, pp. 833–839, May, 2009.

[16] K. Nikiforov, "Multiscale simulation of surface characteristics of field emitter tip," J. Phys.: Conf. Ser., vol. 1124, p. 022023, 2018.

[17] C. Oberdorfer, M. E. Sebastian, and S. Guido, "A Full-scale Simulation Approach for Atom Probe Tomography," Ultramicroscopy,. vol. 128, pp. 55–67, 2013.

Gap in pagination due to withheld papers.

Pages 182-184

Fabrication of ZnO nanowires cold cathode X-ray source with micro patterned transmission anode

Song Kang, Yangyang Zhao, Guofu Zhang, Shaozhi Deng, Ningsheng Xu, Jun Chen*

State Key Laboratory of Optoelectronic Materials and Technologies, Guangdong Province Key Laboratory of Display Material and Technology, School of Electronics and Information Technology, Sun Yat-sen University,
Guangzhou 510275, Guangzhou Province, People's Republic of China
*Corresponding author: stscjun@mail.sysu.edu.cn

Abstract—**Flat-panel X-ray source could realize compact imaging system with low dosage. In order to better control the size of X-ray emission spot from the flat panel X-ray source, we fabricated a flat panel X-ray source with micro-pattern transmission anode. The size of X-ray emission spot can be accurately controlled by modulating the size of the transmission anode pattern, which is expected to improve the imaging performance.**

I. INTRODUCTION

X-ray imaging has been widely used in industry, medical service, security and other fields. At present, most X-ray imaging techniques are based on thermionic cathode X-ray sources. The cold cathode X-ray source is expected to overcome such shortcomings of the traditional thermionic cathode X-ray source as long response time, high power consumption, short lifespan and large volume. Furthermore, the newly-proposed flat-panel cold cathode X-ray source has potential for realizing high resolution, lower dose compact imaging system. Recently, flat-panel X-ray sources (FPXS) using ZnO nanowire cold cathode and transmission anode have been reported [1-5].

The current diode FPXS adopts an unpatterned large area transmission anode. In principle, small-size X-ray emission spot can be obtained by using patterned cathodes. However, due to divergence of the electron trajectory and large cathode to anode distance in the device, it is difficult to achieve micro-size X-ray emission spot, as shown in Fig. 1(a). In this study we control the area emitting X-ray by patterning the metal anode thin film into micro-patterns. As shown in Fig. 1(b) and (c), the size of the X-ray emission spot is decided by the anode pattern. Thus X-ray emission from micro-size spot could be easily achieved. In this work, FPXS devices with patterned transmission anode were fabricated and projection X-ray imaging were realized. This work is expected to provide a route for realizing micro-focus FPXS.

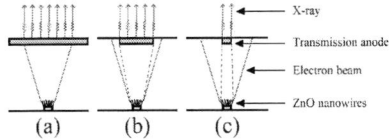

Fig. 1 Schematic showing X-ray emission from unpatterned(left) and patterned transmission anode with different sizes(middle and right).

II. DEVICE STRUCTURE AND FABRICATION

Fig. 2 shows the schematic structure of a diode FPXS device. ZnO nanowires prepared on ITO glass were used as the cathode. Patterned Mo thin film prepared on ITO glass was used as the anode. In the present work, the anode composes of circular patterns, and the diameter of each pattern is 30 μm. The patterns distribute in a 4.8× 4.8 cm^2 area, the spacing between adjacent patterns is 60 μm. The cathode composes of patterned ZnO arrays distributed in a 4.8× 4.8 cm^2 area where the diameter of each pattern was 5 μm and the spacing between adjacent patterns is 25 μm.

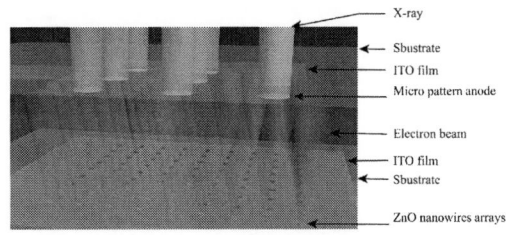

Fig. 2 Schematic of flat-panel X-ray source with patterned transmission anode.

The transmission anode was prepared using magnetron sputtering and patterned by ultraviolet photolithography. Firstly, an ITO film with a thickness of 0.5 μm was deposited on glass substrate by magnetron sputtering, the ITO film serves as the electrode layer. Secondly, micro patterned Al-Mo-Al film were formed on the ITO film using magnetron sputtering and lift-off process. The Al-Mo-Al film has a total thickness of 1.31 μm, where the thickness of each layer is 0.12 μm, 1.07 μm, and 0.12 μm, respectively. The first Al layer is used as a buffer layer to improve adhesion, and the other Al layer is used to prevent oxidation. Patterned ZnO nanowires arrays were prepared on another glass substrate uses e-beam evaporation, liff-off and thermal oxidation. The morphologies of samples were observed with a field emission scanning electron microscope (SEM, SUPRA 60).

To fabricate the diode FPXS device, the anode and cathode are assembled and separated by 6 mm thick spacer. The X-ray source is sealed using low melting-point glass frit. The device was pumped and sealed off at the pressure of 8×10^{-7} Pa. By the above-mentioned process, vacuum-sealed FPXS devices with patterned transmission anode were fabricated.

III. RESULTS AND DISCUSSION

The morphologies of the micro patterned transmission anode are shown in Fig. 3(a) and (b). Fig. 2(c) and (d) present the SEM images of a single ZnO nanowire pattern and its arrays. The length of the ZnO nanowires is about 2-5 μm and the average density is estimated to be 10 μm⁻².

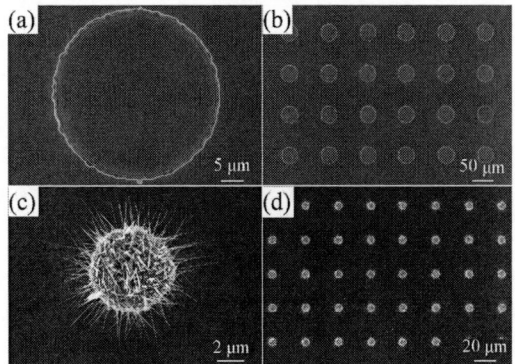

Fig. 3 (a, b) Circular patterned transmission anode with diameter of 30 μm; (c, d) Circular patterned ZnO nanowires arrays with diameter of 5 μm.

The I-V characteristics and corresponding F-N plot measured from one of the FPXS devices is shown in Fig. 4. The maximum voltage and maximum current can reach 42 kV and 229 μA, respectively. The X-ray energy spectrum of X-ray source recorded at the anode voltage of 42 kV is shown in Fig. 5.

The projection imaging performance of the fabricated FPXS was tested using a line-pair testing card. Fig. 6(a) shows the results, which show that a resolution of 2.8 lp/mm can be obtained. In addition, the X-ray images of biological specimen (hippocampus) and chips were obtained as shown in Fig. 6(b) and (c). The results show that the FPXS using micro patterned transmission anode can achieve clear projection imaging.

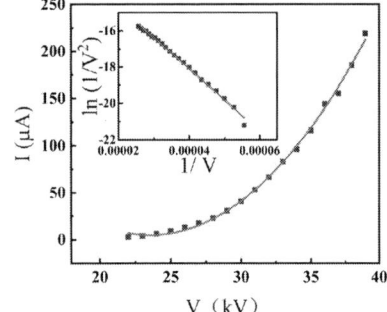

Fig. 4 I-V characteristics of the FPXS. The inset shows the corresponding F-N plot

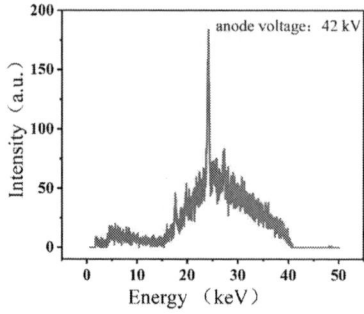

Fig. 5 X-ray energy spectrum of the fabricated FPXS.

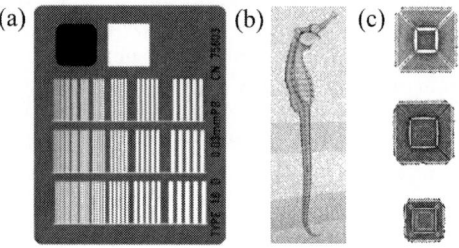

Fig. 6 Projection imaging results. (a) Line-pair-card; (b) hippocampus specimen and; (c) chips.

IV. SUMMARY

We designed and fabricated FPXS with micro-pattern transmission anode and investigated its imaging characteristics. The imaging capability of the fabricated device was verified. Using micro-patterned transmission anode can effectively modulate the X-ray emission spot of FPXS. Our work provides an important route for realizing micro-focus FPXS.

ACKNOWLEDGEMENTS

The authors gratefully acknowledge the financial support from the National Key Research and Development Program of China (Grant No.2016YFA0202000), the Science and Technology Department of Guangdong Province (Grant No.2020B0101020002), Fundamental Research Funds for the Central Universities.

REFERENCES

[1] Daokun Chen, Xiaomeng Song, Zhipeng Zhang, et al, Applied Physics Letters, 107(24):1-5(2015).
[2] Daokun Chen, Yuan Xu, Guofu Zhang, et al, Vacuum, 144, 266-271(2017).
[3] Libin Wang, Yangyang Zhao, Keshuang Zheng, et al, Applied Surface Science, 484,966-974 (2019).
[4] Yangyang Zhao, Yicong Chen, Guofu Zhang, et al, Nanomaterials, 11(240):240 (2021).
[5] Libin Wang, Yuan Xu, Xiuqing Cao, et al, IEEE Transactions on Nuclear Science, DOI 10.1109/TNS.2021.3051008.

Cold Cathode X-Ray Flat Panel Detector Based on Ga$_2$O$_3$ Thin Film Photoconductor

Haojian Huang, Manni Chen, Zhipeng Zhang, Juncong She, Shaozhi Deng, Ningsheng Xu, Jun Chen*

State Key Laboratory of Optoelectronic Materials and Technologies, Guangdong Province Key Laboratory of Display Material and Technology, School of electronics and information technology, Sun Yat-sen University, Guangzhou 510275, People's Republic of China

* E-mail: stscjun@mail.sysu.edu.cn

Abstract— **Large area flat panel detector is the key component of X-ray imaging system. In this study, Ga$_2$O$_3$ thin film photoconductor was prepared using electron-beam evaporation. A cold cathode X-ray flat panel detector was fabricated using Ga$_2$O$_3$ thin film photoconductor as the anode. X-ray response of the detector was measured and the results verified the possibility of using Ga$_2$O$_3$ thin film as the photoconductor for vacuum X-ray detector.**

I. INTRODUCTION

Large area flat panel detector is the key component of X-ray imaging system, which has important applications in medical diagnosis, industrial non-destructive inspection, security screening and scientific research instrument, etc. Highly sensitive flat panel X-ray detectors are key to realize low dose X-ray imaging. The photoconductive cold cathode flat panel X-ray detector consists of cold cathode electron source arrays and photoconductor anode, having the advantages of high-radiation tolerance and high sensitivity [1-2]. Large area cold cathode flat panel X-ray detectors and mechanism for realizing highly sensitive detection have been extensively studied [3-4].

Having a direct bandgap up to 4.5~5 eV and a high absorption coefficient, gallium oxide (Ga$_2$O$_3$) owns advantages such as good stability, easy preparation, and high radiation resistance etc., which makes it an ideal candidate for solar-blind ultraviolet and high-energy radiation detection [5]. Solid state X-ray detectors using single crystalline and thin film Ga$_2$O$_3$ were reported recently [6-9].

In this study, the possibility of using Ga$_2$O$_3$ thin film as photoconductor in a cold cathode vacuum flat panel detector was explored. An X-ray flat panel detector with diode structure was fabricated. Ga$_2$O$_3$ thin film photoconductor prepared using electron beam evaporation was used for the anode and ZnO nanowire field emitter arrays prepared using thermal oxidation technique was used as the cathode. The results verified the X-ray detection capability of the vacuum cold cathode flat panel detector based on Ga$_2$O$_3$ thin film photoconductor.

II. EXPERIMENTAL

Quartz was used as the substrates for the deposition of gallium oxide thin film. Indium tin oxide (ITO) film as the electrode was deposited on the quartz substrate by magnetron sputtering. Then 1.8 μm Ga$_2$O$_3$ thin film was deposited on ITO electrode by electron beam evaporation. The Ga$_2$O$_3$ thin film was characterized using scanning electron microscope (SEM), high-resolution transmission electron microscope (HRTEM), X-ray diffraction (XRD) and ultraviolet-visible (UV-Vis) absorption spectrometer.

Fig.1 shows the device structure of the cold cathode X-ray flat panel detector. Ga$_2$O$_3$ thin film photoconductor was used as the anode. Arrays of patterned ZnO nanowires prepared on ITO glass substrate was used as cold cathode. The distance between anode and cathode is 120 μm. The ZnO nanowires were prepared using thermal oxidation. Patterned ZnO nanowires arrays were used to minimize the screening effect and improve the emission uniformity. The area of each pattern is 25 μm×60 μm.

The X-ray response of the device were tested using a commercial thermionic cathode X-ray source. The detector was placed in a vacuum chamber with a pressure of ~10^{-5}Pa. The response of the device was measured under different dosages of X-ray radiation and anode biases. The emission current and its response to X-ray radiation was recorded using a Keithley 6485 picoammeter.

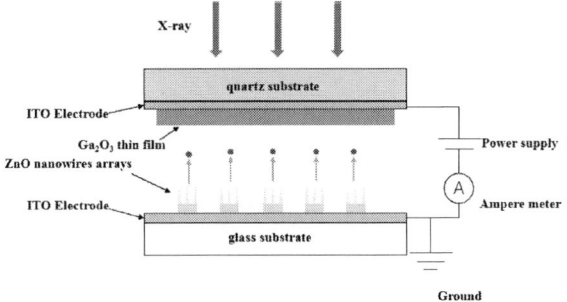

Fig.1 Schematic showing the structure of cold cathode X-ray flat panel detector with Ga$_2$O$_3$ thin film photoconductor anode and ZnO nanowire arrays cold cathode.

III. RESULTS AND DISCUSSION

Fig. 2.(a) shows the SEM images of prepared Ga$_2$O$_3$ thin film. Smooth films were obtained. XRD and HRTEM results indicate the prepared thin film composes of nanocrystalline Ga$_2$O$_3$ grains. Fig.2.(b) shows the top-view SEM images of ZnO nanowire arrays, indicating that the ZnO nanowires uniformly distribute on the patterns with a length of 1~3 μm.

Fig.3 is the ultraviolet-visible transmission spectrum of Ga$_2$O$_3$ thin film. The Ga$_2$O$_3$ thin film has more than 80% average transmission in the visible region and a significant absorption was observed at wavelengths less than 280 nm which is due to the band edge absorption.

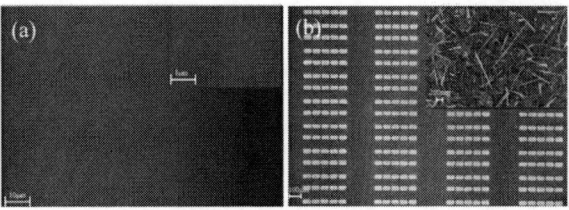

Fig.2 (a) SEM images of Ga$_2$O$_3$ thin filma; (b) Top-view SEM images of ZnO nanowire arrays.

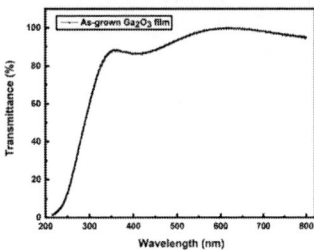

Fig.3 Transmission spectrum of Ga$_2$O$_3$ thin film.

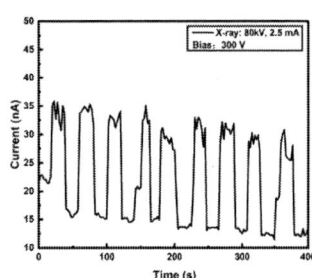

Fig.4 Transient X-ray photocurrent response of the detector with an anode bias voltage of 300 V.

Fig.4 shows the X-ray response of nine consecutive on/off cycles of X-ray exposure (tube current of 2.5 mA and tube voltage of 80 kV), which is manually turned on and off. The emission current distinctly responses to the change of X-ray radiation, suggesting that the repeatability is robust. The dark current is about 12~16 nA and the photocurrent is about 38~35 nA. The average light-dark current ratio for the nine consecutive on/off cycles is about 2.26. The variation of the current might be due to the fluctuation in field emission current.

The X-ray sensitivity was calculated using:

$$S = (I_{x\text{-ray}} - I_{dark}) / (A \cdot D). \qquad \text{(Eq. 1)}$$

Where, A is the active area, $I_{x\text{-ray}}$ is the X-ray radiation generated photocurrent, I_{dark} is the dark current, D is the radiation dose rate. The obtained average sensitivity is about 7.57×10^{-4} μCmGyair^{-1}cm^{-3}.

IV. CONCLUSIONS

In summary, an X-ray flat panel detector was fabricated by using Ga$_2$O$_3$ thin film conductor as the anode and ZnO nanowires field emitter arrays as the cathode. Distinct photoresponse was observed. The results verified the possibility of using Ga$_2$O$_3$ thin film as the photoconductor for vacuum X-ray detector.

ACKNOWLEDGMENTS

The authors gratefully acknowledge the financial support from the National Key Research and Development Program of China (Grant No.2016YFA0202000), National Natural Science Foundation of China (Grant Nos. 91833303 & 62001527), the Science and Technology Department of Guangdong Province (Grant No.2020B0101020002), Fundamental Research Funds for the Central Universities.

REFERENCES

[1] Z. P. Zhang, Z. J. Zhang, et al, "Sensitive and fast direct conversion X-ray detectors based on single-crystalline HgI$_2$ photoconductor and ZnO nanowire vacuum diode," Adv. Mater. Technol. vol.5, pp.1901108, 2020.

[2] Z. P. Zhang, K. Wang, et al, "Electron bombardment induced photoconductivity and high gain in a flat panel photodetector based on a ZnS photoconductor and ZnO nanowire field emitters," ACS Photonics, vol.5, pp.4147-4155,2018.

[3] Z. P. Zhang, K. Wang, et al, "A flat panel photodetector formed by a ZnS photoconductor and ZnO nanowire field emitters achieving high responsivity from ultraviolet to visible light for indirect-conversion X-ray imaging," J. Lightwave Technol. vol.36, pp.5110-5015, 2018.

[4] X. P. Bai, Z. P. Zhang, et al, "Theoretical analysis and verification of electron-bombardment-induced photoconductivity in vacuum flat-panel detectors," J. Lightwave Technol. vol.39, pp.2618-2624, 2021.

[5] M. Higashiwaki, K. Sasaki, et al, "Gallium oxide(Ga$_2$O$_3$) metal-semiconductor field-effect transistors on single-crystal β-Ga$_2$O$_3$(010) substrates," Applied Physics Letters, vol.100, pp.013504, 2012.

[6] Z. P. Zhang, Z. M. Chen, et al, "ε-Ga$_2$O$_3$ thin film avalanche low-energy X-ray detectors for highly sensitive detection and fast-response applications," Adv. Mater. Technol. vol.6, pp.2001094, 2021.

[7] M. N. Chen, Z. P. Zhang,et al, "Fast-response X-ray detector based on nanocrystalline Ga$_2$O$_3$ thin film prepared at room temperature," Appl. Surf. Sci. vol.554,pp.149619, 2021.

[8] H. Liang, S. Cui, et al, "Flexible X-ray detectors based on amorphous Ga$_2$O$_3$ thin films," ACS Photonics, vol.6, pp.351-359, 2018.

[9] X .Lu, L. Zhou, et al, "Schottky X-ray detectors based on a bulk β-Ga$_2$O$_3$ substrate," Appl. Phys. Lett. vol.112, pp.103502, 2018.

Development of gated carbon nanotube cold cathode for miniature X-ray source

Yajie Guo, Junfan Wang, Baohong Li, Yu Zhang, Shaozhi Deng, Ningsheng Xu, Jun Chen*

State Key Laboratory of Optoelectronic Materials and Technologies, Guangdong Province Key Laboratory of Display Material and Technology, School of Electronics and Information Technology, Sun Yat-sen University, Guangzhou 510275, People's Republic of China
* Corresponding author: stscjun@mail.sysu.edu.cn

Abstract—**A gated carbon nanotube cold cathode has been fabricated for miniature X-ray source application. Randomly aligned CNTs were prepared using CVD. Gated CNT cold cathode was fabricated using a metal mesh gate. The current-voltage characteristics, gate transmission rate and the current stability were studied. Current fluctuation less than 1% was achieved.**

I. INTRODUCTION

X-ray source has important applications in the fields of industrial inspection, medical diagnosis and cancer radiotherapy. However, the traditional thermionic cathode X-ray sources have disadvantages of slow response, bulky volume and high-power consumption [1]. The cold cathode X-ray sources have been developed to solve those problems. Meanwhile, it is easy to realize miniaturization or micro-focus X-ray sources with cold cathode technology, which has significant application prospect in industrial inspection and medical radiotherapy [2-4]. Cold cathode X-ray sources have thus become a research hotspot in the field [5-9]. The electron gun is the core part of cold cathode X-ray source. High electron transmission rate and highly stable emission are vital for developing a practical cold cathode electron gun.

In this work, a gated CNT cold cathode electron gun with carbon nanotubes (CNTs) has been fabricated for miniature X-ray source application and gate transmission and current stability have been studied.

II. EXPERIMENTAL

The structure of CNTs cold cathode electron gun is composed of cathode and gate electrode. Fig. 1 shows the structure and measurement circuit of gated CNTs cold cathode electron gun. CNTs were prepared on 1.6 mm diameter circular stainless steel (SS304) substrates by thermal chemical vapor deposition (CVD) [10]. The 0.5 nm thickness Fe thin film has been pre-deposited acting as the catalyst. The growth temperature is 650 °C, while the reaction gas includes acetylene and hydrogen. The distance between cathode and gate electrode is approximately 100 μm. Molybdenum mesh was used as the gate with apertures of diameter 90 μm.

The current-voltage characteristics of the device is measured in a vacuum chamber by using a phosphor screen as the anode. The distance between cathode and anode is 7.5 mm. The anode current (I_a), gate current (I_g) and cathode current (I_c) were recorded by ampere meters.

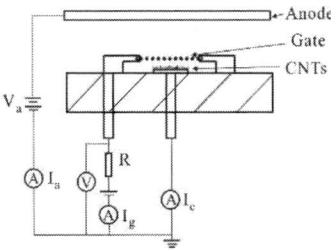

Fig. 1. The structure and measurement circuit of CNTs cold cathode electron gun

III. RESULT AND DISCUSSION

Fig. 2 shows the morphology of CNTs in the device. The CNTs is dense and disorder. The diameter of the CNTS ranges from 20~50 nm and the length could reach several tens microns.

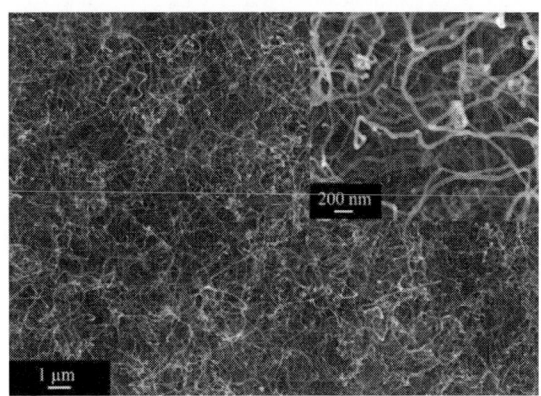

Fig. 2. SEM images of CNTs. Inset shows the magnified morphology of CNTs.

Before the measurement of I-V characteristics, an aging process was first carried out and the cathode continuously operated under certain current. Then, the anode current, gate current, and cathode current versus gate voltage characteristics of the electron gun were measured under 3 kV anode voltage. Typical results were shown in Fig. 3. Current of 130 μA has been achieved at gate voltage of 620 V from cathode of electron

978-1-6654-2590-2/21 $31.00 © 2021 IEEE

gun and the corresponding anode current is 68 μA. The gate electron transmission rate under different gate voltage is shown in the inset of Fig.3. The rate increases sharply and then fall slowly. Maximum gate transmission rate of 59 % was obtained at voltage of 427 V and the gate transmission is about 52 % for the cathode current 130 μA.

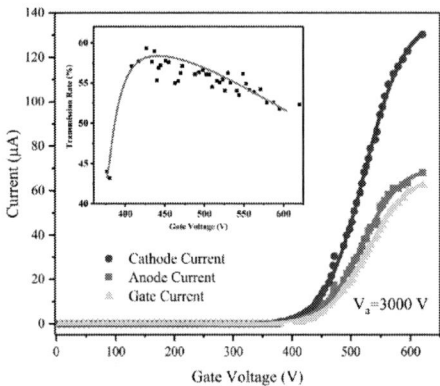

Fig. 3. I-V curves. Inset is the gate transmission rate.

The fluctuation of cathode current was studied. Furthermore, the effect of gate ballast resistor on the current stability was studied. Fig. 4 shows the current stability at ~55 μA over 30 min recorded using three different ballast resistor values, i.e. 500 KΩ, 5 MΩ, 50 MΩ. The corresponding current fluctuation φ_C was calculated by:

$$\varphi_C = \frac{\sum_{i=1}^{n} \left| I_i - I_{average} \right|}{n \cdot I_{average}} . \qquad (1)$$

Where I_i is the cathode current at each moment and $I_{average}$ is the average current.

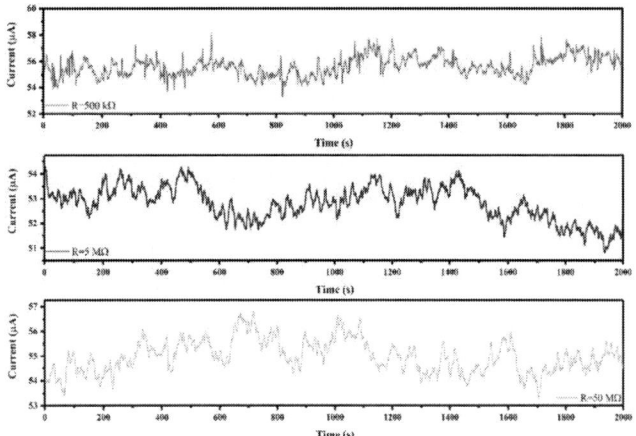

Fig. 4 Current stability of cathode current mesure using different ballast resistor. 500 kΩ (red line), 5 MΩ (bule line), 50 MΩ (green line)

The fluctuations were calculated for different cathode currents and resistor values. The results are summarized in Table.1. Fluctuations of 2.88%@~11 μA, 1.06%@~55 μA, 1.30%@~105 μA were obtained when 500 KΩ resistor was used, indicating stable current can be obtained from our CNTs cathode. The cathode current becomes more stable with increasing resistor value due to the increased regulation effect. Our results show that current fluctuation less than 1% can be achieved by using 5 M and 50 M resistor.

TABLE I The cathode current fluctuation

Resistor	Fluctuation @ cathode current		
500 KΩ	2.88%@~11 μA	1.06%@~55 μA	1.30%@~105 μA
5 MΩ	2.17%@~16 μA	1.06%@~52 μA	0.69%@~132 μA
50 MΩ	2.15%@~15 μA	0.97%@~55 μA	0.78%@~157 μA

IV. Conclusions

A cold electron gun for miniature X-ray source application has been fabricated used CNTs cold cathode. The cathode current of 130 μA and corresponding anode current of 68 μA has been achieved, with a gate transmission rate of approximate 52%. The CNTs cold cathode has good current stability and less than 1% current fluctuation was achieved by using ballast resistor.

Acknowledgments

The authors gratefully acknowledge the financial support from the National Key Research and Development Program of China (Grant No.2016YFA0202000), the Science and Technology Department of Guangdong Province (Grant No.2020B0101020002), Fundamental Research Funds for the Central Universities.

References

[1] H. Sugie , M. Tanemura, V. Filip, et al. Applied Physics Letters, 2001, 78(17):2578-2580.

[2] J. W. Jeong , J. T. Kang, S. Choi, et al. Applied Physics Letters, 2013, 102(2):138-144.

[3] G. Cao , Y. Z. Lee , R. Peng , et al. Physics in Medicine & Biology, 2009, 54(8):2323-2340.

[4] W. Lei , Z. Y. Zhu , C. Y. Liu , et al. Carbon, 2015, 94:687-693.

[5] G. Z. Yue , Q. Qiu , B. Gao , et al. Applied Physics Letters, 2002, 81(2):355-357.

[6] L. B. Wang , Y. Xu , X. Q. Cao, et al. IEEE Transactions on Nuclear Science, 2021, 68(3): 338-345.

[7] Y.Y. Zhao, Y.C. Chen, G. F. Zhang, et al. Nanomaterials, 2021,11(1):240.

[8] D. K. Chen, X. M. Song, Z. Z. Zhang, et al. Applied Physics Letters, 2015, 107:243105.

[9] S. Park , J.-T. Kang, J.-W. Jeong, et al. IEEE Electron Device Letters, 2018, 39(12):1936-1939.

[10] Y Zhang , Y M Tan , L Z Wang , et al. Vacuum, 2020, 172:109071.

Optimization of Focusing Structure for a Micro-Focus X-ray Source

Junfan Wang, Yajie Guo, Haifeng Zhu, Baohong Li, Yu Zhang, Shaozhi Deng, Ningsheng Xu and Jun Chen*

State Key Laboratory of Optoelectronic Materials and Technologies, Guangdong Province Key Laboratory of Display Material and Technology, School of Electronics and Information Technology, Sun Yat-sen University, Guangzhou, People's Republic of China
*Corresponding author: stscjun@mail.sysu.edu.cn

Abstract—**Cold cathode micro-focus X-ray source with a single focusing lens was designed. The focus spot size (FSS) of the electron beam was simulated using CST software. The simulation method is verified by experimental results using an electron gun consisting of gated carbon nanotube cold cathode and focusing electrode.**

I. INTRODUCTION

Micro-focus X-ray source has important applications in industrial and medical imaging, such as micro-computed tomography (micro-CT) [1]. In such application, the X-ray source needs a focused electron-beam, which generates a small focal spot size (FSS). The FSS on the anode target directly determines the imaging resolution[2]. Compared to traditional thermionic cathode X-ray source, cold cathode X-ray source is easy to achieve small FSS due to the narrow energy spreading of the emitting electron. Furthermore, cold cathode X-ray source could be in small size and have fast response. Compact imaging system with high spatial and temporal resolution is expected using such X-ray source[3].

The key part to achieve small FSS is the electrostatic focusing structure of micro-focus X-ray source. In this study, an electrostatic focusing structure with single lens was modeled and simulated using CST software. Various combinations of operation voltages, lens apertures(D_f), lens-cathode distances(Z) revealed the effect on FSS of these variables. Based on the simulation, a focusing electrode was fabricated using 3D printing technique and its focusing performance was characterized experimentally.

II. SIMULATION AND DISCUSSION

An ideal X-ray source model was built in CST STUDIO as shown in Fig. 1(a). Before simulation, following parameters were specified according to the targeted X-ray source device:

(1) Anode-cathode distance: 75 mm, V_{anode}: 60 kV, lens-anode distance: no less than 45 mm.

(2) Cathode emission diameter: 1.6 mm, grid-cathode distance: 0.1 mm, V_{gate} : 1 kV.

(3) Extraction grid mesh diameter: 0.09 mm, center distance: 0.12 mm, grid thickness: 0.02 mm. As shown in Fig. 1(b). gate diameter: 7 mm.

The simulation was divided into two groups according to different independent variables, lens apertures(D_f) and lens-cathode distances(Z). Z varied between 4 and 12 mm in Fig. 2, and Df varied between 10 and 24 mm in Fig. 3. For every Z (or D_f), lens voltage(V_f) was adjusted until that the focus of the electron beam fell exactly on the anode, i.e., the electron beam was about to be overfocused, then the FSS and the corresponding V_f were recorded. Curves in Fig. 2(b) and Fig.

2(a) show the effect of Z on FSS and V_f, respectively. Fig. 3 tells the relationship between D_f (instead of Z) and the same two dependent variables when Z were fixed to be 6 mm here.

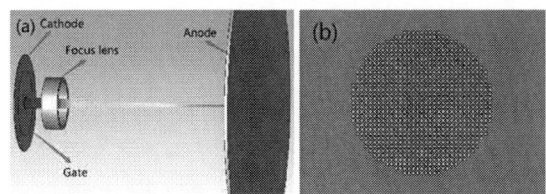

Fig. 1 X-ray source simulation structure. (a) CST model for micro-focus X-ray source. (b) CST model for extraction gate (grid mesh in the center).

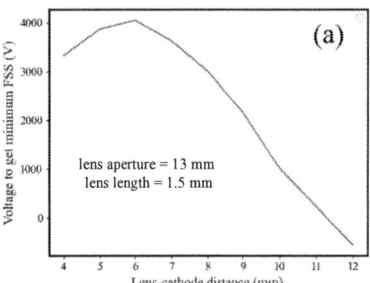

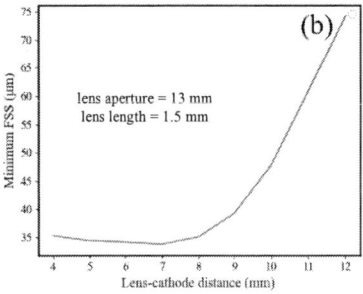

Fig. 2 Minimum FSS and corresponding V_f under D_f = 13 mm and lens length = 1.5 mm. (a) Simulated Voltages to get minimum FSS as a function of Z. (b) Simulated minimum FSS as a function of Z.

Apparently, the larger the voltage difference between Vf and the center voltage of the focusing lens(which means lower V_f in this simulation), the greater the focusing strength will be applied on the electron beam. According to this method, we could succeed in making the focus fall exactly on the anode by adjusting V_f carefully.

Fig. 2(a) and Fig. 3(a) tells us the focusing strength that lens provided changed when the focusing structure(i.e. Z or D_f) changed. When D_f increased, the focusing strength decrease,

978-1-6654-2590-2/21 $31.00 © 2021 IEEE

so lower V_f was needed to make the focus fall on the anode. When Z increased, the focusing strength climbed up and then declined, so the same trend appeared for V_f as shown in Fig. 2(a). From the size of the focus, when another variable was fixed, within a small range of Z or D_f (5~8 mm for Z, 14~18 mm for D_f), minimum FSS approached a minimum value.

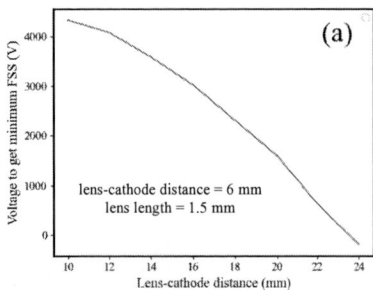

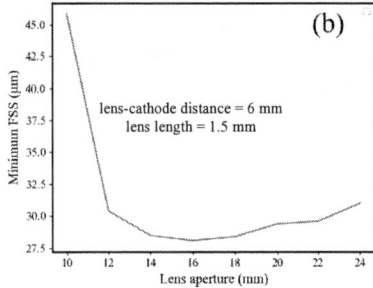

Fig. 3 Minimum FSS and corresponding V_f curves under Z = 6 mm and lens length = 1.5 mm. (a) Simulated voltages to get minimum FSS as a function of D_f. (b) Simulated minimum FSS as a function of D_f.

III. EXPERIMENTAL RESULTS

In order to verify the simulation, we fabricated focus lens and assembled with a gated CNT cold cathode. The focus lens was fabricated using a 3D printing machine. It is difficult for us to measure the focus spot in a real X-ray source operated under high anode voltage. Therefore, we adopted a relative small anode-cathode distance in the experiment. Phosphor screen is used as the anode to visualize the focus spot. The cathode-anode distance was 15 mm and V_{anode} was 5 kV.

The assembled electron gun consisting of the focusing lens and the gated CNT cold cathode is shown in the insert of Fig. 4(a). The cathode is randomly aligned CNTs prepared on round substrate with a diameter of 1.6 mm. The effect of lens voltage on the FSS of the electron beam can be clearly observed which is in consistence with the simulated trends. The recorded images of emission patterns were presented in Fig. 4(a)-(c). Both the images without focus lens and focused under different focusing voltages are given. The minimum FSS observed is about 2.5 mm. The value is larger than the simulated value using the parameters the same as the experimental set-up, which is about 2.1 mm. In the simulation the electrons are assumed to emit in the direction perpendicular to the surface. While in the case of randomly aligned CNTs, the electrons might emitted at large angle. This might cause the different between the simulation and experiment. Overall, the experimental results could verify the validation of the simulation method.

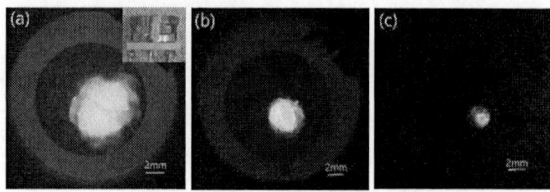

Fig. 4. Picture of the focal spot on the fluorescent screen. (a) Without focusing lens. (b) V_f = -800 V. (c) V_f = -1000 V.

IV. SUMMARY

In summary, we have conducted simulation to explore the influence of variables such as Z, D_f on the FSS in a focusing lens structure designed for a micro-focus X-ray source. The simulation method was validated by experimental results. The results shows that by adopting an optimal parameter, micro-FSS can be achieved by using a simple single focus structure.

ACKNOWLEDGMENTS

The authors gratefully acknowledge the financial support from the National Key Research and Development Program of China (Grant No.2016YFA0202000), the Science and Technology Department of Guangdong Province (Grant No.2020B0101020002), Fundamental Research Funds for the Central Universities.

REFERENCES

[1] M. Dierick et al., "Recent micro-CT scanner developments at UGCT," Nuclear Instruments and Methods in Physics Research Section B: Beam Interactions with Materials and Atoms, vol. 324, pp. 35-40, 2014.

[2] A. V. Avachat, W. W. Tucker, C. H. C. Giraldo, D. Pommerenke, and H. K. Lee, "Looking Inside a Prototype Compact X-Ray Tube Comprising CNT-Based Cold Cathode and Transmission-Type Anode," Radiat Res, vol. 193, no. 5, pp. 497-504, 2020.

[3] S. Park et al., "A Fully Closed Nano-Focus X-Ray Source With Carbon Nanotube Field Emitters," IEEE Electron Device Letters, vol. 39, no. 12, pp. 1936-1939, 2018.

Focal Spot Size Enhancement by Offset control of Triode e-beam Module for High Resolution X-ray Imaging

Yi Yin Yu[1] and Kyu Chang Park[1*]

[1]*Department of Information Display,*
Kyung Hee University, Dongdaemun-gu, Seoul,02447, Korea

*Contact: kyupark@khu.ac.kr, phone +82-2-961-9447

Abstract— **Demands on high quality x-ray imaging techniques are required for early detection of abnormal tissues and quality control of mass production in the field of medical and industrial applications.**

Moreover, smaller focal spot size (FSS) of x-ray sources have been developed for many decades but still their structure and operational scheme are complicated. To optimize electron beam trajectory, at least two types of focusing electrodes are needed.

We demonstrate under 0.5 mm FSS of cold cathode x-ray source triode type e-beam module with vertically aligned carbon nanotubes (VA-CNTs) without any additional focusing electrodes. In addition, gate hole offset effects were investigated to FSS.

The authors anticipate that our simple but sophisticated gate structure will be beneficial for high quality x-ray imaging and cost effective in terms of massive production of cold cathode x-ray generators.

I. INTRODUCTION

Cold cathode based x-ray sources are intensively studied due to their many advantages compare to the conventional thermionic ones [1, 2]. Especially, nano sized electron emitters can remarkably reduce the penumbra area which deteriorates the x-ray image quality in term of focusing. Besides, the instantaneous switching capability of cold cathode x-ray devices diminish the redundant dose [4]. Thanks to the field driven electron emission phenomena the fast switching of cold cathode is achievable.

Many kinds of cold cathodes are studied such as Spindt type Si or metal tips, ZnO NWs and so on. Among them, carbon nanotubes (CNTs) are regarded as a prominent electron source owing to their outstanding electrical and mechanical properties. There are two types of CNTs based electron emitters in accordance with emitter preparation method. One is screen printing process and the other is directly grown free standing VACNTs on the substrate. The latter has superior characteristics in terms of current density and smaller FSS.

Vertically aligned emitter structure is prone to bias electric field to extract electrons from free standing emitter tips. Moreover, due to the high aspect ratio, electron beam represents narrow beam divergence. Meanwhile screen printed CNTs based emitter shows wide beam spread because they are oriented in random direction within the paste nature. Song et. al

reported paste CNTs based x-ray generator in the configuration of triode type but its structure and operation are cumbersome due to the additional focusing electrodes [3].

We studied gate offset effects of triode e-beam module on FSS for the sake of fine resolution x-ray imaging applications without any focus electrodes by employing our VACNTs. Our simple but sophisticated triode e-beam module architecture is regarded as key solution of high quality x-ray imaging and cost effective way for massive cold cathode based x-ray sources fabrication.

II. EXPERIMENT

A. Preparation of VACNTs

Electron emitter arrays of VACNTs were synthesized on Si substrate by home-made plasma enhanced chemical vapor deposition (PECVD) system. The selective positioning emitters were achieved by patterning Ni catalyst thin film before the process. An SEM image of emitter arrays is represented in Fig 1.The diameter and height of individual emitter were 3 and 50 μm respectively. The spacing between adjacent emitters was 15 μm to consider screening effect among the emitters.

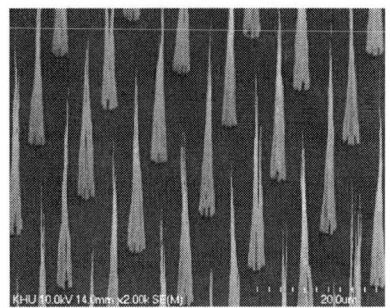

Fig 1. SEM image of vertically aligned carbon nanotubes (VACNTs) on Si wafer. The image was tilted by 45 degree.

B. Triode e-beam module assembly with different gate hole offset

The gate electrodes were prepared to diverse offsets in terms of gate hole arrays. Our emitter array is composed of 2 by 8 emitter matrix and 1 by 1 of gate offset represents additional 1 row and 1 column for its structure. Up to 4 by 4 gate offsets were tested in the triode configuration.

C. X-ray image evaluation with respect to diverse gate offset applied modules

To compare FSS characteristics depending on gate offsets of the triode module, crossed W wires were imaged at the magnification of 6 by changing gate offsets of e-beam module. Both longitudinal and transverse direction of crossed W wire images were evaluated by image process software. For intuitive comparison target phantom images were measured. Both x-ray images were taken at over 60 kV_p of anode bias and same mAs conditions.

III. RESULTS AND DISCUSSION

Gate offset structure of 1 by 1 showed the smallest FSS properties. As increasing gate hole offsets, the sizes of FSS were increased for both longitudinal and transverse directions. The tendency represents linear trends and it might be related to the beam trajectory alteration by gate aperture sizes Theoretical analyse will be conducted for the future study and the offset structure will be optimized later.

IV. CONCLUSIONS

Triode gate structure with smaller offset shows better x-ray imaging qualities in terms of FSS. A linear relationship between gate hole offsets and FSS were observed and it might be related to the beam divergence effect by gate aperture. Smaller than 0.5 mm of FSS was achieved with simple triode e-beam structure without any axillary focal electrodes. The result will be contributed to cost effective and high resolution cold cathode based x-ray fabrication process.

REFERENCES

[1] G. Z. Yue, Q. Qiu, B. Gao, Y. Cheng, J. Zhang, H. Shimoda, S. Chang, J. P. Lu, and O. Zhou, "Generation of continuous and pulsed diagnostic imaging x-ray radiation using a carbon-nanotube-based field-emission cathode," Applied Physics Letters, vol. 81, no. 2, pp. 355–357, Jul. 2002.

[2] A. Basu, M. E. Swanwick, A. A. Fomani and L. F. Velasquez-Garcia, "A portable x-ray source with a nanostructured Pt-coated silicon field emission cathode for absorption imaging of low-Z materials," J. Phys. D. vol. 48, 225501, 2015.

[3] J. W. Jeong, J. W. Kim, J. T. Kang, S. Choi, S. Ahn and Y. H. Song, "A vacuum-sealed compact x-ray tube based on focused carbon nanotube field-emission electrons," Nanotechnology, vol. 24, 085201, 2013.

Outgassing during LAFE operation in the diode system

S.V. Filippov*, A.G. Kolosko, E.O. Popov

Div. of Plasma Physics, Atomic Physics and Astrophysics
Ioffe Institute
St Petersburg, Russia
*s.filippov@mail.ioffe.ru

Abstract—**The paper presents the results of a study of field cathode based on a multi-walled carbon nanotube / polystyrene nanocomposite. Mass spectra and kinetics of the partial pressure of the main volatile products released from the surface of the electrodes were obtained when a constant voltage of different levels was applied to the cathode. The main volatile products during field emission diode operation are H_2, H_2O, CO / C_2H_4 and CO_2. The behavior of H_2O peak intensity is characterized by increased inertia relative to sudden voltage changes.**

Keywords—mass-spectrometry, large area field emitter, residual gas analysis, carbon nanotube, field emission.

I. INTRODUCTION

The stability of large area field emitter (LAFE) operation depends on many factors: external conditions, properties and morphology of the emitter surface, etc. One of the determining factors is the vacuum conditions under which emission occurs, namely, the composition and pressure of the residual atmosphere in the experimental chamber.

Earlier it was shown that the exposure of LAFE based on carbon nanotubes in the atmosphere of various gases has a significant effect on the value of the emission current and its stability [1]. Previous studies have reported a changeable behavior of the emission current (decreasing / increasing tendency) when a constant voltage of different levels is applied [2,3,4]. This behavior can be explained in terms of adsorption-desorption processes occurring on the emitter and opposite electrode surfaces. For a complete understanding of these processes, it is necessary to have an information of the composition and kinetics of the partial pressures of the residual gases during LAFE operation.

In this article, we report on the features of the behavior of the emission current on a multiwall carbon nanotubes / polysterene (MWCNT/PS) LAFE and present the results of a series of experiments to determine the gases evolved during its operation.

II. EXPERIMENTAL DETAILS

LAFE was produced by spin coating the MWCNT/PS suspension on a metal substrate 10 mm in diameter. A metal substrate with the same dimensions as the cathode substrate was used as the anode. The diode system was evacuated to a pressure of $2 \cdot 10^{-7}$ Torr. A time-of-flight mass spectrometer with

its own pumping system based on a getter pump was connected directly to the field emission unit. Other details of experimental setup and measurement technique can be found elsewhere [5, 6].

III. RESULTS AND DISCUSSION

A constant DC voltage was applied to the cathode with a stepwise change in its level up to 3 kV every 30 s (see Fig. 1a). A changeable character of the emission current was observed (see Fig. 1b). Starting from the 2nd step, the current shows a slight decrease over one level up to the 6th step (maximum current level). At decreasing of the applied voltage level from the 7th to 11th step the emission current constantly increases. The observed character of the emission current variation is similar to the results of other studies obtained on carbon nanofibers synthesized using electroplated catalysts [4], on flame-synthesized patterned MWCNTs [3] and single walled carbon nanotubes attached to the looped tungsten wire [2].

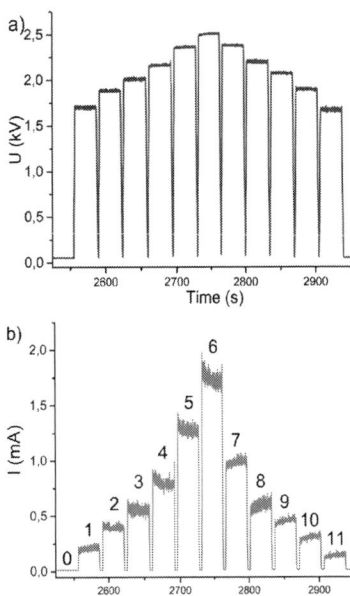

Fig.1. Applied DC voltage levels a) and corresponding changes in the emission current b).

978-1-6654-2590-2/21 $31.00 © 2021 IEEE

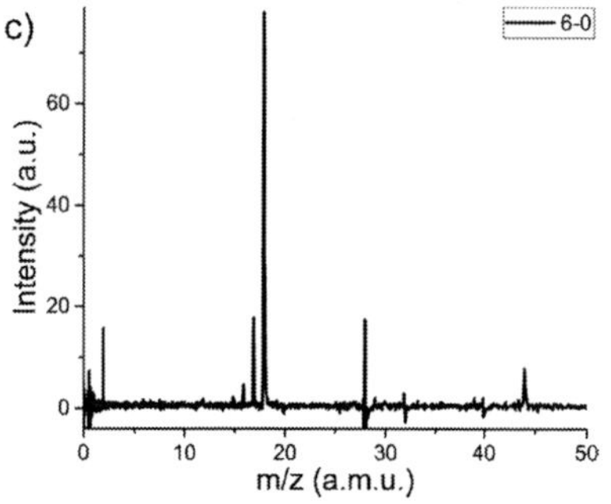

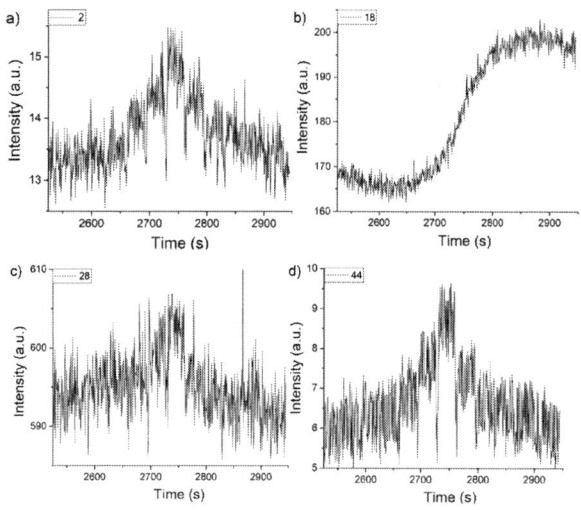

Fig.2. Difference between mass-spectra before (step 0 - background mass spectrum) and during field emission testing (step 6 – maximum current level).

At each current level, the mass spectrum of the residual gases in the experimental chamber was recorded. Fig. 2 shows the difference mass spectrum between LAFE operation (step 6) and in the absence of emission (chamber background – step 0).

The main components released during the stable emission regime are the following peaks: 2 amu (H_2), 17 amu (OH), 18 amu (H_2O), 28 amu (N_2/CO/C_2H_4) and 44 amu (CO_2). It is also worth to note that the absence or insignificant change in the spectrum of 32 amu (O_2), 40 amu (Ar) and 14 amu (N) peaks. The latter fact excludes the contribution of N_2 to the 28 amu, since otherwise atomic nitrogen would be noticeable in the spectrum.

During the entire experiment the kinetics of the peaks corresponding to aforementioned gases were also recorded. Figure 3a-d shows the rate of gas evolution of the emitter-anode system for the main volatile products during the LAFE operation.

Fig. 3. Intensity of gas evolution of the emitter-anode system for the main volatile products during operation LAFE: a) 2 amu - (H_2), b) 18 amu - (H_2O), 28 amu - (CO/C_2H_4) and 44 amu - (CO_2).

Note that H_2, CO/C_2H_4 and CO_2 show an instantaneous and stepwise change similar to the behavior of the emission current. On the other hand, the concentration of H_2O in the chamber

monotonically increases with an increase in the emission current up to the 7th stage, subsequently decreasing due to the evacuation system of the vacuum chamber.

During the sample conditioning two operation modes were found: a gradual decrease and short-term current discharge. With a gradual decrease in the current, peaks with 15 amu and 26 amu appeared in the spectrum corresponding to organic radicals of the C_xH_y, which indicate the possible destruction and shortening of emitting nanotubes. When discharges occur, i.e. of sharp jumps in the emission current, the response of almost all peaks present in the spectrum (including 32 and 40) is observed.

IV. CONCLUSIONS

We have shown that the main volatile products releasing in the chamber during LAFE operation are H_2, H_2O, CO / C_2H_4 and CO_2. The source of such gases is possibly electron-stimulated desorption from the anode. Moreover, concentration of H_2O molecules have a high inertia with respect to voltage and current changes. A mechanism of thermal desorption from the entire surface of the anode is possible during its gradual heating due to electron bombardment. The gradual decreasing in the current level when a high voltage is applied to a as-prepared sample is associated with the burnout of unstable nanotubes (the longest or thinnest).

REFERENCES

[1] A. Wadhawan, R.E. Stallcup II, K.F. Stephens II, J.M. Perez and I.A. Akwani, Appl. Phys. Lett. 79, 1867 (2001).

[2] K.A. Dean and B. R. Chalamala, Appl. Phys. Lett. 76, 375 (2000).

[3] Ch. Li, G. Fang, X. Yang, N. Liu, Y. Liu and X. Zhao, J. Phys. D: Appl. Phys. 41, 195401 (2008).

[4] K.H. Park, S. Lee and K.H. Koh, J. Vac. Sci. Technol. B 24, 1353 (2006).

[5] E.O. Popov, A.G. Kolosko, S.V. Filippov, I.L. Fedichkin and P.A. Romanov, J. of Vac. Sci. Technol. B 33, 03C109 (2015).

[6] A.G. Kolosko, M.V. Ershov, S.V. Filippov and E O. Popov, Tech. Phys. Lett. 39, 484 (2013).

Cathodoluminescent UV Sources for Photocatalytic Disinfection of Air

Evgenii P. Sheshin, Ilya N. Kosarev, Bulat I. Masnaviev, Alexander O. Getman, Ilya A. Savichev and Dmitry I. Ozol[*]

Moscow Institute of Physics and Technology, Dolgoprudny, Moscow Region, Russian Federation

[*]Contact: ozol.mipt@gmail.com

Abstract—**We have created samples of cathodoluminescent sources of 315 nm and 355 nm wavelengths based on commercially available phosphors, and demonstrated their photocatalytic activity. The power density of UV radiation is higher than 10 mW/cm². The photocatalytic activity was tested by measuring the rate of oxidation of acetone vapors under irradiation of a titanium dioxide catalyst, the reaction rate was more than 2 ppm/min.**

I. INTRODUCTION

In the SARS-CoV-19 pandemic, air disinfection issues have become especially topical. There are two methods of air disinfection using UV radiation: direct disinfection using germicidal range UV (250-300 nm) and photocatalytic disinfection by irradiation of 310-370 nm of titanium dioxide TiO_2 [1]. Both methods are quite effective, although each of the methods has pros and cons. Photocatalysis has been shown to be effective of destruction a wide range of microbes - bacteria, fungi, algae, protozoa and viruses as well, include SARS-CoV-19 coronavirus [2].

Cathodoluminescent lamps with field emission cathodes [3],[4] are a new option for environmentally friendly mercury-free UV sources – nowadays a strong tendency exists to eliminate mercury-containing sources especially in household usage as potentially ecologically dangerous [5]. Designing a cathodoluminescent source of 'photocatalytic' range is less difficultthan the source of the germicidal range - in particular, due to possibility of using uviol glass (not quarts), whose transmission in this range is still quite large. The technology of production of such sources is quite simple, well-known since the days of monochrome TV cathode-ray tubes, and with mass production they can be cheap. Effective UV LEDs in the 315-360 nm range, the most suitable for photocatalysis, do not currently exist and are not expected to appear in the coming years [6].

We have created samples of cathodoluminescent sources of 315 nm and 355 nm wavelengths based on commercially available phosphors, and demonstrated their photocatalytic activity.

II. RESULTS

The lamp (Fig. 1,2) with triode scheme comprises cathode, modulator, and anode. The cathode is manufactured of specifically treated carbon fiber. Field emission cathodes on

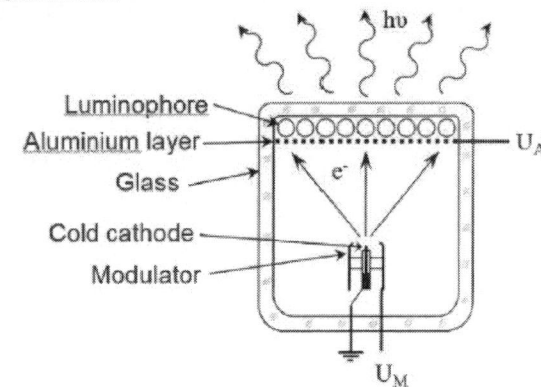

Fig. 1 Schematic view of a cathodoluminescent light source

the basis of carbon fiber exhibit long service life (not less than 10000 hours of continuous operation without marked changes of their parameters). The anode is coated with luminophore. The emission spectrum of cathodoluminescent lamps depends only on the luminophore used. So using the same lamp design, it is easy to obtain sources of different wavelengths, simply replacing the phosphor (Fig.3,4).

Fig. 2 One of the cathodoluminescent UV-lamps (switched on). Note: a very small part of the radiated power is in the visible range

978-1-6654-2590-2/21 $31.00 © 2021 IEEE

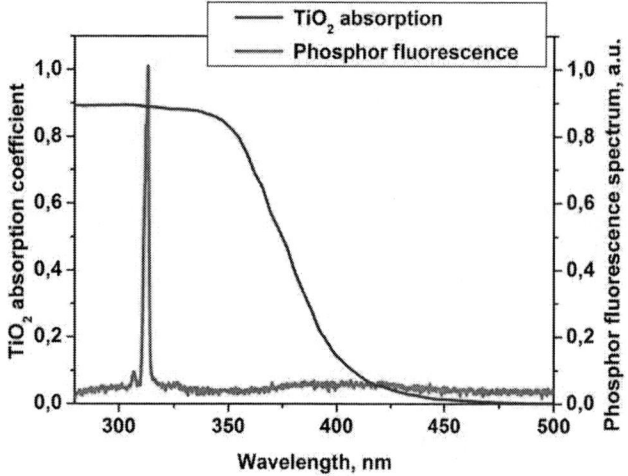

Fig. 3 Cathodoluminescence spectrum of the SrB_6O_{10}:Gd phosphor and its overlapping with the TiO_2 absorption curve

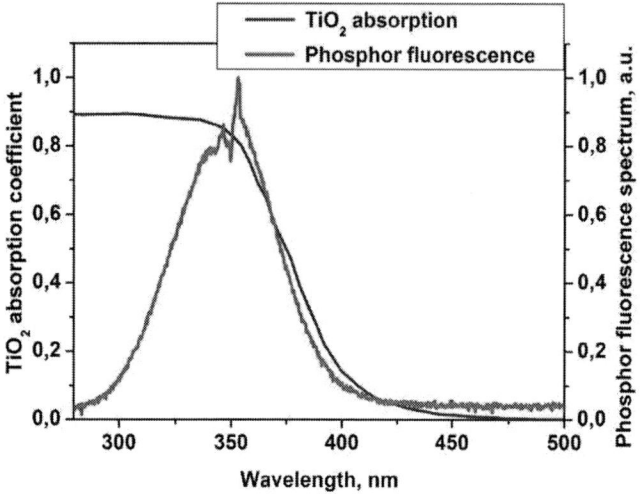

Fig. 4 Cathodoluminescence spectrum of the $BaSi_2O_5$:Pb phosphor and its overlapping with the TiO_2 absorption curve

Many types of UV-band luminophores are already known with an efficiency of cathodoluminescence up to 9% [7] and even 20%.

The power density of CL-lamps UV radiation is higher than 10 mW/cm². The photocatalytic activity was tested by measuring the rate of oxidation of acetone vapors under irradiation of a titanium dioxide catalyst.

The absorption of overband light by a TiO_2 photocatalyst leads to the production of electron-hole pairs. Before the onset of recombination, charge carriers that are on the surface of the photocatalyst can transfer to molecules contained in the air, thereby forming an intermediate reaction product, for example,

an ion of a water molecule or an oxygen molecule. Thus, a chemical reaction is initiated, which is completed after a finite number of transfers. Due to photocatalysis, it is possible to carry out the oxidation of many organic compounds under standard conditions. In this regard, various microorganisms can be destroyed by oxidizing their shells using photocatalysis.

According to [8] the threshold intensity of the overband light for the activation of photocatalysis is about 1 mW/cm².

We carried out an experiment on the photocatalytic oxidation of acetone with a chemical reaction of the process: $C_3H_6O + 4O_2 \rightarrow 3CO_2 + 3H_2O$. A sensor was used to measure the CO_2 concentration. Titanium dioxide is the most studied photocatalyst. Acetone oxidation rate exceeds 2 ppm per minute.

The luminescence intensity of the lamp with a silicate phosphor was about 2 times higher than that of borate phosphor, but due to the worse overlap with the absorption of oxide (compare Fig.2, 3), the photocatalytic efficiency was comparable.

III. SUMMARY

Energy efficiency and power density of tested cathodoluminescent UV-sources are higher than of UV-LEDs of the same range. Thus, the experimental sources had shown their principle suitability for photocatalytic purification and disinfection of air.

It is possible to further increase the power of a single lamp, for example, by using several cathodes [9]. The design of special UV-emitting CRT-phosphors optimized for electron beam excitation will increase the power and efficiency by at least 2–3 times.

REFERENCES

[1] Oppenländer T. Photochemical Purification of Water and Air. WILEY, 2003.

[2] Foster, H., et al. "Photocatalytic disinfection using titanium dioxide: spectrum and mechanism of antimicrobial activity." Applied microbiology and biotechnology 90.6 (2011).

[3] N. Vereschagina, M. Danilkin, M. Kazaryan, D. Ozol, E. Sheshin and D. Spassky "Cathodoluminescent UV-radiation sources", Proc. SPIE 10614, International Conference on Atomic and Molecular Pulsed Lasers XIII, 106141F (16 April 2018)

[4] D. Ozol, E. Sheshin, M. Danilkin and N. Vereschagina "Cathodoluminescent UV Sources for Biomedical Applications" In: Tiginyanu I., Sontea V., Railean S. (eds) 4th International Conference on Nanotechnologies and Biomedical Engineering. ICNBME 2019. IFMBE Proceedings, vol 77. Springer, Cham..

[5] Minamata Convention on Mercury at http://www.mercuryconvention.org

[6] H. Amano, et al. "The 2020 UV emitter roadmap." Journal of Physics D: Applied Physics 53.50 (2020): 503001

[7] M. Broxtermann, et al., "Cathodoluminescence and Photoluminescence of YPO4: Pr3+, Y2SiO5: Pr3+, YBO3: Pr3+, and YPO4: Bi3+," ECS J. Solid State Sci. Technol., vol. 6, pp. R47-R52, April 2017.

[8] A. Fujishima, X. Zhang, and D. A. Tryk, "TiO2 photocatalysis and related surface phenomena," Surf. Sci. Rep., vol. 63, pp. 515-582, 2008.

[9] E. P. Sheshin, V. S. Melekescev, A. Y. Taikin and D. I. Ozol, "Multicathode Field Emission Configurations and their Optimization," 2020 33rd International Vacuum Nanoelectronics Conference (IVNC), 2020

Concept of a Secondary Emission Converter of the Energy of Fast Electrons and γ-Quanta

On the Basis of Carbon Materials (e.g. Graphene)

Dmitry I. Ozol

Moscow Institute of Physics and Technology, Dolgoprudny, Moscow Region, Russian Federation
e-mail: ozol.mipt@gmail.com

Abstract— **Concept of vacuum secondary emission converter of energy of fast electrons and gamma-quanta is proposed. Using thin layers of conductive carbon materials allows theoretically achieve efficiencies of up to 40-50%.**

I. INTRODUCTION

Some kinds of direct (non-thermal) converters of energy of high-energy particles (e.g. β- and γ-particles) by using secondary emission were suggested before [1],[2],[3]. The principle of operation of the most well-thought-out design [2] is based on the alternation of electrodes with different secondary emission coefficients, connected in series, thus, the generator has a low operating voltage of the order of units of volts. Unfortunately, its efficiency η cannot exceed 4% under the most optimistic assumptions – despite the fact that some important factors were not taken into account. Here a new variant is proposed. A key feature of the new secondary-emission converter is the use of very thin electrodes.

II. NEW CONCEPT OF SECONDARY EMISSION CONVERTER

Only a small portion of secondary electrons leaves the substance. Only the electrons generated in a thin (~ 10 nm) near surface layer (moreover, by no means all of them) can escape. A scheme can be imagined where the primary particles - β-particles or γ-quanta – pass through a thin solid film. When the film is sufficiently thin, a considerable part of secondary electrons can escape from such film secondary emission cathode, and a multilayer stack of these film electrodes can serve as an energy converter (Fig.1). The total thickness of the electrodes should be of the order of the path length of the primary particle in this substance (of the order of 10-100 μm for β-particles dependent on the energy and 1000 μm for γ-quanta). Unlike thermal emission converters and β-elements, multilayer secondary emission converter should possess a high operation voltage that depends on the penetration ability of primary particles. The total voltage can amount to hundreds or thousands volts. Another type of commutation of the layers can be suggested that does not demand the equity of secondary emission currents through all interelectrode gaps.

For maximal efficiency it is necessary that the thickness of each layer does not exceed the path length of the secondary electron; i.e. its value is of the order of 10—100 nm dependent on the material; the least values are for metals, and they are greater for semiconductors and dielectrics.

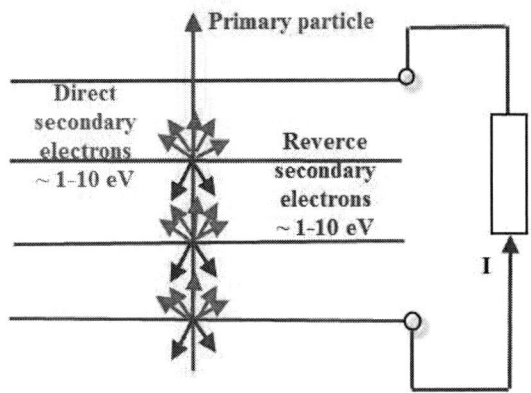

Fig. 1. Scheme of the of the proposed secondary-emission converter.

Manufacturing of sufficiently thin foils and robust multilayer arrangements of them is a very complicated task. Dielectric or semiconductor electrodes can possess the thickness by an order of magnitude greater, however, their manufacturing and durability is also in question. Theoretically, carbon foils, and, especially, graphene, can greatly simplify the problem. First, the path length of secondary and primary electrons in them is considerably greater (because of relatively low Z). Second, it is suggested that even a monoatomic graphene layer can exhibit high mechanical durability. In the context of the secondary emission converter the monoatomic layer is not necessary, even 10-100 layers is enough.

III. EVALUATION OF THE EFFICIENCY

In the first approximation, the total efficiency η is described:
$$\eta = \eta_{em} \cdot \eta_{sp} \cdot \eta_A \cdot \eta_g$$
η_{em} is determined by the ratio of the secondary emission coefficients per shot and per reflection (the difference between the number of forward and reverse secondary electrons - Fig.1), η_{sp} – by the energy spectrum of secondary electrons, η_A – by the work function, η_g –by the spatial distribution of the velocities of secondary electrons. For a preliminary estimate, we assume $\eta_A = 1$ – in particular, because it should be expected that for ultrathin films (in the limit of monatomic layers) work function will be very small. Also we neglect the

978-1-6654-2590-2/21 $31.00 © 2021 IEEE

last factor by putting $\eta_g = 1$ and roughly estimate the first two in a one-dimensional approximation.

η_{em} will depend on the coefficients of secondary emission – ufortunately, for the through secondary emission these value are not exactly registered (especially for very thin layers and for graphene). The two main channels of secondary electron formation are impact ionization and plasmon decay. Their comparative contributions are shown in Fig.2 [4], excitation from the jellium can be identified with impact ionization. It can be assumed that during impact ionization, the resulting secondary electrons fly in the direction of the primary particle. The electrons generated by the decay of plasmons can be considered isotropic, half of them form reverse electrons, half - direct. For thin films, this is the ratio of direct and reverse ~ 6:1, $\eta_{em} \sim 0.85$.

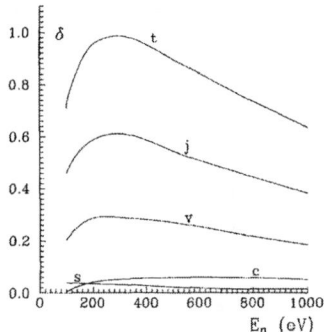

Fig. 2. Total SE yield (t) as a function of primary electron energy and partial yields, corresponding to a different origin of the outgoing electron: (j): individual excitation from the jellium; (v): volume plasmon decay; (s): surface plasmon decay; (c): core level excitation [4].

There are very few works on the through pass secondary emission [5],[6]. In the first approximation one can suggest that the spectra are similar with the some known secondary emission spectra – for example, for metal [7] (Fig.3).

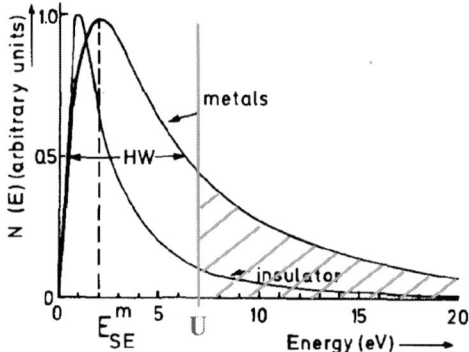

Fig.3. Typical spectra of secondary ion emission from metal and insulator [7]. U is the potential difference between two consecutive film electrodes (acting as the cathode and anode).

Rough estimations can be made in the following way. The secondary electron should do work against electric field, i.e, each electron that reaches the anode should do a useful work **A=q*U** independent on its energy. Electrons of energy E<U will not reach the anode and return to the cathode; therefore, they will not do the useful work. The area shaded with orange color is associated with losses due to anode heating, since the

velocity of these electrons when they arrive to the anode is not equal to zero, and the rest of energy will be lost at their deceleration in the anode material. The useful work is $\int_{U}^{\infty} n(E) \cdot U dE$. The coefficient η_s is defined$_s$ as

$$\eta_{sp} = \frac{\int_{U}^{\infty} n(E) \cdot U dE}{\int_{0}^{\infty} n(E) \cdot E dE}$$ For spectral distribution (Fig.2) the

maximal efficiency for the metal will amount to ~ 50% for $U \sim 2* E_{SE}^{m}$. However, theoretically one can imagine a search for or creation of a substance with the specified as much as possible narrow spectrum of secondary electrons. The less is the scattering over energy, the higher is the upper efficiency limit; in extreme case it amounts to 100% for a monochromatic beam.

All these considerations apply in the case where the layers from which the secondary emission occure, are thick enough so that on the one hand, manages "to form" the spectrum of secondary electrons and on the other hand secondary electrons emitted from the layer number N is completely absorbed in the layer N+1, that is, their energy is not enough to pass through it and perform useful work in the next vacuum gap. However, in the case of very thin layers of the substance with a small atomic number – e.g. graphene – electrons can spend a very small proportion of their own energy to overcome the layer, and in this case, the efficiency increases significantly, because most of the fast electrons from the "orange" region do not spend their excess energy into heating of the anode, and "carry" it in the next interelectrode gap, where do useful work. This allows us to hope for still larger values of efficiency.

IV. CONCLUSIONS

It follows that principally we can expect the efficiency of direct conversion up to 50% or even more. Such a secondary emission converter can be appropriate primarily for conversion of γ-radiation (for hard rays preferably) and then for fast β-radiation. Converters of this scheme are interesting for radioisotopic sources using beta-active isotopes and for fusion reactors where a substantial part of energy losses from hot plasma occur due to the bremsstrahlung X-ray radiation.

REFERENCES

[1] E.Schwarz, "Secondary emission type of nuclear battery," US Patent 2,858,459 A, 1954.

[2] J.Ritter, "Radioisotope photoelectric generator," US Patent 4,178,524.

[3] G.H.Miley, Direct conversion of nuclear radiation energy, University of Illinois, Urbana-Champaign, IL (United States), 1970

[4] Devooght J. et al. Theoretical description of secondary electron emission induced by electron or ion beams impinging on solids // Particle induced electron emission I – 1991. – p. 67-128

[5] H.Bruining, Physics and Applications of Secondary Electron Emission, PERGAMON PRESS, 1962.

[6] U.A.Arifov (ed.), Secondary Emission and Structural Properties of Solids, Consultants Bureau, New York, 1971.

[7] H.Seiler, "Secondary electron emission in the scanning electron microscope," Journal of Applied Physics. 1983 Nov 1;54(11), pp.R1-R18.

Towards a MEMS transmission point X-ray source

Tomasz Grzebyk*, Krzysztof Turczyk, Anna Górecka-Drzazga, Jan A. Dziuban

Faculty of Microsystem Electronics and Photonics, Wroclaw University of Science and Technology,
11/17 Janiszewski St., 50-372 Wrocław, Poland
Tel.: +48 71 320 49 77, tomasz.grzebyk@pwr.edu.pl

Abstract—**The paper presents the concept of a miniature X-ray source fabricated entirely in the MEMS technology. Our construction integrates all the components, including high vacuum micropump, on a single silicon-glass chip. The source generates a conical X-ray beam by using a properly formed transmitting target. In the presented research most attention was paid to determining the optimal target parameters as well as exposure conditions (electron beam energy, exposure time, observation distance and thickness of the target material).**

Keywords- MEMS, X-ray source, field emission elctron source, X-ray target

I. INTRODUCTION

There is a strong research trend towards miniaturization of X-ray sources. Thermal cathodes are being replaced by cold cathodes, power consumption is decreasing, and entire devices are becoming smaller and lighter [1, 2]. Miniature X-ray sources can find applications in X-ray imaging, in spectroscopy or can serve as ionization sources [3, 4]. Miniature X-ray sources already exist, they contain highly developed field emission cathodes made of microstructurized layers or nanomaterials, but the vacuum housing is still made by conventional techniques, often more than 30 years old. Our research team is trying to develop a miniature X-ray source that will be made only in MEMS (Micro-Electro-Mechanical System) technology and which will integrate all its elements on a single multilayer silicon-glass chip.

II. DEVICE CONTRUCTION

The X-ray source, according to the authors' concept, consists of four electrodes: field emission carbon nanotube cathode [5], extracting gate, focusing electrode and a specially designed target (Fig. 1a). The electrodes of the micropump, electron gun and electron-optical microcolumn are formed in anisotropically etched silicon wafers. They are isolated by spacers with via-holes formed in isotropically etched borosilicate glass substrates. All silicon and glass layers are connected together by the anodic bonding process to form a hermetically vacuum tight housing. No classical mechanical processing is applied. A serious problem related to ensuring stable, high vacuum conditions inside the structure was solved by integrating it with the MEMS micropump developed by our team [6].

The source works in a transmission mode. A high-energy electron beam hits the target and excites the radiation. X-rays are generated in all directions, but the shape of the target (a thin, small square membrane) allows them to leave the structure only through a small spot (Fig. 1b). If the X-ray beam is formed at one point, it is conical and the sample placed just above the target is magnified – the source works like a miniature X-ray microscope.

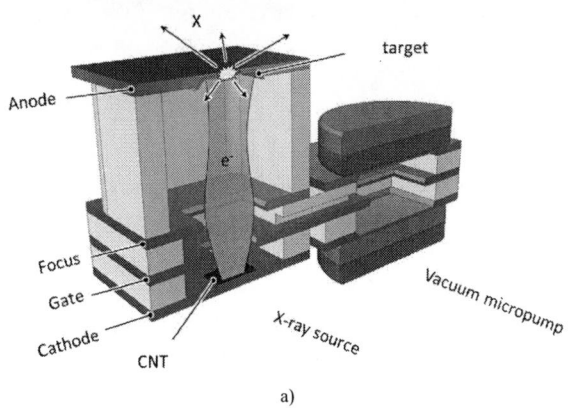

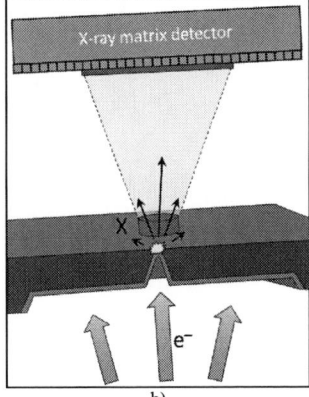

Figure 1. Scheme of the MEMS X-ray source (a) and zoom on the target (b).

III. EXPERIMENT

In our research, we focused on developing of an appropriate target, and examining X-ray images for various parameters of the electron beam.

The first task was to determine the optimal thickness of the target. On the one hand, it must be thick enough to convert as much energy into radiation as possible, but on the other hand, it cannot absorb the generated rays. Moreover, it must ensure high mechanical strength and protection against the penetration of air. In this research the target was made of 400 μm thick, (100) oriented and low resistivity silicon wafer with a square membrane 15 to 90 μm thick and $10 \times 10 \ \mu m^2$ large.

978-1-6654-2590-2/21 $31.00 © 2021 IEEE

Other important parameters are the electron beam energy, exposure time, and distance from target to detector.

Most of the research was done inside a commercial scanning electron microscope (Jeol JSM IT-100), to use a well-defined electron beam (emission current ~1.5 nA, tunable energy). The first X-ray images were obtained on X-ray films (Forma Dentix; Fig. 2). The parameter that was derived in all measured cases was the contrast (contrast = difference in brightness between the exposed and unexposed parts expressed in gray scale, Fig. 3).

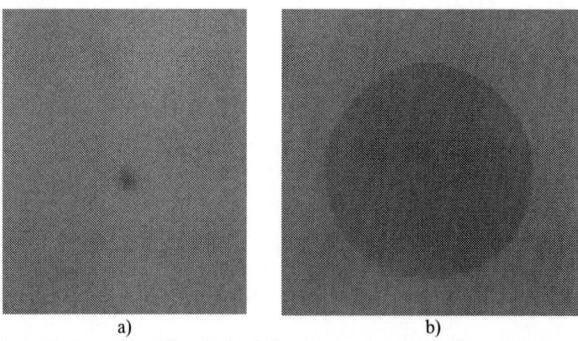

a)   b)

Figure 2. Examples of the obtained X-ray images: a) with a X-ray film placed directly above the target, b) with a film placed d = 7 mm from the target; E = 9 keV.

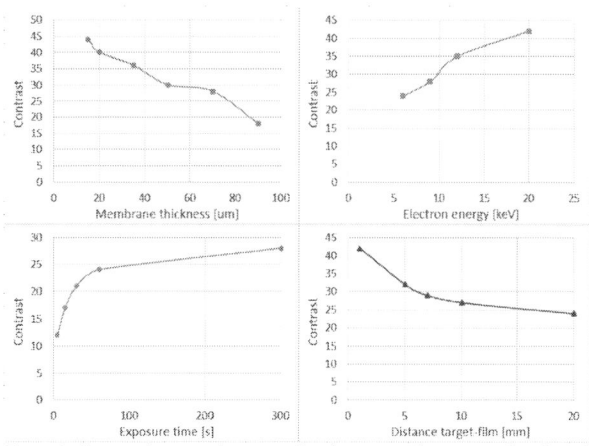

Figure 3. Contrast between the exposed and unexposed areas as a function of: membrane thickness, electron beam energy, exposure time and distance between the target and the X-ray film.

It is clearly visible that as the thickness of the target membrane increases, the image contrast decreases (from 45 at 15 μm to 20 at 90 μm). So, if the target is thinned in the central part to 10–15 μm, most of X-rays will be transmitted. If the rest of the target is thicker than ~120 μm, it will absorb the radiation – therefore even if the electron beam is not perfectly focused – a point X-ray beam can be obtained.

What is not surprising – the higher beam energy resulted in higher image contrast, but even 5 keV was enough to record clear images. Contrast increases significantly for exposure times up to 60 s, longer exposure did not cause visible changes. The exposure time can be significantly reduced if the beam current is higher (the relation should be proportional). The greater the distance (d) between the target with the sample and the X-ray film, the radiation intensity decreases, but even at 20 mm the pattern of the film is visible. And the greater the distance, the greater the magnification of the observed object.

After parametrization of the target and exposure conditions, the first image of a biological object (bee's leg) was obtained on the X-ray film, showing a satisfactory contrast and resolution (Fig. 4).

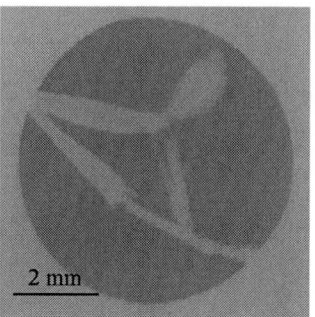

2 mm

Figure 4. Image of the bee leg obtained on a X-ray film; E = 7 keV, d = 12 mm.

IV. CONCLUSSIONS

The miniature X-ray source was made as a multilayer silicon-glass structure. Measurements confirmed that to obtain soft X-rays one can use monocrystalline silicon targets, with a small thin membrane formed in its center. The Si wafer which is at least 120 μm thick does not allow the radiation to pass through, but its thin part which is less than 30 μm - does. A voltage of 5 keV is sufficient to expose the film. Time necessary to irradiated X-ray film varies for different energies, membrane thickness and distance from a target, it could be as short as 1 s, but to get a clear image 1 minute was more proper. When the geometrical and exposure parameters are correctly chosen, clear and magnified images can be obtained.

ACKNOWLEDGMENT

The work was partially financed by the statutory grant of The Wrocław University of Science and Technology.

REFERENCES

[1] A. Górecka-Drzazga, Miniature X-Ray sources, Journal of Microelectromechanical Systems, vol. 26, 1 (2017) 295-302.
[2] J-W. Jeong, J-W. Kim, J-T. Kang, S. Choi, S. Ahn and Y-H. Song, A vacuum-sealed compact X-ray tube based on focused carbon nanotube field-emission electrons, Nanotechnology, vol. 24, no. 8, pp. 085201-8, 2013.
[3] S. Cornaby, et al., "Simultaneous XRD/XRF with low-power X-ray tubes," Proc. Adv. X-Ray Anal., vol. 45. 2002. pp. 34–40.
[4] J. Kawai, High-sensitivity small-size X-ray fluorescence spectrometers, document Guest Forum 12, 52-57, 2008
[5] K. Laszczyk, M. Krysztof, Electron beam source for miniaturized electron microscope on chip, Vacuum, vol. 189, 110236, 2021
[6] T. Grzebyk, A. Górecka-Drzazga, J. A. Dziuban, "High vacuum micropump for miniature nanoelectronics devices," Technical Digest of IVNC 2013, July 8-12, Roanoke, VA, USA, pp. 1-2.

Optimization of Gated ZnO Nanowire Field-Emitter Arrays by Tuning Pixel Density

Songyou Zhang, Xiuqing Cao, Guofu Zhang, Shaozhi Deng, Juncong She, Ningsheng Xu and Jun Chen*

State Key Laboratory of Optoelectronic Materials and Technologies
Guangdong Province Key Laboratory of Display Material and Technology
School of Electronics and Information Technology, Sun Yat-sen University, Guangzhou, People's Republic of China
* E-mail: stscjun@mail.sysu.edu.cn

Abstract— **Gated nanowire field emitter arrays (FEAs) have potential applications in vacuum microelectronic devices. In this study, a coaxis-gated ZnO nanowire FEAs structure with an in-plane focusing gate was optimized by tuning the pixel density using simulation. Gated ZnO nanowire FEAs with optimal parameters was designed and fabricated. Their emission characteristics were studied.**

I. Introduction

Large area gated field emitter arrays (FEAs) have important applications in vacuum microelectronic devices such as addressable flat panel X-ray sources and photodetector [1-2]. ZnO nanowires are ideal candidate material for large area FEAs due to their feasibility of large area uniform controllable preparation. Early studies show that the emission current can be effectively modulated in coaxis-gated ZnO nanowire FEAs and focusing capability has been demonstrated with an in-plane focusing gate electrode [3-5].

In this study, in order to further lower the operation voltage and increase the emission current of the coaxis-gated ZnO nanowire FEAs, we optimize the design by tuning the pixel density while proportionally changing the geometric parameters. The emission current was simulated for FEAs with different pixel densities and the FEAs with optimal size were fabricated and characterised.

II. Device Structure And Simulation

Fig. 1 shows the structure of the coaxis-gated ZnO nanowire FEAs with an in-plane focusing gate electrode [6]. In this structure, the cathode, gate and focusing electrode are arranged from inside to outside on the same top layer plane. The bottom electrodes of the cathode and gate are perpendicularly arranged to achieve addressable pixel. The cathode and the gate at the top layer are connected to the corresponding electrodes at the bottom layer through via holes.

A simulation was carried out using COMSOL multiphysics software based on a 2D model to investigate the field emission current of ZnO nanowires FEAs with different pixel densities. In the simulation, the spacing (D) between adjacent pixels was adjusted, and the corresponding device geometric size of the pixel changed in the same proportion as the pixel spacing. Fig. 2 illustrates the relationship between the simulated current density and spacing (D), where the maximum emission current density could be observed at D = 100 μm. When the D is small, there are more field emitter arrays per unit area. In the meantime, the cathode has a smaller pattern, which means less nanowires around the edge of the pattern are modulated by the gate. Therefore, an optimal D will occur.

Based on the simulation results, FEA device was designed with 200 × 200 arrays of patterned ZnO nanowire field emitters and the size of each pixel is 100 μm× 100 μm. For each pixel, the cathode pattern has a radius of 18 μm and is surrounded by a ring electrode acting as the control gate. The inner and outer diameters of the ring-shaped gate electrode were 24 and 32 μm, respectively.

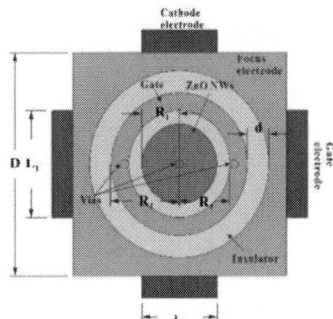

Fig. 1 Top view of one pixel in the coaxis-gated ZnO nanowire FEAs with an in-plane focusing gate electrode

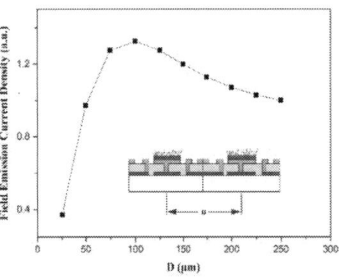

Fig. 2 Simulated field emission current density as a function of pixel to pixel distance(D). Inset shows the schematic of simulation model.

III. DEVICE FABRICATION AND CHARACTERISTICS

A four-step-mask microfabrication process is used to prepare the coaxis-gated ZnO nanowire FEAs which is similar to early report [6]. First, a 120-nm-thick chromium layer was deposited on the glass substrate by magnetron sputtering, and the patterned bottom electrode was obtained by a lift-off process. Then a layer of 1.5-µm-thick silicon oxide was deposited by PECVD as the insulating layer between the bottom electrode and the upper electrode. After this, reactive ion etching was used to etch the via holes to connect the upper and bottom electrodes. Then, the patterns of the upper cathode, gate and focus electrode were obtained through one-step mask, and a 320-nm-thick indium tin oxide (ITO) layer was deposited by magnetron sputtering. Latter, the Zn film was deposited by electron beam evaporation deposition and patterned by lift-off process. Finally, ZnO nanowire arrays were synthesized by thermal oxidation in air at 470℃ and maintained for 3 h.

Fig. 3 shows the SEM images of the fabricated ZnO nanowire FEAs. The population density of ZnO nanowires was estimated to be 8×10^7 cm^{-2}, and the morphologies of nanowires at different pixels were similar, with the length of about 1-2 µm and the tip diameter of about 10-30 nm.

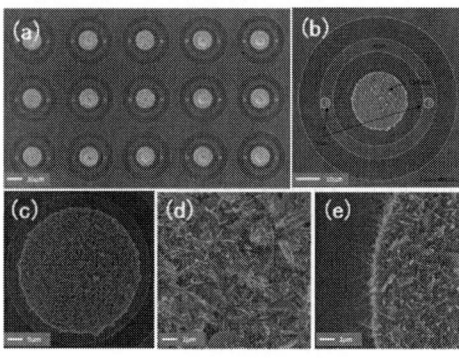

Fig. 3 SEM images of the fabricated ZnO nanowire FEAs. (a) ZnO nanowire FEAs. (b) Single pixel of ZnO nanowire FEAs. (c) The cathode pattern. (d) ZnO nanowires at the inner part of the cathode pattern. (e) ZnO nanowires at the edge of the cathode pattern.

The field emission characteristics were measured in a vacuum chamber under a base pressure of 5×10^{-5} Pa. To verify the addressing performance and gate-control capability, we used green phosphor as the anode and assembled it with the FEAs. The distance between anode and cathode was 0.25 mm. In this way, the image of the emitted electrons could be observed on the anode screen with a digital camera. The anode current under different gate voltages is illustrated in Fig. 4, which is recorded from 200 × 2 pixels. The gate voltage was varied from 0 to 120 V while the anode voltage was kept at 1300 V. When the gate voltage is larger than 100 V, the anode current rises rapidly. The inset in Fig. 4 shows the field emission

images which were recorded under applied gate voltage of 80, 100 and 120 V, respectively. A maximum current density of ~1 mA/cm^2 was obtained.

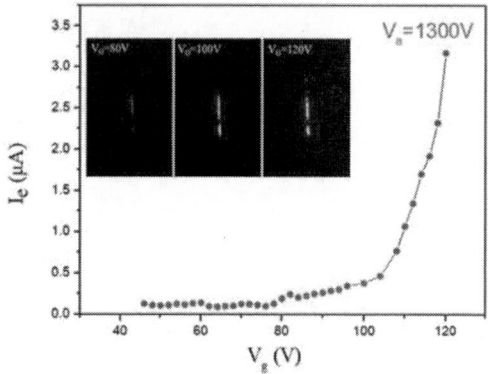

Fig. 4. Plot of anode current versus gate voltage for 200 × 2 pixels. Inset shows the corresponding field emission images.

IV. CONCLUSIONS

Simulation results shows that higher field emission current can be obtained from coaxis-gated ZnO nanowire FEAs by optimizing the distance between adjacent pixels. The ZnO nanowire FEAs with optimal pixel density were fabricated, and good field emission characteristics were obtained.

ACKNOWLEDGMENT

The authors gratefully acknowledge the financial support from the National Key Research and Development Program of China (Grant No.2016YFA0202000), National Natural Science Foundation of China (Grant Nos. 91833303 & 62001527), the Science and Technology Department of Guangdong Province (Grant No.2020B0101020002), Fundamental Research Funds for the Central Universities.

REFERENCES

[1] D. K. Chen, X. Song and Z. P. Zhang, et al., "Transmission type flat-panel X-ray source using ZnO nanowire field emitters," Applied Physics Letters, vol. 107(24), pp. 243105, 2015.

[2] Z. P. Zhang, S. Zheng and K. Wang, et al., "Electron bombardment induced photoconductivity and high gain in a flat panel photodetector based on a ZnS photoconductor and ZnO nanowire field emitters," ACS Photonics, vol. 5(10), pp. 4147-4155, 2018.

[3] L. Zhao, Y. X. Chen and Y.M. Liu, et al., "Integration of ZnO nanowires in gated field emitter arrays for large-area vacuum microelectronics applications," Current Applied Physics, vol. 17(1), pp. 85-91, 2017.

[4] Y. M. Liu, L. Zhao, and Z.P. Zhang, et al., "Fabrication of ZnO nanowire field-emitter arrays with focusing capability," IEEE Transactions on Electron Devices, vol. 65(5), pp. 1982-1987, 2018.

[5] X. Q. Cao, J. Yin, and L. B. Wang, et al., "Fabrication of coaxis-gated ZnO nanowire field-emitter arrays with in-plane focusing gate electrode structure," IEEE Transactions on Electron Devices, vol. 67(2), pp. 677-683, 2019.

Study of Nanoscale Cathodes for Gas Discharge Devices

Sergey M. Karabanov

Ryazan State Radio Engineering University

pvs.solar@gmail.com

Abstract— **This paper presents the effect of the magnesium oxide film on the characteristics of a glow discharge with a metal film cathode based on the Cr-Cu-Cr composition. It has been found that for the mixture of Ne + 3% Xe at the pressure of 133 hPa the minimum value of the cathode fall of the glow discharge is (32.5±2) V. The onset of relaxation oscillations of the discharge current is established. The observed phenomena can be explained by the generation and dynamics of charging the magnesium oxide film.**

I. INTRODUCTION

A glow discharge is an independent electric discharge in gas with a cold cathode at the currents of 10^{-5}-1 A. It is observed at relatively low gas filling pressures between metal electrodes. The distinctive features of the normal glow discharge are the constant cathode fall (U_{cf}) and glow discharge current density (j_d), depending only on the electrode material, composition and the gas filling pressure. The more the work function of the cathode material, the more the normal cathode fall. The area of the discharge adjacent to the cathode is called the cathode dark space, where the main processes determining the discharge properties occur.

The main processes determining the glow discharge properties are the secondary ionic-electron emission and photoemission caused by the discharge plasma radiation [1,2].

In glow discharge devices various metals (W, Mo, Ni, etc.) are used as cathode materials. Inert gases and their mixtures (Ne + 1% Ar, Ne + 0,3% Xe, Ne + 3% Xe) are used as the gas filling [1,2].

To create various gas-discharge devices, film cathodes (a metal base with films deposited on the surface) are also used. This makes it possible to increase the cathode emission capacity [1].

Dielectric MgO, La$_2$O$_3$ films are used in gas-discharge AC displays to increase their efficiency. MgO films have a high coefficient of the secondary ionic-electron emission of 0.57 and a minimum operating voltage of about 90 V. The MgO films are highly stable in the discharge [3].

It has been found that oscillatory processes are observed in a discharge with oxide films on a metal cathode [4].

This paper presents the results of studying the effect of the MgO film deposited on the surface of a film cathode on the parameters of a glow discharge in inert gas.

II. RESULTS AND DISCUSSION

A magnesium oxide film with the thickness of 20-25 nm was applied by the electron-beam evaporation. The film had an island structure. The Cr-Cu-Cr composition was used as the film electrodes of the gas-discharge gap. The layers thicknesses were 50 nm, 500 nm and 150 nm, respectively.

The structure of the metal composition was determined by the copper film structure. In the case under study the grain size in the copper film did not exceed 0.5 μm, and the grain size of the chromium film was less than 10 nm. Fig. 1 shows the discharge gap design.

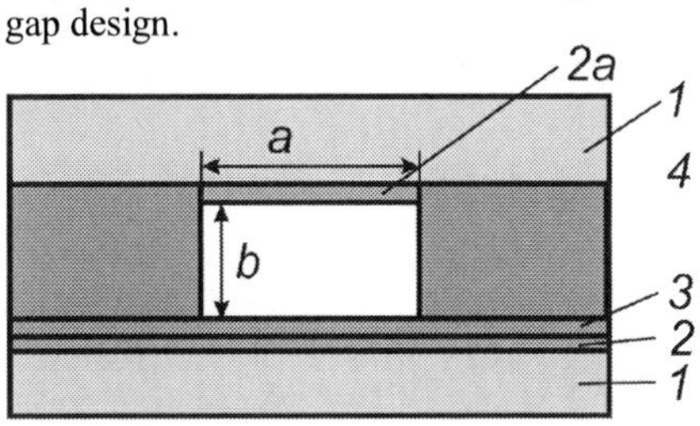

Fig. 1 Discharge gap design: 1 – glass substrate, 2 – film cathode, 2a – anode, 3 – MgO film, 4 – dielectric matrix (a=0.5 mm; d=0.5 mm).

Inert gases and their mixtures (Ne+1% Ar, Ne+0.3% Xe, Ne+3% Xe) were used as gas filling. In all cases the gas mixture pressure was 133 hPa. The cathode fall (U_{cf}) and the discharge current were measured.

Table I shows minimum values of the cathode fall (U_{cf}) for the investigated gas fillings.

TABLE I
MINIMUM U_{cf} (V) FOR CATHODES WITH DIFFERENT GAS FILLINGS

Cathode material	Gas filling			
	Ne	Ne+1% Ar	Ne+0.3% Xe	Ne+3% Xe
Cr-Cu-Cr-MgO	65	54	57	32.5
Cr-Cu-Cr	200	115	140	190

The U_{cf} had minimum values after the treatment of the cathode surface with a discharge. At the initial moment, regardless of the gas filling type, unstable combustion of the discharge was observed, then, in the course of the cathode surface formation, the U_{cf} stabilized and then increased again. Fig. 2 shows the typical dependence of U_{cf} on the discharge processing time.

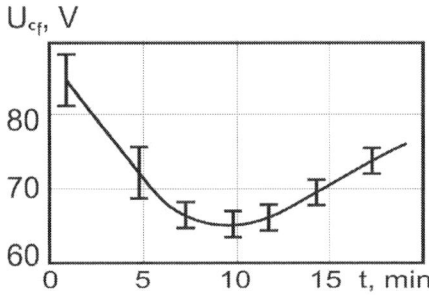

Fig. 2 Experimental dependence of the U_{cf} value on the cathode processing time with the discharge for neon (the discharge current is 1.5 mA).

978-1-6654-2590-2/21 $31.00 © 2021 IEEE

The U_{cf} for the gas filling of Ne+0.3% Xe is 14 V lower than the previously obtained value for a cathode based on barium oxide films [2].

Fig. 3 shows the time dependence of the U_{cf} for the Ne+1% Ar gas filling. Relaxation oscillations of the U_{cf} and discharge current are observed.

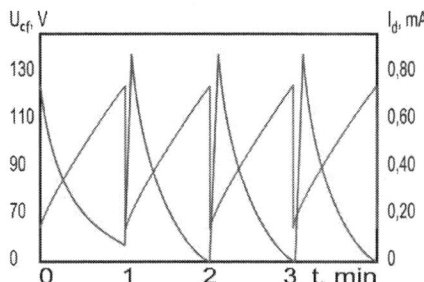

Fig. 3 Experimental dependence of the U_{cf} (blue line) and discharge current (red line) on time (for gas filling of Ne+1%Ar).

The oscillation period of the discharge current is about 1 min. At the initial moment, non-periodic oscillations of the discharge current occur, which then turn into periodic ones. Electron-graphic studies of the cathode surface have shown that the observed relaxation processes are accompanied by change of the topography of the MgO film. From the beginning, the MgO film is continuous, then, as the discharge is applied, an island structure with the grain size from several microns to several tens of microns (up to 50 μm) is formed. As the duration of the discharge treatment increases, the MgO film is sputtered, that is accompanied by the U_{cf} increase (Fig. 2).

The observed phenomena can be explained as follows. After the discharge ignition its emission properties change while the cathode surface forms. At the initial moment, when the MgO film covers the entire cathode surface, the discharge combustion is unstable; then, when the film island structure is formed, the U_{cf} decreases and, in some cases, periodic oscillations of the discharge current appear.

The significant U_{cf} decrease for a cathode covered with a MgO film is connected with the anomalously high secondary emission from the cathode surface caused by film charging [5,6]. The field density value in the MgO film at the moment of discharge combustion can reach $10^6 – 10^7$ V/cm. Charging occurs as a result of ionic-electron and photoemission. The increase of the secondary emission coefficient of a cold cathode for a cathode with the MgO film in comparison with a metal cathode reaches $5 \cdot 10^2 – 10^3$ [5]. The maximum value of the secondary emission coefficient is achieved for the Ne+3Xe gas filling, where the minimum value of U_{cf} is observed. This gas filling is characterized by the high current density [7] and increase of the ultraviolet radiation fraction in the discharge emission spectrum [8].

The observed discharge relaxation oscillations are probably connected with periodic changes of the value of the accumulated charge on the MgO film. This is caused by the generation of the autoelectronic emission from the metal cathode surface, which is caused by a strong electric field formed by the accumulated charge on the MgO film surface. The dynamics of charging the dielectric films is determined by the charge relaxation time, which is 10^2 s. for MgO films.

Similar charging time values are given in [6]. The experimentally observed oscillation period of 1 min. coincides with charge relaxation time.

For the detailed explanation of the observed phenomena, it is necessary to study the emission at continuous ion bombardment and discharge combustion under the given conditions.

III. Conclusions

Thus, in this work, the minimum value of U_{cf} of the glow discharge on a metal film cathode covered with the MgO film is obtained. It is 32.5 ± 2 V for the Ne+3%Xe gas filling. The generation of relaxation oscillations of the discharge current has been established.

The observed phenomena can be explained by the formation of an accumulated charge on the MgO films surface, which may result in the autoelectronic emission from the metal cathode surface.

References

[1] E.V. Zykova, E.T. Kucharenko, V/L/ Aivazov, 'Study of a glow discharge with cold cathodes coated with dielectric films', Radio Engineering and Electronics, 1979, no. 7. p. 1464.

[2] T. Takeishi, "Normal cathode fall of glow discharge", J. Phys. Soc. Japan, 1956, no. 11, p. 676.

[3] Display devices, ed. by J. Pankove, Springer, 1980.

[4] A.R. Shulman, S.A. Fridrikhov, Secondary emission methods for solid state research, Moscow, Science, 1979.

[5] I.N. Bronstein, B.S. Freiman, "Secondary electron emission, Moscow, Science, 1969.

[6] L.N. Dobretsov, M.V. Gomoyunova, Emission electronics, Moscow, Science, 1966.

[7] F.M. Yablonsky, Gas discharge devices for displaying information, Moscow, Energy, 1979.

[8] Gas discharge panels for color TV display, NHK Technical Monograph., 1979.

UV lighting with carbon nanotube based cold cathode electron beam (C-beam) and its characteristics

Sung Tae Yoo[1], and Kyu Chang Park[1,*]

[1]*Department of Information Display, Kyung Hee University, Dongdaemun-gu, Seoul, 02447, Korea*
Contact: kyupark@khu.ac.kr, phone +82-02-961-9447

Abstract— **Here, we investigated the generation of ultraviolet (UV) light in various wavelength ranges with carbon nanotube (CNT) based cold cathode electron beam (C-beam) irradiation technology. The wide band gap anode materials were excited by electrons emitted by C-beam, confirming the generation from UVA to UVC. The properties of UV light are affected by the anode materials and the electron emission properties. UV light source using C-beam irradiation has the advantage of flat panel UV light source and can be applied to various fields.**

I. INTRODUCTION

Currently, the world is threatened by the viral infection caused by covid-19. Ultraviolet (UV) is a tool that sterilizes viruses and bacteria, and is widely used for various purposes in various fields, as well as preventing the spread of infectious diseases. UV light is divided into UVA (400 ~ 320 nm), UVB (320 ~ 280 nm), UVC (280 ~ 200 nm), VUV (200 ~ 120 nm), and EUV (120 ~ 10 nm). UVA, which has a relatively low photon energy, causes chemical reactions used for UV curing and organic decomposition. UVB can be used for phototherapy, vitamin D synthesis, and plant lighting. UVC is used in many applications such as disinfection, deodorization, detection and optical cleaning [1]-[2]. EUV light lead the latest semiconductor advances and expand photolithography technology to a few nano-meter node areas. Research on UV light sources is becoming more and more important.

UV generation technology using electron beam (e-beam) has been studied since the early 2000s. In 2009, K. Watanabe et al. irradiated e-beam on hexagonal boron nitride to obtain UVC light with a wavelength of 225 nm [3]. Since then, research on UV generation by e-beam irradiation has been conducted using AlGaN multi-quantum wells (MQWs) and various wide bandgap materials as an anode. However, most of these studies are on the generation of UV light based on thermal electron sources [2].

Cold cathode-based e-beams are being explored for next vacuum nano-electronic applications such as X-ray [4], microscope [5], and UV lighting [6]-[10]. Thermal electron sources including tungsten filament need to be heated to around 1000 degrees to emit electrons, but the cold cathode-based e-beam does not require heating, so heating time is not required, and it has the advantage that it can be driven immediately. Among these cold cathode-based e-beams, carbon nanotube (CNT) emitters are the most promising electron source candidates. CNT is a very good material electrically,

chemically, and mechanically, and it is suitable for use as an electron emission source when making vertically aligned CNT (VACNT) by direct growth using plasma enhanced chemical vapor deposition (PECVD). Since VACNTs are fabricated using traditional photolithography and PECVD, electron emission sources can be fabricated in a desired location and in a desired shape. Based on these advantages, the electron emission area can be adjusted according to the required UV emission area, and it is also advantageous for large-area UV source development.

We synthesized VACNTs and made carbon nanotube based cold cathode electron beam (C-beam), and conducted research on UV generation by irradiating electrons emitted from the C-beam on various anode materials.

II. EXPERIMENT

A. Fabrication of VACNTs and C-beam

VACNTs were grown via PECVD after Ni deposited on a Si wafer was patterned using photolithography at desired locations. The inset scanning electron microscope image in Fig. 1 shows the grown VACNTs. VACNTs were well grown vertically with a dot size of 3 μm and a height of about 40 μm. VACNTs were used as a cold cathode, and a metal mesh was used as a gate to control electron emission from the VACNTs. C-beam is made by insulating the gate and cathode with ceramic for modularity.

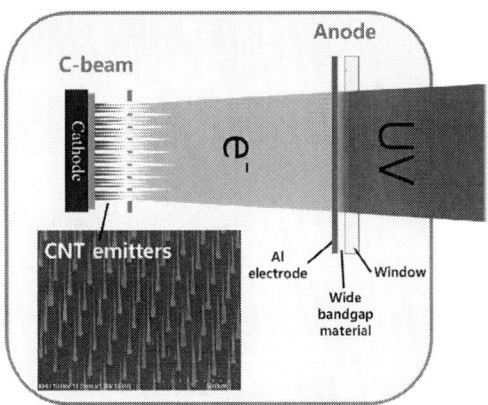

Fig. 1 Schematic of UV lighting with carbon nanotube-based cold cathode electron beam (C-beam) irradiation.

B. UV lighting system

As shown in Fig. 1, the electrons emitted from the C-beam are accelerated by the voltage applied to the Al electrode and collide with the wide bandgap material to generate UV light through excitation and relaxation. The inside of the UV lighting system is in a 10^{-7} Torr vacuum. When electrons are emitted and accelerated, a positive voltage is injected into the anode and gate.

III. RESULTS AND DISCUSSION

Using electrons emitted from the C-beam, UV characteristics change depending on the anode materials and the anode current and voltage. We irradiated electrons to anode materials such as SrB_4O_7:Eu phosphor, Zn_2SiO_4, Sapphire, and AlGaN MQWs, and it was confirmed that the wavelength from UVA to UVC was changed with the anode materials [6]-[9]. According to the C-beam irradiation conditions, changes in UV intensity and properties were confirmed, and it was proven that *E. coli* was sterilized with UV light with a peak wavelength of 230 nm [10]. Details regarding properties of UV light and C-beam irradiation conditions will be announced.

IV. CONCLUSIONS

In summary, we made UV lighting using a C-beam that uses VACNTs as field emitters. By irradiating C-beam to wide band gap anodes, UV light was successfully generated from UVA to UVC, and their characteristics were confirmed. UV characteristics are changed depending on the irradiation conditions of the C-beam and the anode materials, and the sterilization effect was confirmed with the generated UVC light.

ACKNOWLEDGMENT

This work was supported by the Technology Innovation Program (20013595, Extreme ultraviolet light source using nano electron beam) funded By the Ministry of Trade, Industry & Energy (MOTIE, Korea) and the BK21 FOUR Program funded by the Ministry of Education (MOE, Korea) and National Research Foundation of Korea (NRF) (21A20130000018).

REFERENCES

[1] M. Kneissl, T.-Y. Seong, J. Han, and H. Amano, "The emergence and prospects of deep-ultraviolet light-emitting diode technologies," Nat. Photonics, vol. 13, no. 4, pp. 233–244, Mar. 2019.

[2] D. Li, K. Jiang, X. Sun, and C. Guo, "AlGaN photonics: recent advances in materials and ultraviolet devices," Adv. Opt. Photon., vol. 10, no. 1, p. 43, Jan. 2018.

[3] K. Watanabe, T. Taniguchi, T. Niiyama, K. Miya, and M. Taniguchi, "Far-ultraviolet plane-emission handheld device based on hexagonal boron nitride," Nat. Photonics, vol. 3, no. 10, pp. 591–594, Sep. 2009.

[4] X. Qian, A. Tucker, E. Gidcumb, J. Shan, G. Yang, X. Calderon-Colon, S. Sultana, J. Lu, O. Zhou, D. Spronk, F. Sprenger, Y. Zhang, D. Kennedy, T. Farbizio, and Z. Jing, "High resolution stationary digital breast tomosynthesis using distributed carbon nanotube x-ray source array," J. Med. Phys., vol. 39, no. 4, pp. 2090–2099, Mar. 2012.

[5] H. R. Lee, D. W. Kim, O. J. Hwang, B. Cho, and K. C. Park, "Scanning electron imaging with vertically aligned carbon nanotube (CNT) based cold cathode electron beam (C-beam)," Vacuum, vol. 182, p. 109696, Dec. 2020.

[6] S. T. Yoo, H. I. Lee, and K. C. Park, "363 nm UVA light generation with carbon nanotube electron emitters," Microelectronic Eng., vol. 218, p. 111142, Oct. 2019.

[7] S. T. Yoo, H. I. Lee, and K. C. Park, "Optimization of Zn_2SiO_4 Anode Structure for Deep Ultraviolet Generation With Carbon Nanotube Emitters," IEEE J. Electron Devices Soc., vol. 7, pp. 735–739, 2019.

[8] S. T. Yoo and K. C. Park, "Sapphire Wafer for 226 nm Far UVC Generation with Carbon Nanotube-Based Cold Cathode Electron Beam (C-Beam) Irradiation," ACS Omega, vol. 5, no. 25, pp. 15601–15605, Jun. 2020.

[9] S. T. Yoo, B. So, H. I. Lee, O. Nam, and K. C. Park, "Large area deep ultraviolet light of $Al_{0.47}Ga_{0.53}N/Al_{0.56}Ga_{0.44}N$ multi quantum well with carbon nanotube electron beam pumping," AIP Adv., vol. 9, no. 7, p. 075104, Jul. 2019.

[10] S. T. Yoo, J. Y. Lee, A. Rodiansyah, T. Y. Yune, and K. C. Park, "Far UVC light for E. coli disinfection generated by carbon nanotube cold cathode and sapphire anode," Curr. Appl. Phys., vol. 28, pp. 93–97, Aug. 2021.

978-1-6654-2590-2/21 $31.00 © 2021 IEEE

Microscope equipped with graphene-oxide-semiconductor electron source

Yukino Kameda [1,2+], Katsuhisa Murakami [2], Masayoshi Nagao [2], Hidenori Mimura[1] and Yoichiro Neo [1]

[1] Research Institute of Electronics Shizuoka University,
Johoku 3-5-1, Naka-ku, Hamamatsu, Shizuoka 432-8011, Japan
[2] National Institute of Advanced Industrial Science and Technology,
AIST Tsukuba Central2, Umezono 1-1-1, Tsukuba, Ibaraki 305-8568, Japan
[+]Contact: kameda.yukino.16@shizuoka.ac.jp

Abstract— **We propose the scanning electron microscope (SEM) using the graphene-oxide-semiconductor type planar electron emission device as an electron source. Graphene/Oxide/Semiconductor (GOS) structure realizes very high electron emission efficiency of over 30 % and it can be operated in low vacuum condition and low applied voltage. Therefore, SEM using the GOS electron source can eliminate several condenser lenses from the electron optics which are needed in the conventional SEM due to the very small divergence angle of the electron beams emitted from the flat surface. In this study, we installed the GOS type electron source into the conventional SEM and successfully obtained the stable SEM image with a probe current of 20 pA without noise using GOS electron source. In addition, we found the optimal electron optics for the parallel electron beams with a very small divergence angle of approximately 0 degree by the simulation of the electron beam trajectory.**

Keywords— *Electron source, graphene, metal/oxide/semiconductor structure*

I. Introduction

A planar type electron emission device based on a Metal/Oxide/Semiconductor (MOS) structure can be operated in low vacuum conditions and low applied voltage. However, its very low electron emission efficiency below 1 % is the critical issue for its practical applications. We demonstrated very high electron emission efficiency of over 30 % using a Graphene/Oxide/Semiconductor (GOS) structure since the graphene gate electrode can suppress the electron inelastic scattering within the topmost gate electrode of the planar type electron emission device. This result will open the planar type electron emission devices up to several applications such as an electron microscope, an ion neutralizer for space propulsion, a flat panel X-ray source.

In the case of the application of the electron guns for the scanning electron microscope (SEM), the electron optics and vacuum system of SEM become very simple if a GOS electron source is used as the electron gun. The GOS electron source can eliminate several condenser lenses from the electron optics which are needed in the conventional SEM due to the very small divergence angle of the electron beams emitted from the flat surface. In addition, there are no need for a turbo molecular pump, which is also needed in the conventional SEM, since the GOS electron source can operate under a low vacuum condition of around 10 Pa.

In this study, we installed the GOS type electron source into the conventional SEM and demonstrated the observation of the SEM image. We further calculated the trajectory of the electron beams emitted from the GOS type electron source.

II. Experimental

The fabrication processes of the GOS devices are as follows. The starting substrate is the n-Si substrate with a 300 nm-thickness of SiO_2 layer. At first, electron emission area with a diameter of 200 μm was formed by a conventional photolithography and BHF wet etching. Then, the thin SiO_2 layer with a thickness of around 10 nm was grown by thermal oxidation on the emission area after RCA cleaning. Next, graphene was directly synthesized on the SiO_2 surface by plasma-enhanced chemical vapor deposition (CVD) using Ar and CH_4. Then, the graphene electrode was patterned by photolithography and oxygen plasma ashing. Finally, Ti and Ni was deposited on the graphene electrode of the outside of the electron emission area by electron beam evaporation as the contact electrode. These are thoroughly described in ref [1], [2] and [3].

We installed the GOS type electron source into the conventional W-filament type SEM without condenser lenses. Figure1 shows a schematic of the experimental setup. The sample stage connecting the picoammeter (R8240, Advantest) for a stage current measurement is grounded. The GOS electron

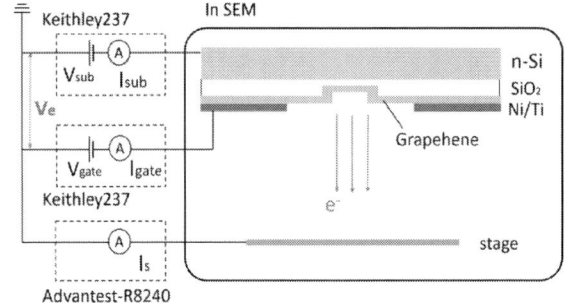

Figure1. Circuit diagram of operating SEM installed the GOS electron source.

978-1-6654-2590-2/21 $31.00 © 2021 IEEE

source was operated at the floating bias of -1kV. The bias of the substrate of the GOS electron source was fixed at -1kV. The operation bias of the GOS electron source was controlled to 10 V by the adjusting the gate bias voltage (i.e, the actual gate bias is -990V). The emission current was calculated by the differential of the gate current (I_{gate}) and substrate current (I_{sub}) measured by the source measure units (237, Keithley). We used Cu wire mesh as an observation sample.

We further calculated the trajectory of electrons emitted from the GOS electron source including the electron optics of the SEM under the experimental conditions by SIMION 8.2.0.5 (Adaptas Solutions, LLC).

III. RESULT AND DISCUSSIONS

Figure 2 shows the SEM image of Cu wire mesh taken by the GOS electron source with a probe current of 20 pA. The stable SEM images without noise were obtained using GOS electron source, which indicate that emission current of the GOS electron source is very stable.

Figure2. SEM image using the GOS electron source

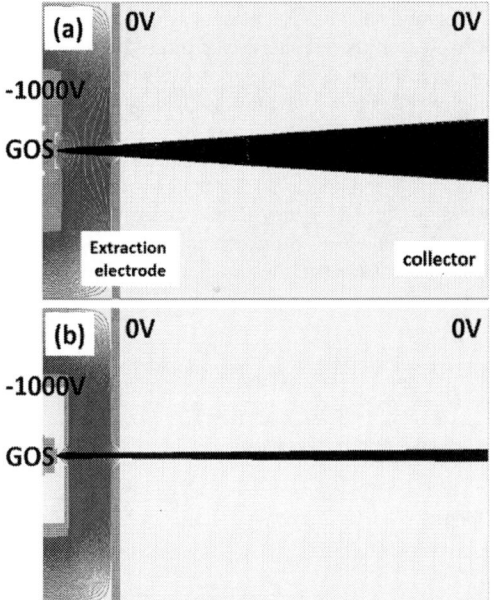

Figure3. calculation result of electrons beam emitted from the device. (a)The potential of Wehnelt electrode is same as GOS electron source (i.e., -1kV). (b) Position of Wehnelt electrode is away from the GOS device.

Figure3 (a) and (b) show the calculated trajectory of electron beams emitted from the GOS electron source. The black lines and red lines indicate the trajectory of electrons and the equipotential lines, respectively. In the case that the Wehnelt electrode is same potential as the substrate of the GOS electron source, electron beams largely spread with its divergence angle of 7.4 degree, as shown in Fig.3 (a). While the Wehnelt electrode was positioned away from the GOS device, the spread of the electron beam was suppressed due to the very small divergence angle of approximately 0 degree, as shown in Fig.3(b). This result will lead to increase the probe current and improve the resolution of the SEM image by the optimal electron optics for the GOS electron sources.

IV. CONCLUSIONS

We can successfully obtain the stable SEM image with a probe current of 20 pA without noise using GOS electron source. In addition, the divergence angle of the electron beams of the GOS electron sources was found to be very small by the optimal placement of the Wehnelt electrode in the electron trajectory simulations. The parallel electron beams with very small divergence angle emitted from the GOS electron sources would improve the resolution of the SEM. We have a future plan to fabricate the GOS electron source with the focusing electrode to further decreasing the size of the electron beams.

ACKNOWLEDGMENT

This work was supported by JSPS KAKENHI Grant Numbers JP18H01505, JP18K18910, 19K04516, and JP21H01401.

REFERENCES

[1] K. Murakami, J, Miyaji, R. Furuya, M. Adachi, M. Nagao, Y. Neo, Y. Takao, Y. Yamada, M. Sasaki, and H. Mimura, "High-performance planar type electron source based on a graphene-oxide-semiconductor structure", Appl. Thys. Lett. 114, 213501 (2019).

[2] K. Murakami, M. Adachi, J. Miyaji, R.Furuya, M. Nagao, Y. Yamada, Y. Neo, Y. Takao, M. Sasaki, and H. Mimura, "Mechanism of Highly Efficient Electron Emission from a Graphene/Oxide/Semiconductor Structure", ACS Appl. Electron. Mater 2020, 2, 2265-2273.

[3] K. Murakami, S. Tanaka, T. Iijima, M. Nagao, Y. Nemoto, M. Takeguchi, Y. Yamada, and M. Sasaki, "Electron emission properties of graphene-oxide-semiconductor planar-type electron emission devices", J. Vac. Sci. Technol. B 36(2), Mar/Apr 2018, 02C110

[4] H. Mimura, Y. Neo, H. Shimawaki, Y. Abe, K.Tahara, and K.Yokoo, "Improvement of emission current from a cesiated matal-oxide-semiconductor cathode", Appl. Phys. Lett.88, 123514(2006).

[5] K. Yokoo, G. Koshita, S. Hanazawa, Y. Abe, and Y. Neo, "Experiments of highly emissive metal-oxide-semiconductor electron tunneling cathode", J. Vac. Sci. Technol. B 14, 2096(1996).

Nanosphere Lithography to Enhance the Field Emission Properties of a Self Aligned Nanocarbon Based Field Emitters

Nirupama M.P*[1] , Satyanarayana B.S.[2], O.S. Panwar[3]

*BML Munjal University, Sidhrawali, Gurugram-122413, Haryana, India

+Corresponding author: dr.nirupamamp@gmail.com

Abstract— **An interesting application of nanocarbon based Field Emitters are in Field Emission Electrical Propulsion System (FEEP) as Field Emitters (FE). The nanocarbon is grown at room temperature to 250°C using indigenously designed and developed cathodic arc system. The best and stable electron emitting nanocluster material process technology is identified by studying its various properties like composition, morphology, electrical and field emission properties. To enhance the field emission properties of a nanocluster carbon-based field emitters a Nanosphere Lithography technique is used, and the process growth parameters are repeated for these samples and tested for its field emission properties. A nanosphere lithography technique-based nanocluster Field Emitters showed better field emission properties compared to as grown nanocluster based FE's.**

Keywords: Nanosphere Lithography, Nanocluster carbon-based Field Emitters, Cathodic Arc, Field Emission Electrical Propulsion (FEEP) Vacuum Nanoelectronics

I. INTRODUCTION

Nanocarbon based electron emitters are in the forefront now. It finds application in devices based on cold cathodes. Nanocarbon exists in various forms: nanocluster carbon, tetrahedral amorphous carbon [ta-C], nanostructured graphite, nano horns, diamond like carbon [DLC]. Among all of them nanocluster carbon has exhibited low field emission. The various forms of nanocarbons can be grown with different fabrication processes at high temperature, indicating a need for low temperature process technology, which reduce the cost and emit electrons at low fields [1, 2].

It is crucial to develop an electron emitter at low temperatures, which will exhibit high current, low power. So, there is a need for developing a cold cathode-based field emitter. Nanosphere lithography (NSL) is a simple, cost effective, high throughput, and self-assembly nanofabrication technique which used production of periodic particle array (PPA) in nanometer scale. This process can be used to demonstrate the production of large variety of nanostructures and ordered nanoparticle arrays on various substrates.

The Self-Assembled Mask Less Nanoparticle arrays fabricated using Nanosphere lithography (NSL) to create nanocarbon emitter arrays. By varying various deposition parameters, the nanocarbon and its various facets are grown using cathodic arc system. Nanocluster carbon is an interesting and novel material in this research work. It is a unique form of nanocarbon with mixed phased material properties having sp^2

and sp^3 carbon bonds. Considering its morphology, from smooth to various dimension clusters and fibrous properties, it can be used in various applications such as in spacecraft propulsion system, large area flexible microelectronics and many more. As such there are no reports related to nanocluster carbon in space related applications, although the carbon nanotubes are extensively studied and demonstrated in space related applications. The research work focuses on demonstrating NSL technique to synthesize self-aligned metal nanoparticle arrays and to fabricate nanocarbon based field emitters. The various properties of these nanocarbon based field emitters were studied using various characterization equipment's like: OLUMPUS, Optical Microscope, Scanning Electron Microscope (SEM) (Hitachi-SU 3500), Near Field Scanning Optical Microscope (NSOM, WiTec ALPHA 300, 100RAS) for AFM (Atomic Force Microscopy) measurements [3-5].

This study offers the scope to understand and control the material from diamond like to graphite like and also have mixed phase materials, which can help to create nanoclusters to nanotubes, fibers and even graphene like materials. Hence an effort has been made to develop an indigenous process equipment, and characterization equipment's to study the material effectively.

II. EXPERIMENTAL DETAILS

A monolayer of uniformly distributed polystyrene spheres of various diameter 100 nm, 200 nm and 500 nm were fabricated using dip coating, capillary force, controlled evaporation, sputtering, dip and dry process. Figure 1 (a, b, c) describe the various fabrication process of nanocarbon based field emitter array. A thin layer of metal was grown by sputtering as shown in figure 1 (a). The thickness of the metal layer varied from 5-7 nm. The nanospheres were lifted off by ultrasonic agitation for about 30 minutes. The nanocluster carbon films were grown on the patterned substrates using cathodic arc deposition system as shown in the figure 1(b, c). The nanocarbons are grown at various process conditions. Depending on the understanding of the emitter material, the emitter is developed with a clear awareness of the emission site density, ability to control the emitter current, the breakdown condition of the emitter.

III. RESULT AND DISCUSSION

Samples were grown at various growth parameters: (a) As grown nanocluster carbon, (b) As grown nanocluster carbon grown at 250℃, (c) Nanocluster carbon samples grown with various partial pressure of nitrogen with zero bias conditions, (d) Nanocluster carbon samples grown with various bias conditions, (e) Nanocluster carbon samples grown with various partial pressure of nitrogen with 60V bias conditions, and (f) gives the Raman spectra of nanocluster carbon samples with fixed $N_2=10^{-4}$ Torr and He= 2.1E-04 and 5.0E-03 Torr. Near Field Scanning Microscope (NSOM) with an excitation laser of 532nm wavelength was used to carry out the Raman analysis of the nanocluster carbon. Figure 2(a) Depicts Raman spectra of as grown nanocluster carbon, Figure 2(b) Gives the Raman spectrum of nanocluster carbon samples grown with various bias conditions.

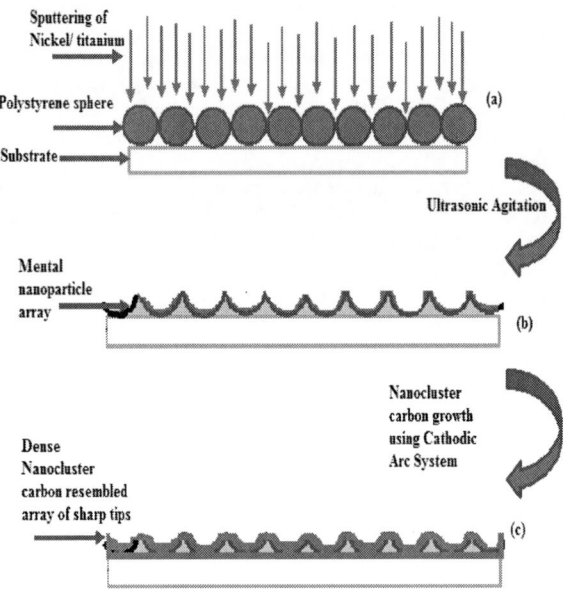

Figure 1 (a, b, c): Fabrication process of nanocarbon based field emitter array.

IV. CONCLUSION

This work is a result of an effort to develop an indigenous facility and an affordable process to study a stable low temperature grown self-aligned nanocarbon based field assisted electron emitter and their patterned arrays for vacuum nanoelectronics applications. Field emission measurements were conducted in vacuum at 10^{-7} Torr, in diode configuration. The field emission measurements from these films showed an enhanced field assisted electron emission property, but the emission sites on the films were uncertain. The emission current was low at low-voltage and increased with the high voltage. The emission from the nanocluster carbon grown on a self-aligned metal patterned substrate, showed shallow slope, indicating they are amorphous or mixed phase carbon.

V. ACKNOWLEDGEMENT

The authors would like to acknowledge BML Munjal University, Gurgaon, Haryana, for providing the research facility to carry out various aspects of the work.

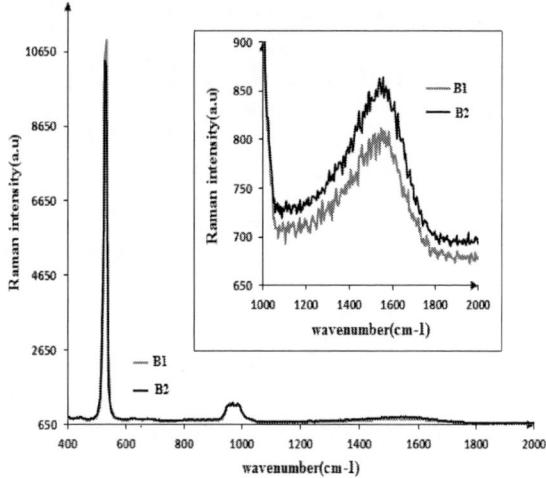

Figure 2 (a) Depicts Raman spectra of as grown nanocluster carbon.

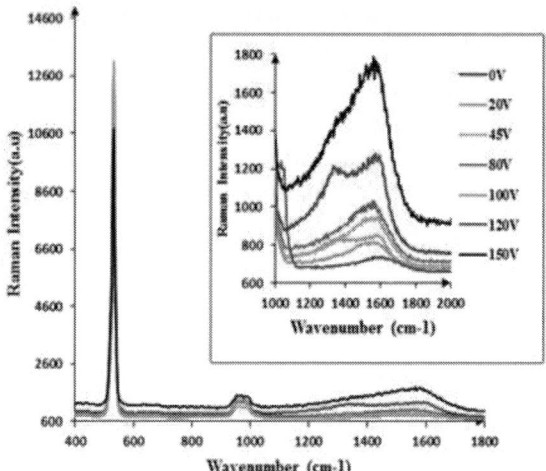

Figure 2 (b) Shows the Raman spectrum of nanocluster carbon samples grown with various bias conditions.

VI. REFERENCES

[1] Satyanarayana, B.S., Nishimura, K. and Hiraki, A. "Field emission from novel room temperature grown carbon based multilayered cathodes", Mat. Res. Soc. Symp. Proc., 685E, D15.5.1–D15.5.6 (2001)

[2] M.P. Nirupama, SB Gandla, A Bhattacharya, B.S. Satyanarayana "Indigenous design and development of the cathodic arc system for the growth of nanocarbon thin Films" Materials Today: Proceedings 5 (1), 3121-3129, 2018

[3] Li L, Zhai T, Zeng H, Fang X, Bando Y, et al. (2011) Polystyrene sphere-assisted one-dimensional nanostructure arrays: synthesis and applications. Royal Society of Chemistry 1: 40-56.

[4] Nirupama M.P, Amarnath Bheemaraju, O. S. Panwar, B. S.Satyanarayana, "Nickel Nanoparticle Arrays Prepared Using Nanosphere Lithography", (IJRASET)ISSN: 2321-9653, Volume 6 Issue II February 2018,pp,75-78.

[5] Minh Nhat Dang, Minh Dang Nguyen, Nguyen Khac Hiep, Phan Ngoc Hong, In Hyung Baek and Nguyen Tuan Hong, "Improved Field Emission Properties of Carbon Nanostructures by Laser Surface Engineering", Nanomaterials 2020, 10, 1931

Field Emission Characteristics of ZnO Nanowire Driven by Pulsed Voltage

Deyi Huang, Yangyang Zhao, Shuai Wang, Guofu Zhang, Juncong She, Shaozhi Deng, Ningsheng Xu and Jun Chen*

State Key Laboratory of Optoelectronic Materials and Technologies, Guangdong Province Key Laboratory of Display Material and Technology, School of Electronics and Information Technology, Sun Yat-sen University, Guangzhou 510275, People's Republic of China

* E-mail: stscjun@mail.sysu.edu.cn

Abstract— **The pulsed field emission characteristics of ZnO nanowires were studied. The current response was analysed under pulsed high voltage driving. Emission currents higher than the value obtained under direct current voltage were recorded. The possible mechanism was discussed.**

I. INTRODUCTION

Quasi-one-dimensional ZnO nanowires are excellent cold cathode materials, which have low turn-on field, good emission uniformity and stability over a large area [1]. Large area flat panel X-ray source have been reported using ZnO nanowire filed emitter arrays [2-4]. By making use of the fast response of cold cathode, X-ray sources could operate under pulsed mode, which can effectively eliminate the blur during the imaging of moving object [5]. Besides, low X-ray dosage is expected which lessens the harm caused by X-ray irradiation to human body in medical applications [6]. Studying the pulsed field emission properties of ZnO nanowires is important for their applications in pulsed X-ray source.

In this study, the field emission characteristics of ZnO nanowire were studied when driven by pulsed voltage. The field emission current was measured under pulsed voltage with different voltage and frequency. The experimental results show that the emission current obtained under pulsed voltage is higher than that obtained under direct current (DC) voltage. The observed phenomena were discussed by a pulsed voltage induced plasma assisted emission mechanism.

II. EXPERIMENT

The experimental set-up of the measurement is shown in Fig. 1. The sample is ZnO nanowires prepared on ITO glass with an area of 4.8 cm×4.8 cm. Arrays of ZnO nanowires patterns with diameter 5 μm were prepared using thermal oxidation method [7]. Metal thin film coated on glass substrate was used as the anode. The distance between cathode and anode was set as 5 mm.

The measurements were carried out in a vacuum with a pressure of ~10^{-6} Pa. The emission current of ZnO nanowires were obtained by measuring the voltage drop on a 500 kΩ sampling resistor with an oscilloscope when applying DC voltage or pulsed voltage. The current under different peak voltage value and frequency of the pulsed voltage were recorded. The morphologies of the ZnO before and after measurements were observed using scanning electron microscope (SEM).

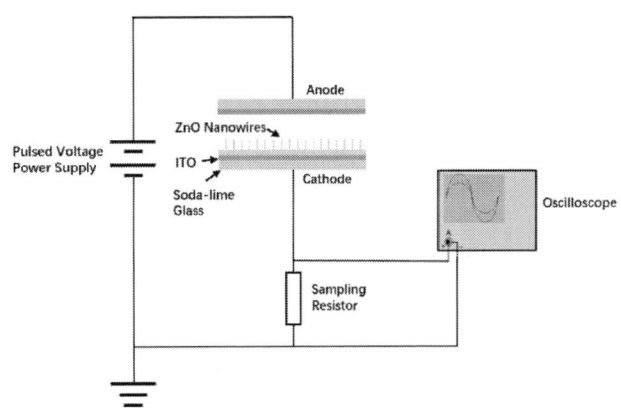

Fig. 1 Schematic of the experiment set-up

III. RESULTS AND DISCUSSION

Fig. 2 shows a typical current response recorded driven by a 1 Hz pulsed voltage. The pulsed voltage has a peak voltage of 26 kV and the duration is 10 ms. The current increases sharply with when the pulsed voltage is on while decays at a slow speed. The peak current is about 386 μA. For comparison, when applied with DC voltage, the current is about 27 μA for the same voltage. The relative increase in the current is about 1330%. Furthermore, higher relative current increase was observed when increasing the peak voltage. We also observed that when the frequency of the pulsed voltage increased, the emission current decreased. Furthermore, the current response under reversed pulsed voltage (i.e. the cathode is applied with positive voltage and the anode is grounded) is measured as well, showing neglectable pulsed current response.

978-1-6654-2590-2/21 $31.00 © 2021 IEEE

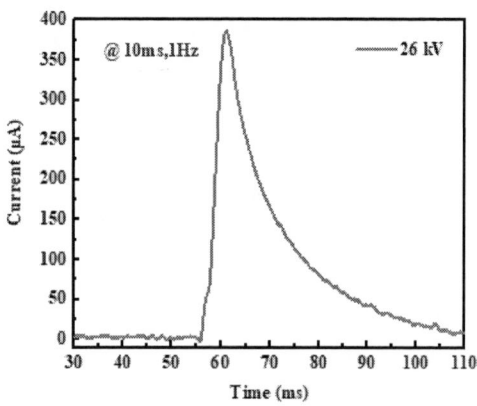

Fig. 2 The current response to a 10 ms 1Hz high voltage pulse.

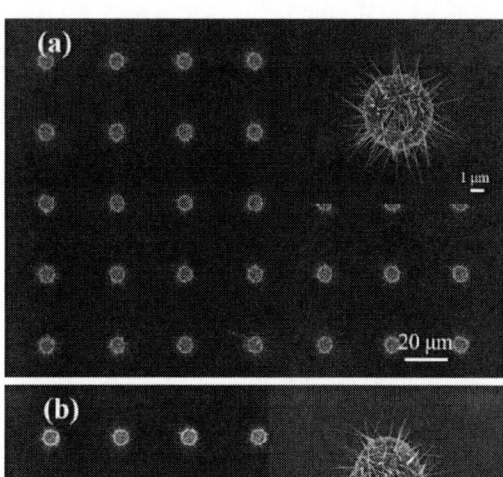

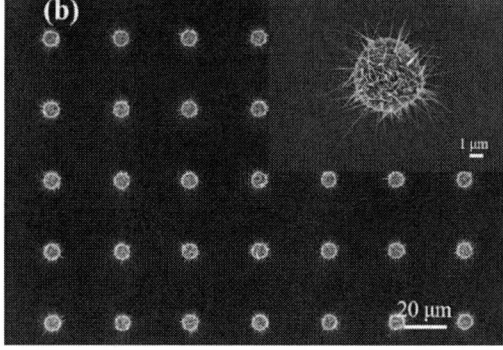

Fig. 3 SEM images of the sample. (a) before and (b) after measurement.

In order to inspect if the morphology of ZnO nanowires changed after the measurement, we observed the morphology of the sample using SEM. Fig. 3 shows the images of ZnO nanowires before and after the measurement. It can be seen from the picture that no obvious changes in the morphology of the nanowires occur.

The observed phenomena can be tentatively explained by a plasma induced field emission enhancement mechanism. When the emitter is driven under high voltage in pulsed mode, the sudden release of adsorbed gas molecules on the anode and cathode surface will be ionized. Higher anode current than that driven by DC voltage can be observed. When the voltage pulse repeats or the pulse frequency increases, the responding current will decrease. However, the exact mechanism needs more careful study.

IV. CONCLUSIONS

In summary, ZnO nanowires were prepared by thermal oxidation, current characteristics were tested under DC voltage and pulsed voltage. The result shows that the emission current of ZnO nanowires under pulsed voltage is significantly larger than that under DC voltage. The exact mechanism for the phenomenon needs further study.

ACKNOWLEDGMENT

The authors gratefully acknowledge the financial support from the National Key Research and Development Program of China (Grant No.2016YFA0202000), the Science and Technology Department of Guangdong Province (Grant No.2020B0101020002), Fundamental Research Funds for the Central Universities.

REFERENCES

[1] C. X. Zhao et al., "Large-Scale Synthesis of Bicrystalline ZnO Nanowire Arrays by Thermal Oxidation of Zinc Film: Growth Mechanism and High-Performance Field Emission," *Crystal Growth & Design*, vol. 13, no. 7, pp. 2897-2905, Jul 2013.

[2] D. K. Chen *et al.*, "Transmission type flat-panel X-ray source using ZnO nanowire field emitters," (in English), *Applied Physics Letters,* vol. 107, no. 24, p. 5, Dec 2015.

[3] L. B. Wang et al., "Fabrication of large-area ZnO nanowire field emitter arrays by thermal oxidation for high-current application," *Applied Surface Science*, vol. 484, pp. 966-974, Aug 2019.

[4] Y. Y. Zhao et al., "High Current Field Emission from Large-Area Indium Doped ZnO Nanowire Field Emitter Arrays for Flat-Panel X-ray Source Application," *Nanomaterials*, vol. 11, no. 1, p. 16, Jan 2021.

[5] W. Lei, Z. Y. Zhu, C. Y. Liu, X. B. Zhang, B. P. Wang, and A. Nathan, "High-current field-emission of carbon nanotubes and its application as a fast-imaging X-ray source," *Carbon*, vol. 94, pp. 687-693, Nov 2015.

[6] C. R. Inscoe et al., "Tomosynthesis imaging of the wrist using a CNT x-ray source array," in *Medical Imaging 2019: Physics of Medical Imaging*, vol. 10948, T. G. Schmidt, G. H. Chen, and H. Bosmans, Eds. (Proceedings of SPIE, Bellingham: Spie-Int Soc Optical Engineering, 2019.

[7] Z. P. Zhang et al., "Controllable preparation of 1-D and dendritic ZnO nanowires and their large area field-emission properties," *Journal of Alloys and Compounds*, vol. 690, pp. 304-314, Jan 2017.

Efficient fabrication of vertical carbon nanotube array cold cathode using laser cutting

Chuyang Liao, Jiupeng Li, Xiaoyu Qin, Qi Bo, Baohong Li, Shaozhi Deng, Yu Zhang*

State Key Lab of Optoelectronic Materials and Technologies,
Guangdong Province Key Lab of Display Material and Technology,
School of Electronic and Information Technology, Sun Yat-Sen University
Guangzhou 510275, People's Republic of China
*Corresponding author: stszhyu@mail.sysu.edu.cn,

Abstract— **A vertical carbon nanotube (CNT) sheet array was fabricated using a laser cutting method which has the advantages of simple, efficient, clean and low cost. A CNT sheet array with 3 μm sheet width, 20 μm sheet height, 20 μm spacing and 3.5 mm² overall size can emit a current of 50 mA, a corresponding current density of 1.43 A /cm² in pulse field emission. The uniform emission site distribution showed that the electrostatic shielding effect can be eliminated in the array pattern. The performance of the CNT sheet array was promising for cold cathode device. The laser cutting method opens up a new way to fabricate high quality field emitter in simple process.**

Keywords—*carbon nanotube; sheet array; laser cutting; field emitter*

I. INTRODUCTION

High current cold cathode is the heart of the vacuum electronic devices. In recent decades, several kinds of cold-cathode based devices, such as X ray tube, microwave tube and lighting element etc. had been fabricated and proved the feasibility and advantage of nanomaterial cold cathode. However, approaching to application, there are still several hampers on the way, mainly in fabrication, performance and reliability [1-2].

Carbon nanotubes (CNTs) are one of the best cold cathode materials with high field enhancement factor, high electrical conductivity and high emission current. The preparation of CNT films using thermal chemical vapor deposition (TCVD) has become mature and low-cost. For the necessity of device and avoiding the electrostatic shielding effect, CNT arrays fabrication is important, in the meantime, is a key difficulty. Pattern lithography is the most popular way, while it has several drawbacks such as complicated steps, photoresist pollution and growth angle bias etc.

In our previous work, we proposed a 2D sheet type CNT field emitter using laser cutting which demonstrated a good field emission performance [3]. The advantage of the laser cutting is high resolution in micrometer scale, direct writing without chemical pollution and simple process with low cost. The method is also suitable for array fabrication. Thus, in this work, we went on develop the laser cutting method to fabricate CNT sheet array. The laser-cutting CNT sheet array showed a promising large current field emission characteristic.

II. EXPERIMENTAL

The laser cutting system is home-made which comprises a high intensity focused pulsing laser, a precise mobile platform and an automatic control interface. The focal laser spot is as small as 2 μm in diameter. The step-by-step precision is 10 nm.

The CNT array fabrication process is a simple two-step process briefly described below. First, a vertical CNT film was grown on silicon substrate using TCVD. Second, the CNT thin film was placed on the platform and irradiated by the focused laser spot. The intensive heat induced by the laser burns the CNTs in a second. Then moving the platform according to the designed path, a cutting pattern was fabricated on the CNT film, such as sheet array, point array and random pattern. The cutting resolution is the same as the diameter of the laser spot. The cutting depth is around 20 μm depending on the robust quality of CNT.

The morphology of CNT sheet array was characterized by scanning electron microscopy (SEM). The field emission characteristics were measured in an ultra-high vacuum chamber using a diode structure with a gap of 150 μm. The DC and pulse field emission characteristics, emission uniformity and stability were measured.

III. RESULTS AND DISCUSSION

The morphology of a laser-cutting CNT sheet array is shown in Fig. 1. The array area is 3.5mm² comprising 96 CNT sheets (Fig. 1a). The dimension of a single CNT sheet is 3 μm in width, 20 μm in depth. (Fig. 1b&c). The CNT sheets are closely aligned with each other on the substrate and are well bonded to the substrate. The space between two sheets is 20 μm. The designed pattern parameter followed the previous theoretical prediction that the ratio of height and space should be large than 1 to avoid the electrostatic shielding effect. On top of a sheet (Fig.4d), the CNTs uniformly distributed which acts as the main emission site during field emission.

The DC and pulsing field emission characteristics of the CNT sheet array are shown in Fig.2. In DC mode (Fig. 2a), the

978-1-6654-2590-2/21 $31.00 © 2021 IEEE

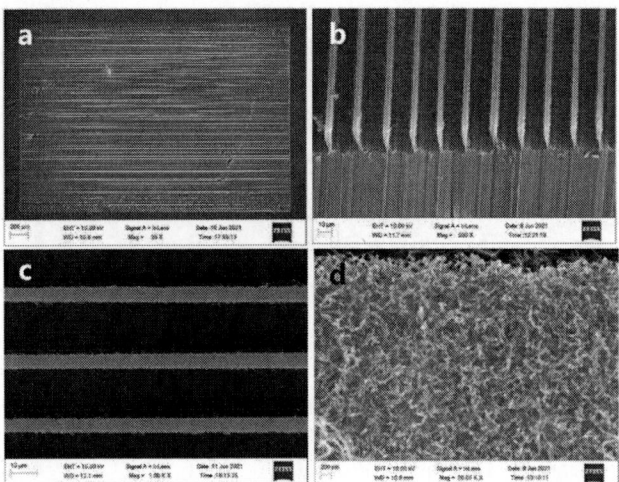

Fig. 1 Morphology of a CNT sheet array, (a) full view on top; (b) side view; (c) magnified top view of three sheets; (d) details of CNTs on top of a sheet.

turn-on field and threshold field were 2.2 V/μm and 3.3 V/μm respectively. The maximum DC emission current was 7 mA at electric field of 9.7 V/μm. The corresponding current density was 0.2 A/cm². According to vacuum breakdown mechanism by Joule heating [4], it is preferred to adopt pulse mode to achieve larger field emission current. In pulse mode (pulse width 1 μS, duty ratio 0.5%, frequency 5 kHz) (Fig. 2b), a maximum pulse current of 50 mA was achieved at 20 V/μm. The corresponding current density was 1.43 A/cm². Both of Fowler-Nordheim (FN) plot showed a linear slope which confirms a traditional field emission behaviour.

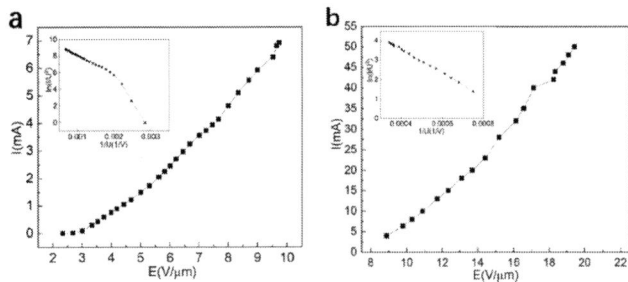

Fig. 2 field emission characteristics of a CNT sheet array, emission current vs. electric field curve and FN plot (inset) (a) in DC mode, (b) in pulse mode.

The current stability and field emission site uniformity are also showed in Fig. 3. The current fluctuation was only 2% during 1-hour maintaining in 1 mA current which demonstrated a pretty good stability during field emission. The emission site image (Fig. 3b) showed the emission sites distributed uniformly on the whole sample area without an edge effect. The sample morphology after measurement was observed in SEM. The surface structure was not seriously damaged, indicating that the CNT sheet could withstand such large current. In order to exhibit its advantage, a same CNT thin film without array was also measured to see the differences. In the same measurement condition, the current fluctuation was 15.7% with an increasing current drift and several strike sparks during emission (Fig. 3c). The emission site mostly distributed on the edge of the sample (Fig. 3d). The results clearly showed that the thin film does not

bring most of the CNTs into field emission and restricts its efficiency in field emission.

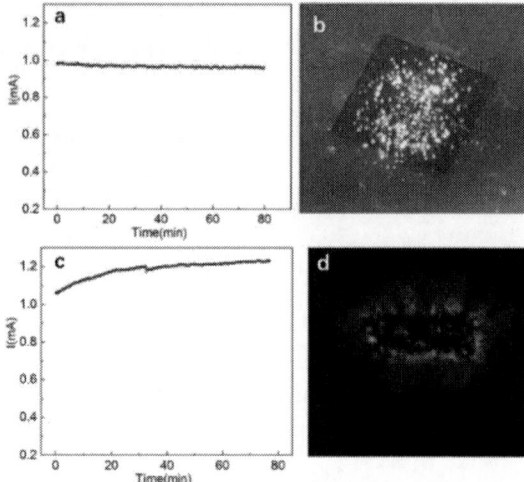

Fig. 3 (a) current stability curve in 1-hour and (b) emission site image of a CNT sheet array. (c) current stability curve in 1-hour and (d) emission site image of a CNT thin film without array.

IV. CONCLUSIONS

A CNT sheet array was fabricated using a focused laser cutting method which is a simple, clean, efficient and low-cost process. The laser-cutting CNT sheet array demonstrated the overwhelming field emission characteristics than the CNT thin film. A 50 mA emission current, 1.43 A/cm² current density and 2% current fluctuation were achieved in a 3.5 mm² CNT sheet array which is promising for cold cathode device application. The field emission performance would be further improved through increasing the laser cutting resolution and optimizing the array structure. This method opens up a new way to fabricate high quality field emitter.

ACKNOWLEDGMENT

This work was supported by the National Natural Science Foundation of China (grant no. 61874142), the National Key Basic Research Program of China (grant no. 2019YFA0210201 and 2019YFA0210202), the Science and Technology Department of Guangdong Province, and the Fundamental Research Funds for the Central Universities.

REFERENCES

[1] Chung S , Huang B , Chiang C , et al. "Carbon nanotubes on flattened tin alloy spheres in a Ball Grid Array (BGA) for cold cathode applications", IEEE-NANO IEEE Conference on Nanotechnology IEEE, 2008.

[2] Kang J S, Kim J H , Park K C , et al. "Fabrication of highly stable electron beams with CNT cold cathode", 30th International Vacuum Nanoelectronics Conference (IVNC) , 2017.

[3] Zhang W M , Deng S Z , Hong T Z , Zhang Y . "Design and characteristics of a vertical carbon nanotube sheet field emitter", 33th International Vacuum Nanoelectronics Conference (IVNC),2020.

[4] Zhang Y , Deng S , Du J , et al. "Effects of pulsewidth and area of carbon nanotube films on their pulsed field emission characteristics", IEEE transactions on electron devices, 60(8): 2677-2681, 2013.

Functionalize of vertically aligned CNTs emitter (C-beam) for surface modification and patterning of self-assembled monolayers (SAM)

Alfi Rodiansyah[1], Kyu Chang Park[1,*]

[1]*Department of Information Display, Kyung Hee University, Dongdaemun-gu, Seoul, 02447, Korea*
Contact: kyupark@khu.ac.kr, Phone +82-02-961-9447

Abstract—**In this study, we report the process of surface modification and patterning of SAMs (self-assembled monolayers) using vertically aligned CNTs as emitter sources (C-beam lithography). We demonstrated the simply patterning process of C-beam lithography by using a diode configuration to transfer pattern onto silicon substrate. Through water contact angle measurement, we also investigated the surface modification of the OTS (octadecyl trichlorosilane) as SAMs. Using the small regime of current exposure at 0.28mA/cm^2 and 60 s time exposure, the desirable patterns were appears after lift-off process. By utilizing the small power consumption bellow 1.5kV, the pattern can be easily formed on silicon substrate. By this study, we accomplished the new approach C-beam lithography as pre elementary study for alternative electron beam lithography.**

I. INTRODUCTION

Patterning and surface modification are the key factor for technological development in the semiconductor industry [1]. The need of patterning devices that can achieve a desirable size with high efficiency are fundamental in micro-nano fabrication. Electron beam lithography (EBL) is one of promising techniques to achieve pattern down to 2 nm [2]. However, EBL techniques have limits due to require sensitive resist with multi step process, utilize high power and complexity structure as well. Principally, the patterning by electron beam lithography is utilizing the focused electron beam that interacted with substrate. Therefore, the patterning at nano-scale can be achieved easily. However, utilize the conventional EBL to achieve into nanoscale feature size patterning by EBL require some high cost and more complexity structure [3][4]. Conventional EBL has a lacking due to low efficiency, which limits producing a small number of nano structure for application purposes.

In this paper, we introduced the C-beam for direct patterning Self-assembled monolayers as a new alternative electron lithography technique. The basic principle of this technique is fabrication the pattern by utilize vertically aligned CNTs as electron sources (C-beam). As well known that CNT as emitter material has outstanding properties that possess as new candidate for electron emission sources. In this study, we have been demonstrated and developed a new method of patterning using electron beam lithography based on C-beam as electron sources. Our efforts are developing new compatible system EBL by utilizing CNT emitters with lower energy consumption

bellow 1.5kV. We accomplished to design small line pattern as well as for micro fabrication process. The main purposes of C-beam lithography are to create a specific pattern in a resist layer of SAMs and subsequent transfer of that pattern into or onto the underlying silicon substrate.

II. EXPERIMENT

A. Fabrication VACNT emitters and SAMs Sample

Vertically aligned Carbon nanotubes (VACNTs) emitters were formed on the silicon wafer substrate through resist assisted patterning process. In this study, we grew line emitters show Fig.1 inset that we made by photolithography process. The CNTs emitters were grown by DC-PECVD (direct current Plasma enhance chemical vapor deposition). Furthermore, we conduct I-V measurement to characterize the emitters as show in Fig 1.

The SAMs was prepared onto silicon substrate using octadecyl trichlorosilane (OTS). The SAMs were formed by immersion the silicon water into OTS/toluene 1:200 v/v solution for 4 hours. To avoid the contamination, the clean silicon wafer was prepared by pre-soaking into acetone and isopropyl alcohol in ultrasonic bath for 10 minute for each. After immersed in OTS/toluene solution, the samples rinsed with pure toluene solvent and blown with nitrogen.

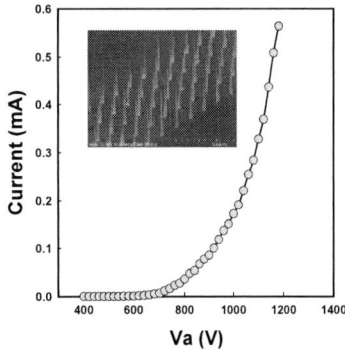

Figure 1. I-V characteristic VACNT emitter (inset:SEM of VACNT 40x magnification).

978-1-6654-2590-2/21 $31.00 © 2021 IEEE

B. Self-assembled Monolayers Patterning

Fig. 2 shows the steps to make patterns on silicon substrate by C-beam lithography. The OTS sample was placing directly on the anode with the gap 250 µm. The exposure time was conducted for 60 s. While for current exposure, we used 0.28 mA/cm². To investigate the effect of current density and time exposure on SAMs surface change, we did water contact angle measurement. Finally, the OTS sample after exposure by C-beam was etched by KOH for 30 minute to form the patterns.

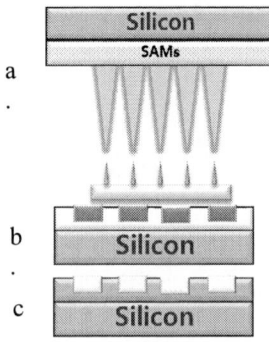

Fig 2. Schematic of C-beam exposure on SAMs-Silion (a) SAMs/silicon after exposure (b) Silicon substrate after etching (c)

III. RESULT AND DISCUSSION

Direct patterning SAMs on the silicon substrate using C-beam as the electron sources was demonstrated. The basic principle of this technique is utilized the electron that emitted form C-beam to change the structure of SAMs without any additional electrode. The effect of electron irradiation from C-beam on SAMs can be identified by the change of properties and structures of SAMs. To analyze the changing of SAMs surface after C-beam irradiation, we carried out the water contact angle measurement. Fig. 3(a)(b)(c) shows the comparison of water contact angle between bare silicon substrate, silicon substrate as deposited by OTS and silicon substrate after exposed by C-beam. The change of water contact angle has been indicated the surface change of SAMs.

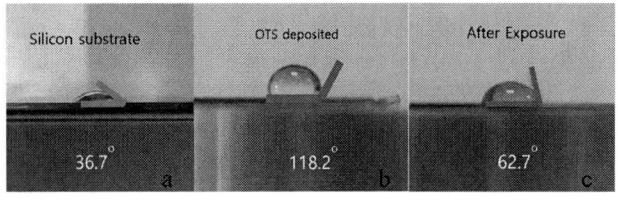

Fig 3. Water contact angle measurement (a) silicon substrate (b) as deposited by OTS (c) after exposure by C-beam

As shows in Fig. 4, the irradiated region appears some pattern after etching. There are the differences surface roughness between area that exposed and the area that non-exposed by C-beam. Apparently, the areas are exposed by C-beam seems rougher due to some pattern appears. While for area non-expose has smoother surfaces. In this case, the OTS acted like positive resist for pattern formation on SiO₂ substrate. Therefore, the patterns appear on the region that has exposed by C-beam.

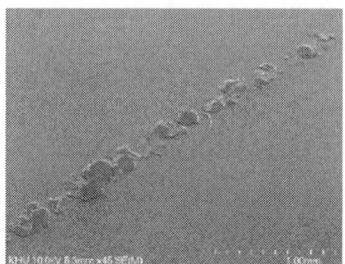

Fig 4. SEM of silicon substrate after etching by KOH.

IV. CONCLUSIONS

In this study, C-beam lithography have been demonstrated to generate pattern on silicon substrate using OTS molecule as Self-assembled monolayers. Through investigation of water contact angle measurement, we obtained the surface modification of OTS structure. In addition, we conclude that there are surface modification of OTS due to electron that emitted from C-beam. In the end, the pattern can be accomplished by lift-off technique; the pattern appeared on silicon substrate after etched by KOH solution.

ACKNOWLEDGMENT

This work was supported by the Technology Innovation Program (20013595, Extreme ultraviolet light source using nano electron beam) funded By the Ministry of Trade, Industry & Energy (MOTIE, Korea) and the BK21 FOUR Program funded by the Ministry of Education (MOE, Korea) and National Research Foundation of Korea (NRF) (21A20130000018).

REFERENCES

[1] Gentili, M.; Giovanella, C.; Selci, S. Nanolithography: A Borderland between STM, EB, IB and X-Ray Lithographies; Kluwer Academic Publishers: Dordrecht, The Netherlands, 1994

[2] Tseng, A., Kuan Chen, Chen, C. and Ma, K., 2003. Electron beam lithography in nanoscale fabrication: recent development. *IEEE Transactions on Electronics Packaging Manufacturing*, 26(2), pp.141-149.

[3] I. S. Jacobs and C. P. Bean, "Fine particles, thin films and exchange anisotropy," in Magnetism, vol. III, G. T. Rado and H. Suhl, Eds. New York: Academic, 1963, pp. 271–350.

[4] Pimpin, A. and Srituravanich, W., 2012. Review on Micro- and Nanolithography Techniques and their Applications. *Engineering Journal*, 16(1), pp.37-56

Gap in pagination due to withheld paper.

Pages 219-220

Field emission behaviour of fresh and aged Sb_2Te_3 nanosheets

Somnath R. Bhopale[1,] and Mahendra A. More[1*]

[1]*Center for Advanced in Material Science and Condensed Matter Physics, Department of Physics, Savitribai Phule Pune Univeristy, Pune-411007, India.*
*Corresponding Author: mam@physics.unipune.ac.in

Abstract— **We have synthesized nanosheets of Sb_2Te_3 by a simple hydrothermal method. Fresh and aged Sb_2Te_3 nanosheets were characterized by X-ray Diffraction, Raman spectroscopy and Field emission scanning electron microscopy (FESEM), prior to field emission (FE) studies. It has been observed that most of the physicochemical properties of the Sb_2Te_3 nanosheets are not changed upon aging of nearly two years. Field emission behaviour of fresh and aged Sb_2Te_3 nanosheets have been studied at base pressure of ~1 x 10^{-8} mbar. It is interesting to note that the overall FE behaviour of fresh and aged emitters are identical in terms of values of turn-on, and threshold fields 4.8, 4.65 and 5.6, 5.25 V/μm along with the maximum emission current density 97 and, 92 μA/cm^2 extracted at applied fields ~ 7.4, 6.75 V/μm, respectively. It clearly implies such long environmental exposure does not affect the conducting surface states of Sb_2Te_3 nanosheets due to its topological behaviour.**

I. INTRODUCTION

Field electron emission characteristics of nanomaterials mainly depend on their aspect ratio, electronic and mechanical properties. Carbon based materials have been proven better field emitters due to very high aspect ratio and excellent electronic and mechanical properties [1]. A new family of materials is introduced in condensed matter physics, known as Topological Insulators (TIs). Owing to its unique electronic properties many researchers are attracted. On their surfaces, highly conducting states are present, but in bulk it is insulting. TIs surface conducting states are very robust against chemical passivation or non-magnetic defects [2].The TIs have been realized for potential materials in different applications; Bi_2Se_3 and Sb_2Te_3 nanosheets have shown interesting field emission characteristics [3] [4]. We have studied the FE behaviour of fresh and aged Sb_2Te_3 nanosheets synthesized using simple hydrothermal method.

II. SYNTHESIS AND CHARACTERIZATION

For hydrothermal synthesis, antimony chloride ($SbCl_3$), tellurium dioxide (TeO_2), hydrazine hydrate ($N_2H_4:H_2O$) as a reducing agent, Polyvinyl alcohol (PVA) as a surfactant, and deionized water (DI) water as a solvent, were used. The details of hydrothermal synthesis are explained elsewhere [4]. The X-ray diffraction (XRD) patterns were recorded using Bruker D8 advance diffractometer (40 kV, 40 mA, Cu-Kα, λ = 1.54 Å). The morphological analysis was done using FESEM (FEI Nova NanoSEM 450, operating voltage 15 kV). The FE measurements were carried out at base pressure ~1x10^{-8} mbar

in a planar diode configuration. The emitter was prepared by sprinkling small quantity of as-synthesized Sb_2Te_3 powder onto a piece of carbon tape (1 cm x 1cm). A semitransparent phosphor screen was used as an anode. The cathode and anode were held parallel to each other, with separation of 1 mm.

III. RESULTS AND DISCUSSIONS

Figure 1 (a) shows XRD patterns of fresh and two-years aged Sb_2Te_3 nanosheets. The presence of well-defined diffraction peaks, indexed to (015), (1010), (110), (205), (0210) and (1115) planes of Rhomohedral crystal structure of Sb_2Te_3 (JCPDS card no.72-1990). Interestingly, there is hardly any change seen in the XRD pattern of aged sample as compared to the freshly prepared one. The XRD analysis clearly reveals no effect of aging on crystalline phase of Sb_2Te_3. Figure 1 (b) shows Raman spectra of fresh and aged Sb_2Te_3 nanosheets. According to group theory, Sb_2Te_3 exhibits three Raman active vibrational modes, A^1_{1g}, A^2_{1g} and E^2g. Both the samples show characteristic peaks around ~120, 139, and 165 cm^{-1}. From the observed bands, Sb_2Te_3 phase is confirmed, surface oxidation was not seen in the aged Sb_2Te_3 nanosheets.

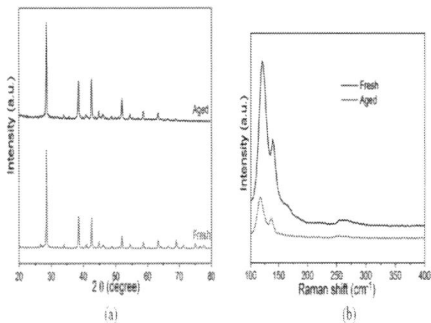

Fig. 1 (a) XRD patterns and (b) Raman spectra of fresh and aged Sb_2Te_3 nanosheets

Figure 2 shows typical FESEM images of hydrothermally synthesized of Sb_2Te_3 nanosheets recorded at different magnifications. In low magnification image (fig 2(a)), agglomeration of nanostructures is observed, whereas the large magnified image (fig. 2(b)) shows formation of sheets like nanostructures under the prevailing experimental conditions. The sheets have varied sizes (length and breadth ~100 to 500 nm), however their thickness is less than 20 nm.

978-1-6654-2590-2/21 $31.00 © 2021 IEEE

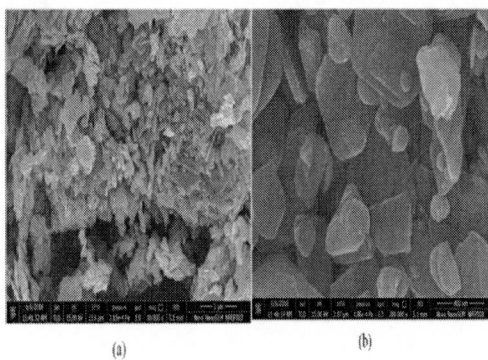

Fig. 2 Typical FESEM images of Sb_2Te_3 nanosheets recorded at (a) low and (b) high magnifications.

Figure 3 shows the FE characteristics of fresh and aged Sb_2Te_3 nanosheets emitters in terms of current density (J) versus applied field (E) plots, and (b) corresponding $\ln(J/E^2)$ versus (1/E) plots.

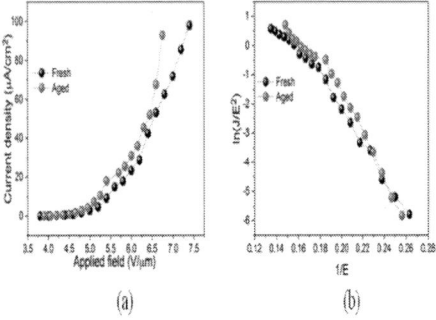

Fig. 3 FE characteristics of fresh and aged Sb_2Te_3 nanosheets emitters (a) current density (J) versus applied field (E) plots, and (b) corresponding $\ln(J/E^2)$ versus (1/E) plots

The emission current density initially increases slowly and then rapidly, with increase in the applied field. The exponential nature of the J-E plot is indicative that the emission characteristic is as per the Fowler-Nordheim (F-N) theory [5]. A careful observation of the J-E plots of fresh and aged samples reveals no noticeable difference. From the observed J-E plots, the values of turn-on and threshold fields (emission current densities 1 and 10 $\mu A/cm^2$, respectively) are estimated at 4.8, and 5.6 V/μm (for fresh sample) and 4.65, 5.25 V/μm (for aged sample), respectively. Furthermore, the fresh emitter delivered maximum current density $\sim 97\mu A/cm^2$ at an applied field of 7.4 V/μm, whereas from the aged sample emitter, maximum emission current density of $\sim 92 \mu A/cm^2$ was extracted at an applied field of 6.75 V/μm. These observed values clearly imply that the FE characteristics of Sb_2Te_3 nanosheets emitter do not change upon two years of aging. It can be speculated such a long exposure to environment does not noticeably affect the conducting surface states of Sb_2Te_3 nanosheets. Moreover, the

Sb_2Te_3 nanosheets are very robust against morphological as well as electronic alterations upon very long environmental exposure. It means that Sb_2Te_3 nanosheets emitter protected conducting surface edges states from environmental doping. Figure 1 (b) shows the F-N plots, i.e. plot of $\ln(J/E^2)$ versus (1/E). Both the F-N plots are nearly identical and exhibit linear nature. A careful observation shows deviation from the linearity in high field region, with a tendency towards saturation.

IV. CONCLUSIONS

We have successfully synthesized Sb_2Te_3 nanosheets by hydrothermal method. Physico-chemical properties of fresh and aged Sb_2Te_3 nanosheets have been studied. It is observed that physico-chemical properties do not change with aging. FE behaviour of fresh and aged Sb_2Te_3 nanosheets is similar, revealing no signification effect of environmental exposure (aging). The FE parameters, namely values, the turn and threshold fields and maximum current density and corresponding strength of applied field remain invariant upon aging; imply robustness of Sb_2Te_3 nanosheets, protecting its morphological and electronic properties against environmental exposure. .

ACKNOWLEDGMENT

Mr. Somnath R. Bhopale would like to thank CSIR (09/137(0629)/2020-EMR-I) for financial assistance.

REFERENCES

[1] Dwivedi Neeraj, Dhand Chetna, J. David, Anderson, Erik C,Kumar, Rajeev, Srivastava, A. K., Malik, Hitendra K., Saifullah, M. S. M., Kumar, Sushil, Lakshminarayanan, Rajamani, Ramakrishna, Seeram, Bhatia, Charanjit S, "The rise of carbon materials for field emission," Journal of Materials Chemistry C, vol. 9, pp. 2620-2659, Feb 2021.

[2] Moore, Joel E, "The birth of topological insulators," Nature, vol. 464, pp. 194-198, March 2010.

[3] Huang, Huihui Li, Yuan Li, Qi Li, Borui Song, Zengcai Huang, Wenxiao Zhao, Chujun Zhang, Han Wen, Shuangchun Carroll, David Fang, Guojia "Field electron emission of layered Bi_2Se_3 nanosheets with atom-thick sharp edges," Nanoscale, vol. 6, pp.8306-8310, May 2014.

[4] Somnath R. Bhopale, M. A. More, "Field emission behaviour of hexagonal Sb2Te3 micron sized platelets of nanometric thickness," AIP Conference Proceedings, vol. 2335, pp.080004, March 2021.

[5] Fursey, G.N. "Field Emission in Vacuum Microelectronics," Springer US, 2007.

$PtSe_2$ Nanosheets as Efficient Field Emitter

Mahendra S. Pawar[1,2], Mahendra A. More[3], and Dattatray J. Late[1,4,*]

[1]Physical and Material Chemistry Division, CSIR - National Chemical Laboratory, Pune, 411008, Maharashtra, (India)
[2]Academy of Scientific and Innovative Research (AcSIR), Ghaziabad 201002, India
[3]Centre for Advanced Studies in Materials Science and Condensed Matter Physics, Department of Physics, Savitribai Phule Pune University, Pune 411007, (India)
[4]Centre for Nanoscience & Nanotechnology Amity University Maharashtra, Mumbai-Pune Expressway, Bhatan, Post − Somathne, Panvel, Mumbai, Maharashtra 410206, (India)
*Contact: datta099@gmail.com

Abstract— **We have grown $PtSe_2$ nanosheets on Si substrate using chemical route followed by thermal annealing in inert atmosphere. The structural and morphological investigations were carried out with the help of Raman spectroscopy and Transmission Electron Microscopy (TEM). The field emission (FE) studies have been reported for the $PtSe_2$ nanosheets in planar diode configuration under ultrahigh vacuum conditions. The turn on field was found to be 5.4 V/µm and FE current density of 505 µA/cm^2 was drawn at an applied electric field of 10.8 V/µm. The emission current-time plot recorded at the pre-set value of emission current of 5 µA over a period of more than 5 h exhibits an initial increase and subsequent stabilization of the current.**

I. INTRODUCTION

In the recent years, layered two dimensional (2D) materials have attracted great deal of attention in nanoelectronic devices due to their unique physicochemical properties.[1,2] The 2D materials include graphene, transition metal dichalcogenides (TMDC's), Black Phosphorous, transition metal oxides and hexagonal boron nitride. Among these graphene is the most widely studied 2D material over the past decade.[3,4]The TMDC's have a general formula MX_2 where, M is the transition metal (Mo, W, V, Sn) and X be the chalcogen (S, Se and Te). These TMDC's possesses tunable bandgap i.e. direct bandgap in monolayer and indirect bandgap in bulk which makes these materials potential candidate in future electronic and optoelectronic devices. There are many reports available on TMDC's such as MoS_2, WS_2, $MoSe_2$, and $SnSe_2$ for various device applications such as field effect transistor, photodetector, gas sensor, and energy storage.[5-7] The other noble transition metals such as Pd and Pt based TMDC's are less explored till date. It is reported that bulk $PtSe_2$ is semimetalic in nature with zero band gap.[8] The theoretical study showed that $PtSe_2$ can afford high charge carrier mobility with a direct band gap of 1.2 eV.[9] To support this results, Wang et al. demonstrated and verified growth of monolayer $PtSe_2$ by direct selenization of Pt crystals under ultra-high vacuum conditions.[9] This property of $PtSe_2$ i.e. transition from semimetal to semiconductor proposes that it could be the encouraging entrant in electronics device applications.

In this report, we present the synthesis of $PtSe_2$ nanosheets using simple chemical route followed by thermal annealing in the Ar atmosphere. We confirm the structural and morphological information of the $PtSe_2$ using Raman spectroscopy and TEM. We have performed FE investigations on $PtSe_2$ nanosheets and extracted emission current density of 505 µA/cm^2 with turn on field found to be 5.4 V/µm.

II. EXPERIMENTAL SECTION

The $PtSe_2$ nanosheets were obtained using two step method and the details are as follows: Initially we have prepared PtSe complex using chemical route, the precursors used are 0.5 ml of 0.015M solution of H_2PtCl_6, 0.5 ml of 0.5M hexamethylenetetramine, 0.8 mg of Se powder and 0.1M $NaBH_4$. The as prepared PtSe complex was then transferred onto the Si substrate followed by thermal annealing at 500°C for 5h in Ar atmosphere. The details about the synthesis procedure has been described in earlier report.[10]

III. RESULTS AND DISCUSSION

The structural and morphological investigations are carried out with the help of Raman spectroscopy and TEM. The Fig. 1(a) shows the Raman spectra for the $PtSe_2$ nanosheets. The two Raman active modes are identified at 172 and 201 cm^{-1} assigned to E_g and A_{1g} modes respectively. The Fig. 1(b) presents the low magnification TEM image of the $PtSe_2$ nanosheets. From Fig. 1(b) it is clear that few layer nanosheets are formed on the surface of Si substrate.

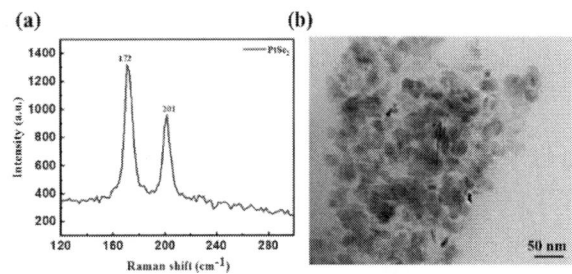

Fig. 1 (a) Raman spectra and (b) low magnification TEM image of the $PtSe_2$ nanosheets.

978-1-6654-2590-2/21 $31.00 © 2021 IEEE

Further, we used these nanosheets for the FE investigations and the measurements are performed in planar diode configuration under ultra-high vacuum (10^{-8} mbar) and depicted in Fig.2. The emission current density as a function of applied electric field plot shown in Fig. 2(a). The turn on field is defined as the field required to withdraw a current density of 1 μA/cm^2 which is found to be 5.4 V/μm.

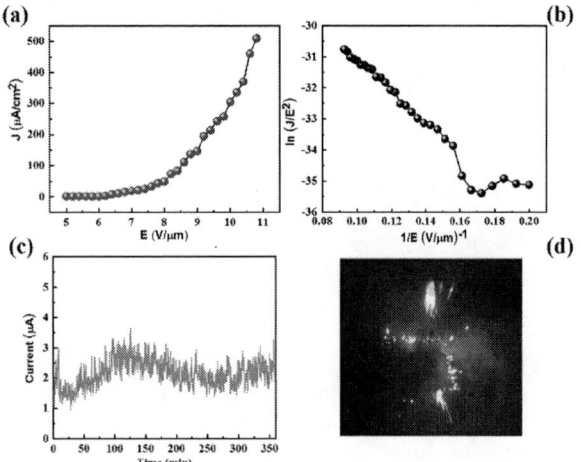

Fig. 2 FE characteristics for PtSe$_2$ nanosheets (a) J-E plot, (b) F-N plot, (c) I-t plot and (d) FE image captured at current density of 50μA/cm^2.

We achieved emission current density of 505 μA/cm^2 at an applied electric field of 10.8 V/μm. The Fowler-Nordheim plot is shown in Fig. 2(b) and the F-N equation is given below:

$$J = \lambda_M a \phi^{-1} E^2 \beta^2 \exp\left(-\frac{b\phi^{\frac{3}{2}}}{\beta E} v_F\right)$$

The non-linear behaviour of the plot suggests that the emission is happening due to semiconducting nature of the PtSe$_2$. We also performed stability experiment upto 6h shown in Fig.2(c). We observed stable emission current over the duration of the measurement and in addition to this some noise was seen which may be due to the adsorption and desorption of the residual gas molecules on the emitter surface. The Fig. 2(d) shows the FE image recorded at emission current density of ~50 μA/cm^2. Our results shows that few layer PtSe$_2$ nanosheets could play the key role in future electronic devices.

IV. Conclusions

In conclusion, we have employed simple chemical route for the synthesis of Ptse$_2$ nanosheets. The structural and morphological analysis was done with Raman spectroscopy and TEM.

Through FE measurements, we have achieved emission current density of 505 μA/cm^2 with turn on field found to be 5.4 V/μm. Our study will accelerate further device research of PtSe$_2$ nanosheets in next generation nanoelectronic devices.

Acknowledgment

M. S. Pawar acknowledges CSIR-Delhi for the financial assistance and this research work was supported by the SERB Government of India under the SERB Research Scientist scheme provided to Dr. D. J. Late.

References

[1] G. Fiori, F. Bonaccorso, G. Iannaccone, T. Palacios, D. Neumaier, A. Seabaugh, S. K. Banerjee and L. Colombo, "Electronics based on two-dimensional materials," Nat. Nanotechnol., vol. 9, pp. 768-779, 2014.

[2] F. H. L. Koppens, T. Mueller, P. Avouris, A. C. Ferrari, M. S. Vitiello, and M. Polini, "Photodetectors based on graphene, other two-dimensional materials and hybrid systems," Nat. Nanotechnol., vol. 9, pp. 780-793, 2014.

[3] S. Wang, P. K. Ang, Z. Wang, A. L. L. Tang, J. T. Thong, and K. P. Loh, "High mobility, printable, and solution-processed graphene electronics," Nano Lett., vol. 10, pp. 92-98, 2010.

[4] J. Zhu, D. Yang, Z. Yin, Q. Yan, and H. Zhang, "Graphene and graphene-based materials for energy storage applications," Small, vol. 10, pp. 3480-3498, 2014.

[5] Q. H. Wang, K. Kalantar-Zadeh, A. Kis, J. N. Coleman, and M. S. Strano, "Electronics and optoelectronics of two-dimensional transition metal dichalcogenides," Nat. Nanotechnol., vol. 7, pp. 699-712, 2012.

[6] Y. Liu, X. Duan, Y. Huang, and X. Duan, "Two-dimensional transistors beyond graphene and TMDCs," Chem. Soc. Rev., vol. 47, pp. 6388-6409, 2018.

[7] H. Li, Y. Shi, M. H. Chiu, and L. J. Li, "Emerging energy applications of two-dimensional layered transition metal dichalcogenides," Nano Energy, vol. 18, pp. 293-305, 2015.

[8] G. Y. Guo, and W. Y. Liang, "The electronic structures of platinum dichalcogenides: PtS$_2$, PtSe$_2$ and PtTe$_2$," J. Phys. C: Solid State Phys., vol. 19, pp. 995, 1986.

[9] Y. Wang, L. Li, W. Yao, S. Song, J. T. Sun, J. Pan, X. Ren, C. Li, E. Okunishi, Y. Q. Wang, and E. Wang, "Monolayer PtSe$_2$, a new semiconducting transition-metal-dichalcogenide, epitaxially grown by direct selenization of Pt," Nano Lett., vol. 15, pp. 4013-4018, 2015.

[10] A. Ali Umar, S. K. Md Saad, and M. Mat Salleh, "Scalable mesoporous platinum diselenide nanosheet synthesis in water," ACS Omega, vol. 2, pp. 3325-3332, 2017.

Electron emission from a solvothermally synthesized ZnS-RGO nanocomposite field emitter

Sanjeewani R. Bansode, Mahendra A. More, Rishi B. Sharma *

Centre for Advanced Studies in Materials Science and Condensed Matter Physics, Department of Physics, Savitribai Phule Pune University, Pune 411007, India.

email id: rbsharma111@gmail.com

Abstract- **A simple solvothermal method was used to synthesize reduced graphene oxide (RGO) nanocomposite which was found to be decorated with zinc sulphide (ZnS) microspheres. The field emission studies were carried out in UHV environment (1×10^{-8} mbar) to evaluate the stability of electron emission and the current density levels. The results of XRD and RAMAN spectra confirmed the formation of hexagonal phase ZnS and the reduction of GO to RGO. The SEM and TEM images showed that ZnS microspheres were deposited on RGO sheets. The turn-on field value, defined at an emission current density of 1 μA/cm^2, is found to be 2.32 V/μm for the ZnS-RGO emitter. Furthermore, the ZnS-RGO emitter delivers a maximum emission current density of 1050 μA/cm^2 at an applied field of $\sim$ 4.1V/μm. The results obtained herein propose the ZnS-RGO nanocomposite emitter to be a potential electron source for practical applications in vacuum nanoelectronic devices.**

Keywords: - ZnS, RGO, Field Emission, XRD, SEM, TEM

I INTRODUCTION

In recent times, single layer, few layers (reduced graphene oxide (RGO)) and multilayers of graphene have attracted attention due to their unique electrical and mechanical properties, large surface area, high chemical stability[1]. The excellent electronic properties of graphene makes it a promising candidate for field emission (FE) based applications [2]. Graphene has also been used as a host material for the homogeneous growth of desired nanostructures [3]. The FE studies of metal chalogenides-graphene system such as WS$_2$-RGO, MoS$_2$-RGO and CdS-RGO have been reported earlier[4-6]. Amongst the metal chalcogenides ZnS is the widely studied material for its application in optical devices, flat panel displays, light emitting diodes (LEDs), field emitters, sensors and lasers [7,8]. As ZnS–few layered graphene (ZnS–FLG) forms an important metal chalcogenide–graphene system, it is considered worthwhile to investigate its field emission properties in detail.

In this paper, we report field emission studies from ZnS-RGO nanocomposite synthesized by one-step solvothermal method.

II EXPERIMENTAL

The ZnS-RGO nanocomposites were synthesized by one step solvothermal method. In typical synthesis process, 1.5 g of Zinc Chloride (ZnCl$_4$) and 0.75 g of Thiourea were dissolved in 40 ml of ethanol with stirring and followed by addition of 50 mg of GO. The mixture was transformed into 100 ml Teflon-lined container. The autoclave was maintained at 180°C for 16 hrs and then air cooled at room temperature. The resultant suspension was filtered and washed 3-4 times by ethanol and finally dried under IR lamp.

III RESULTS AND DISUCSSION

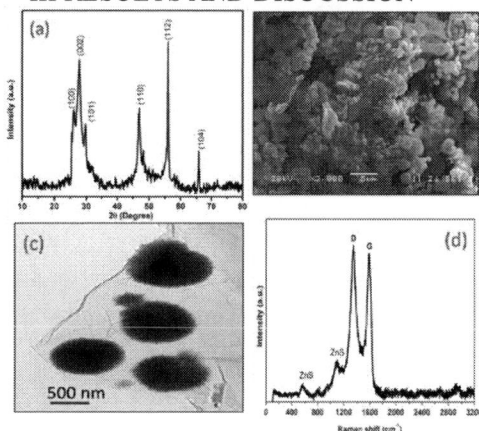

Fig. 1 Characterization of ZnS-RGO nanocomposite (a) XRD spectrum (b) SEM (c) TEM (d) RAMAN spectrum.

The XRD spectrum of ZnS-RGO nanocomposite, shown in Fig.1(a), exhibits a set of well defined diffraction peaks, which could be indexed to crystalline phase of ZnS (Hexagonal phase) (JCPDS No. 75-1547). The Fig.1 (b and c) depict that S EM and TEM images of ZnS-RGO nanocomposite showing ZnS microspheres distributed on RGO sheets. As can be seen from Fig. 1(d), the Raman peaks of RGO at 1351 cm^{-1} and 1583cm^{-1} corresponding to D and G bands respectively. The peaks at 573 cm^{-1} and 1082 cm^{-1}

related to pure ZnS are due to longitudinal optical (LO) phonons [9].

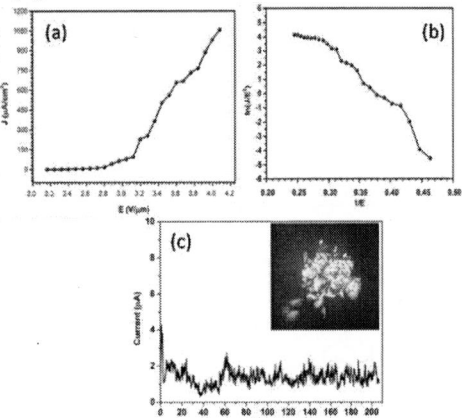

Fig.2 Field emission from ZnS-RGO nanocomposite (a) Field emission current density verses applied electric field (J-E) characteristics. (b) Corresponding Fowler–Northeim (F-N) plot. (c) Field emission current density (I-t) stability (inset shows Field emission micrograph)

The field emission current density versus applied field (J-E) characteristic of the ZnS-RGO nanocomposites is shown in Fig. 2(a). The values of turn-on and threshold fields, defined as the fields, required to draw emission current densities of 1 and $10 \mu A /cm^2$, respectively, are found to be 2.32 and 2.72 for an anode- cathode separation of ~1mm. The emissions current density was observed to be 1050 $\mu A/cm^2$ at an applied field of ~4.1 V/μm. The low turn-on and threshold field values for ZnS-RGO nanocomposite emitter may be attributed to the modifications in the surface morphology and the electronic structure of the nanocomposite. As shown in Fig.2(b),the F-N plot of the ZnS-RGO nanocomposites non-linear in high field region, indicative of semiconducting behavior. The emission current stability at the pre-set value 1uA over a duration of more than 3h is shown in Fig. 2(c). The emission current is found to be stable at the pre-set value of1uA/cm² with small fluctuations (≤ 20%).The fluctuations may be attributed to the switching on/off of nanoprotrusions on the emitter surface. In Fig. 2(c), the inset shows a typical field emission micrograph of ZnS-RGO nanocomposite.

IV CONCLUSIONS

The ZnS-RGO nanocomposite has been synthesized by a one step solvothermal method. The XRD analysis indicates formation of the crystalline nature of the ZnS-RGO nanocomposite. The surface morphology of the ZnS-RGO nanocomposite is characterized by the presence of randomly distributed ZnS microspheres on the RGO sheets. The FE characterstics of nanocomposite gives turn on and threshold field values are 2.32 and 2.72 V/μm. The ZnS-RGO nanocomposite emitter is found to deliver a current density 1050 μA/cm² at an applied electric field of ~4.1V/μm. The ZnS-RGO nanocomposite exhibits good field emission stability and may find useful applications in various nanoellectronic devices.

ACKNOLEDGEMENT

RBS wishes to thank the Council of Scientific and Industrial Research (CSIR), Government of India, New Delhi for granting the Emeritus Scientist scheme. SRB thanks CSIR for the Research Associateship. Authors thank the Head of the Department, Department of Physics, Savitribai Phule Pune University, Pune for providing the experimental facilities

REFERENCES

[1] S. K. Tiwari, S. Sahoo, N. Wang, A. Huczko, "Graphene research and their outputs: Status and prospect"*Journal of Science: Adv. Mater. Devices*, vol. 5, pp 10-29 Jan 2020.

[2] X. Shao, A. Khursheed,"A Review Paper on "Graphene Field Emission for Electron Microscopy" *Appl. Sci.,*vol. 8, pp 868,May 2018.

[3] A.K.M. A. Iqbal. Md. , N. Sakib , A.K.M. P. Iqbal ,D. M. Nuruzzaman, Graphene-based nanocomposites and their fabrication, mechanical properties and applications, *Materialia,* vol.12, pp. 100815, August 2020.

[4] C.S.Rout, P. D. Joshi, R. V. Kashid, D. S. Joag, M.A. More, A.J. Simbeck, M.Washington3, S. K. Nayak, D. J. Late, Superior Field Emission Properties of Layered WS2-RGO Nanocomposites,*Sci. Rep.* vol. 3, pp. 3282, Nov. 2013.

[5] S.R.Bansode, K. Harpale, R. T. Khare, P. S. Walke, M. A.More,Morphological,structural and field emission characterization of hydrothermally synthesized MoS2-RGO nanocomposite, *Mater. Res. Express*, vol.3, pp 115023, Nov 2016

[6] S.R.Bansode, R.T. Khare, K. Harpale, M. A. More,Facile, single step synthesis of CdS—RGO heterostructure and its photo enhanced field emission investigations, *Mater. Res. Express*, vol.3, pp 035602, Nov 2016.

[7] M. Indhumathy, A. Prakasam, ZnS-Reduced Graphene Oxide Nanohybrid Materials as Photoanodes with Improved Photovoltaic Performance, *Journal of cluster science,*vol.31,pp.1, Jan 2020.

[8]A. Jha,S. K. Sarkar, D. Sen, K. K. Chattopadhyay, Carbon fiber-ZnS nanocomposite for dual applications as an efficient cold cathode as well as luminescent anode for display technology, *Nanoscale*, vol. 7, pp 6, Dec. 2014.

[9] Y. Ding, T. Cong, X. Chu, Y. Jia, X.Hong, Y. Liu, Magnetic-bead-based sub-femtomolar immunoassay using resonant Raman scattering signals of ZnS nanoparticles, *Anal Bioanal Chem*, vol. 408, pp. 5013–5019, May 2016.

Low-Macroscopic-Field Electron Emission from Metal Thin Films

I.S. Bizyaev, P.G. Gabdullin*, M.A. Chumak, V.Ye. Babyuk, S.N. Davydov, A.V. Arkhipov, O.E.Kvashenkina

Peter the Great St. Petersburg Polytechnic University, St. Petersburg, Russia
*Contact: gabdullin_pg@spbstu.ru

Abstract— **Films of Mo, Zr, W, Ni and Ti with thickness 2–20 nm were deposited by magnetron sputtering on Si substrates. After conditioning by the combined action of temperature (300–600°C) and electric field, films of Mo and Zr showed the capability of cold electron emission in an electric field 1.5-5 V/μm. Microscopic studies of the emissive samples revealed a partial reconstruction of the initially continuous films to the island form.**

I. INTRODUCTION

Various species of electrically heterogeneous nano-structured (EHN) carbon are known to emit electrons at room temperature in an electric field with a low macroscopic magnitude (LMF) – of the order of 1 V/μm [1–3]. However, the physical mechanism of emissivity for some of the EHN carbons remains unclear [2–4]. In our previous studies [5, 6], we investigated a structurally simple type of EHN emitters, namely, thin films of sp^2-bonded carbon consisting of nanoscale islands. Their LMF emissivity was explained in [7] by a model which took into account peculiar electronic properties of sp^2 carbon. In this work, we studied the emission properties of thin films of metals with similar EHN morphology.

II. EXPERIMENTAL METHODS

Thin films of Mo, Zr, W, Ni and Ti were deposited on flat Si substrates with a native oxide layer by magnetron sputtering in a Mantis HEX deposition system. Film thickness, growth rate and substrate temperature were varied in the range of 2–20 nm, 0.1–1.0 Å/s and 100–150°C, respectively.

Emission characteristics of the manufactured samples were measured in the parallel-plate geometry with a field gap width of 0.6 mm. The surface of each sample was studied before and/or after the emission tests with AFM, SEM and STM methods. Local electronic properties were determined by scanning tunneling spectroscopy (STS) with the use of the Omicron VT AFM XA 50/500 universal tool.

III. EXPERIMENTAL RESULTS

Many (~20) of the investigated metal film samples showed a more or less pronounced LMF emissivity. Near-threshold parts of a few typical I–V characteristics presented in Fig. 1 appear to be roughly exponential, which implies involvement of electron tunneling in the emission mechanism. The threshold field values for the best samples were as low as 1.5–4 V/μm. Among the tested film materials, Mo and Zr showed the best LMF emissivity. W and Ni film samples produced an emission current at field magnitudes 5–7 V/μm, but the current was unstable. With Ti films, no emission was ever obtained. Mo films with the best emission properties had thickness in the range 6–10 nm. It is important to note that the tested films usually did not possess the LMF emissivity in the as-grown

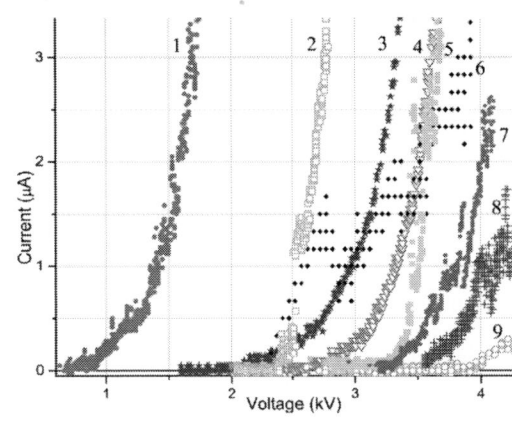

Fig. 1. Room-temperature emission I-V characteristics measured in planar geometry (cathode-anode distance 0.6 mm, anode diam. 6 mm) with film samples: 1 – Mo 10 nm; 2 – Mo 4 nm; 3, 4 – Mo 6 nm; 5 – Zr 8 nm; 6 – Zr 6 nm; 7 – Mo 8 nm; 8 – Mo 2 nm; 9 – Ni 6 nm.

state – this property was developed in the course of their conditioning. The samples were treated by the combined action of an electric field (~1 V/μm) and heating to 300–600°.

No whiskers or other high-aspect elements were found on the samples by microscopic studies, either before or after the emission experiments. Typical and most noticeable features that appeared after sample conditioning and/or emission testing represented circular breaches in the metal coating up to several microns in size (Fig. 2*a*). The structure of such film defects implies that it may have appeared in a process known as "dewetting" [8]. Size distribution of such holes in the emissive films was continuous, starting from tens of nanometers. The smallest ones were found over the entire area exposed to the electric field (Fig. 2*b*) – but not on the film margins outside the

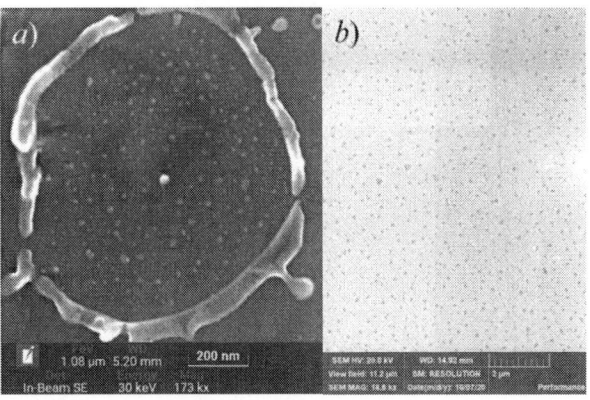

Fig. 2. SEM images of film structure defects on a Mo 10 nm sample after emission tests: a) a large, micron-scale hole; b) uniformly distributed small holes (dark dots) – the image is contrast-enhanced, scale bar is 2 μm.

anode "footprint" (the film samples were wider than the anode size); they were also not seen on pristine, as-grown films. Therefore, we related the smallest defects (Fig. 3a) to the combined action of the electric field and temperature during the conditioning. Presumably, they represent nuclei developing into the larger holes (and LMF emission centers) as a result of dewetting stimulated by emission-induced local heating and ion bombardment [9].

Tunneling current characteristics measured at different points inside and near the smallest hole (Fig. 3b–d) reveal dissimilar local electronic properties. The one in Fig. 3b shows a finite density of states (DOS) at the Fermi level, which is inherent in conductors. The graph in Fig. 3c displays a bandgap ca. 1 eV wide – apparently, the probe was in a contact with the semiconductor substrate at the hole bottom. Other spectra acquired at the hole bottom were staircase-like and included several approximately equidistant steps (Fig. 3d). Such shapes are usually associated with nanoparticles having isolated electron systems; they are explained by either dimensional quantization [3, 10] or the Coulomb blockade effect [11]. Thus, it can be concluded that at least some nanoislands directly observed in microscopic images (Fig. 2a, 3a) are insulated from their environment – as it is required by several emission models [1, 2, 4], including the model proposed in [7].

IV. DISCUSSION AND SUMMARY

We associate the observed LMF emissivity of metal films with the presence of nanoparticles in the electrically heterogeneous system created as the result of field or thermo-field conditioning of the emitter. Earlier, similar conclusion was made with respect to cold electron emission from MOS (metal/oxide/semiconductor) sandwich structures [12] and from island metal films [13]. In those cases, the emission was induced by an electric current passed across or along the system, or by some other energizing factors (e.g. irradiation with an IR laser [13]). These factors generated hot electrons in the nanoparticles where the most efficient relaxation channel was suppressed by dimensional effects [14]; such electrons were easily emitted. In contrast to those studies, the cold emission in our experiments was induced only by the action of an electric field. In [7], we

proposed a model of LMF emission from carbon films, which can also be adopted to the case of metallic nanoisland films. According to this model, electron emission is facilitated by electric field of insulated islands charged to electric potentials of the order of Volts. Their positive charge can be acquired and maintained by nanoscale thermoelectric phenomena. Details of the model will be disclosed elsewhere.

Concluding, the phenomenon of cold electron emission from metallic (Mo, Zr, W, Ni and Ta) thin films on oxidized Si was investigated in the reported experiments. Achievement of LMF emissivity was associated with formation of regions where the film lost its integrity and separate metal nanoislands appeared. Film disintegration was promoted by heating the sample to 300–600°C with simultaneous application of an electric field. Even though the achieved parameters of cold emitters were modest, the data obtained can be useful for a better understanding of the LMF emission mechanisms realized in electrically heterogeneous nanostructures.

REFERENCES

[1] A. Evtukh, H. Hartnagel, O. Yilmazoglu, H. Mimura, and D. Pavlidis, Vacuum Nanoelectronic Devices: Novel Electron Sources and Applications. Chichester, UK: John Wiley & Sons, 2015.

[2] R. Forbes, "Low-macroscopic-field electron emission from carbon films and other electrically nanostructured heterogeneous materials: hypotheses about emission mechanism," Sol. St. Electron., vol. 45, pp. 779–808, June 2001.

[3] V. Filip, L.D. Filip, H. Wong, "Review on peculiar issues of field emission in vacuum nanoelectronic devices," Sol. St. Electron., vol. 138, pp. 3–15, December 2017.

[4] E.D. Eidelman and A.V. Arkhipov, "Field emission from carbon nanostructures: models and experiment," Physics-Uspekhi, vol. 63, pp. 648–667, July 2020.

[5] A. Andronov, E. Budylina, P. Shkitun, P. Gabdullin, N. Gnuchev, O. Kvashenkina, and A. Arkhipov, "Characterization of thin carbon films capable of low-field electron emission," J. Vac. Sci. Technol. B, vol. 36, 02C108, March 2018.

[6] P. Gabdullin, A. Zhurkin, V. Osipov, N. Besedina, O. Kvashenkina, and A. Arkhipov, "Thin carbon films: correlation between morphology and field-emission capability," Diam. Relat. Mater., vol. 105, 107805, May 2020.

[7] A.V. Arkhipov, E.D. Eidelman, A.M. Zhurkin, V.S. Osipov, and P.G. Gabdullin, "Low-field electron emission from carbon cluster films: combined thermoelectric/hot-electron model of the phenomenon," Fuller. Nanotub. Car. N., vol. 28, pp. 286–294, April 2020.

[8] C.V. Thompson, "Solid-state dewetting of thin films," Annu. Rev. Mater. Res., vol. 42, pp. 399–434; August 2012.

[9] M.S. Tuzhilkin, P.G. Bespalova, M.V. Mishin, I.E. Kolesnikov, K.V. Karabeshkin, P.A. Karaseov, and A.I. Titov, "Formation of Au nanoparticles and features of etching of a Si substrate under irradiation with atomic and molecular ions," Semiconductors, vol. 54, pp. 137–143, January 2020.

[10] L.D. Filip, M. Palumbo, J.D. Carey, and S.R.P. Silva, "Two-step electron tunneling from confined electronic states in a nanoparticle," Phys. Rev. B, vol. 79, 245429, June 2009.

[11] N. Bagraev, A. Bouravleuv, L. Klyachkin, A. Malyarenko, W. Gehlhoff, Yu. Romanov, and S. Rykov, "Local tunnelling spectroscopy of silicon nanostructures," Semiconductors, vol. 39, pp. 685–696, June 2005.

[12] R.E. Thurstans and D.P. Oxley, "The electroformed metal-insulator-metal structure: a comprehensive model," J. Phys. D, vol. 35, pp. 802–809, April 2002.

[13] R.D. Fedorovich, A.G. Naumovets, and P.M. Tomchuk, "Electron and light emission from island metal films and generation of hot electrons in nanoparticles," Phys. Rep., vol. 328, pp. 73–179, April 2000.

[14] Y. Bilotsky and P.M. Tomchuk, "Peculiarity of electron–phonon energy exchange in metal nanoparticles and thin films," Surf. Sci., vol. 602, pp. 383–390, January 2008.

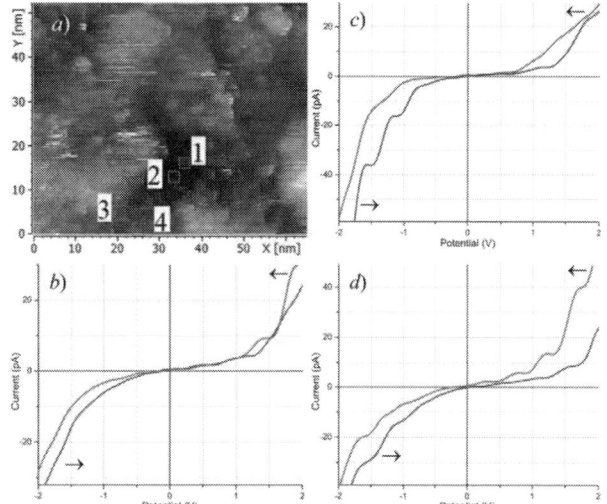

Fig. 3. a) High-magnification STM image of an activated 10 nm Mo film sample; b),c),d): STS spectra measured at positions 3, 2 and 1, respectively.

978-1-6654-2590-2/21 $31.00 © 2021 IEEE

Cold cathode electron gun based on single wall carbon nanotubes field emitters for THz traveling wave tube

Ruirui Jiang[1], Baoqing Zeng[1*], Jianlong Liu[1], Kaiqiang Yang[1], and Jing Zhao[1]

[1]*School of Electronic Science and Engineering, University of Electronic Science and Technology of China, Chengdu, Sichuan, China*

*Corresponding author: bqzeng@uestc.edu.cn

Abstract— **A cold cathode electron gun for THz traveling wave tube (TWT) was designed. The single wall carbon nanotubes (SWCNTs) cold cathode was fabricated by screen printing with the radius of 0.1mm. The turn-on field of the SWCNTs cathode was 0.78V/um, and the field enhancement factor was 8500. The maximum emission current density exceeded 10A/cm^2. Based on the field emission test, we designed a cold cathode electron gun with the beam waist radius of 0.03mm. And the area compression ratio of the electron gun is about 11. The emission current density and beam radius of the designed SWCNTs cold cathode electron gun have met the design requirements of THz TWT.**

Keywords—*field emission, cold cathode, electron gun, THz TWT*

I. INTRODUCTION

Field emission cathode or cold cathode has the advantages of anti-radiation, instantaneous start-up, high current density and miniaturization, etc., which are ideal for the electron sources of traveling wave tubes (TWTs) [1]. Field emission cathodes apply the strong external electric field to extract electrons from various materials via tunnelling effect [2]. Carbon nanotubes-based field emission cathode is an ideal new type of electron source for TWTs, which has the characteristics of high aspect ratios, high stability, high field emission current, room-temperature working and low cost [3].

The terahertz (THz) region of the electromagnetic spectrum has enormous potential for communications, space research, medicine, biology, and remote sensing [4]. TWTs are high power, high gain microwave amplifiers that continue to play an important role in THz region. The electron gun, which forms an electron beam for effective RF wave to beam interaction, is the most critical component of any TWT [5, 6]. It plays a vital role in determining the high electronic efficiency, high linearity and high reliability which are major considerations for a THz device.

II. EXPERIMENTS

The single wall carbon nanotubes (SWCNTs) cathode used in this work was prepared by screen printing, as we reported elsewhere. The field emission test was carried out in high vacuum chamber and the base pressure was kept at about 1×10^{-4} Pa. The SWCNTs cathode and a metal anode formed a diode structure and the space was remained at 300μm. The radius of the SWCNTs cathode is 0.1mm. For characterization of the field emission properties, the electric field was applied using the dc drive or the pulse (frequency: 10 Hz, pulse width: 1ms) drive. The turn-on field (corresponding to the emission current density of 10 μA/cm^2) of the SWCNTs cathode was tested under dc drive. In the case of the pulse drive, the maximum emission current was measured.

The surface morphology of the SWCNTs cathode was characterized by scanning electron microscopy (SEM) and transmission electron microscopy (TEM). The electromagnetic and particle simulation software CST was used to research the electron gun.

III. RESULTS AND DISCUSSION

A. Field emission test of SWCNTs cathode

The field emission characteristics were analyzed with the Fowler-Nordhiem (*F-N*) theory described by [2]

$$J = A\frac{(\beta E)^2}{\varphi}e^{-\frac{B\varphi^{3/2}}{\beta E}} \tag{1}$$

Where *J* is the emission current density, *E* is the electric field of the cathode surface, *β* is the field enhancement factor, *A* and *B* are constants ($A = 1.54\times10^{-6}$ A·eV·V^{-2} and $B = 6.83\times10^{3}$ V·eV$^{-3/2}$·μm^{-1}), and *φ* is the work function of the cathode.

Fig. 1 (a) is the SEM image of the SWCNTs cathode. It can be seen that there are plenty of emission points on the cathode surface, which indicates excellent field emission performance. Fig. 1 (b) is the TEM image of the SWCNTs, from which we can see that the SWCNTs have ultrahigh aspect ratio and hence the SWCNTs cathode has very low turn-on field.

Fig. 2 (a) is the field emission test of the SWCNTs cathode at dc mode. The turn-on field is as low as 0.78MV/m, and the field enhancement factor reaches 8500. After many repeated measurements, the cathode has good repeatability and stability. Under the pulse condition, the measured maximum emission current is 3.92mA, and the corresponding current density exceeds 10A/cm^2, as shown in Fig. 2 (b).

978-1-6654-2590-2/21 $31.00 © 2021 IEEE

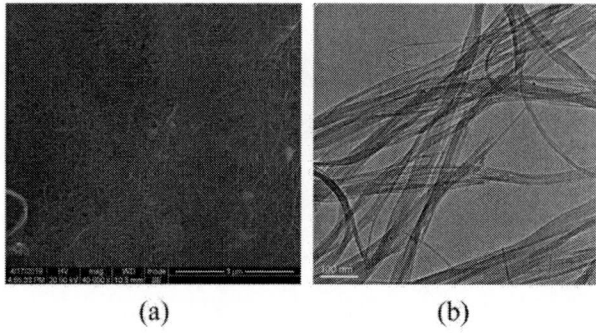

Fig. 1 (a) The SEM image and (b) The TEM image of the SWCNTs cathode.

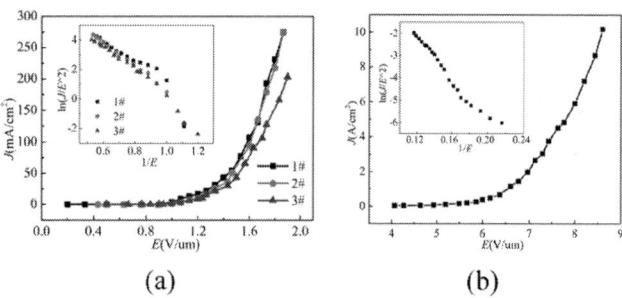

Fig. 2 The J-E plots and corresponding F-N plots (a) under dc mode and (b) under pulse mode.

B. Simulation design of the cold cathode electron gun

Fig. 3 is the simulation schematic diagram of SWCNTs cold cathode electron gun. The electron beam is focused by carefully adjusting the position and voltage of electrodes. The emission current is controlled by the grid voltage.

The designed parameters are listed in Table I.

TABLE I
THE DESIGNED PARAMETERS OF THE COLD CATHODE ELECTRON GUN

Parameters	Values
The radius of cathode	0.1 mm
The aperture of the grid	0.5 mm
The aperture of focusing electrode	0.8 mm
The aperture of anode	0.3mm
The distance between the cathode and grid	1 mm
The distance between grid and focusing electrode	0.5 mm
The distance between focusing electrode and anode	1 mm
The voltage of cathode	0 V
The voltage of grid	4000 V
The voltage of focusing electrode	500 V
The voltage of anode	10000 V

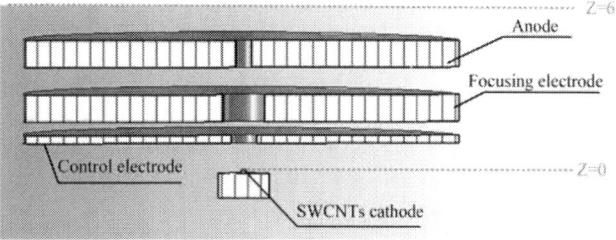

Fig. 3 The schematic diagram of SWCNTs cold cathode electron gun.

Fig. 4 is the electron trajectories of the SWCNTs cold cathode electron gun. The electron beam is well focused. From the electron distribution at the cross section of beam waist (Z=6mm), we can see that the radius of the electron beam waist is 0.03mm, which has met the design requirements of some THz TWT. And the area compression ratio of the electron gun is about 11.

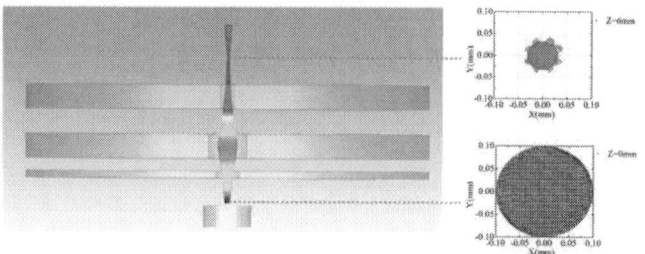

Fig. 4 The electron trajectories of the SWCNTs cold cathode electron gun and the distribution of electron beam on the cross section of the cathode (z=0) and the beam waist (z=6).

IV. CONCLUSIONS

In conclusion, we designed a cold cathode electron gun for THz TWT based on the SWCNTs field emission. The vacuum electronic devices using cold cathode can work at room temperature without the cathode heater, which greatly decrease the volume and power consumption. The low turn-on field of the SWCNTs cathode also reduce the operation voltage of the electron gun. THz TWT with cold cathode can realize miniaturization and instantaneous start-up.

ACKNOWLEDGMENT

This work was supported in part by National Natural Science Foundation of China under Grant 61921002, Grant 61571103, Grant 61531010, Grant 61601101 and Grant 61941110, and in part by Fundamental Research Funds for the Central Universities, and National Key Laboratory of Science and Technology on Vacuum Electronics under Grant 6142807190205.

REFERENCES

[1] H. M. Manohara, R. Toda, R. H. Lin, A. Liao, M. J. Bronikowski, and P. H. Siegel, "Carbon nanotube bundle array cold cathodes for THz vacuum tube sources", J. Infrared, Millimeter, Terahertz Waves, vol. 30, no. 12, pp. 1338–1350, 2009.

[2] Z. L. Lin, X. J. Wang, "Field electron emission" in *Cathode Electronics*. Beijing, China: National Defense Industry Press, 2013, pp. 98-115.

[3] X. F. Shang, J. J. Zhou, P. Zhao, Z. H. Li, S. Qu, Z. Q. Gu, Y. B. Xu, and M. Wang, "The enhanced field-emission properties of screen-printed single-wall carbon-nanotube film by electrostatic field", Appl. Surf. Sci., vol. 256, no. 7, pp. 2005–2008, 2010.

[4] P. H. Siegel, "Terahertz technology", IEEE Transactions on Microwave Theory and Techniques, vol. 50, no. 3, pp. 910-928, 2002.

[5] R. K. Sharma, A.K. Sinha, S. N. Oshi, "An improved method for the synthesis of anode aperture for Pierce guns", IEEE Transactions on Electron Devices, vol. 48, no. 2, pp. 395-397, 2001.

[6] Y. Cheng, Z. Chen, Y. Wang, H. Yin, "Low perveance confined-flow Pierce gun for a 0.14 THz broadband folded waveguide traveling wave tube", Vacuum, vol. 86, no. 3, pp. 335-339, 2011.

First-Principle Model of the Electron Field Emission From Silicon Nano-Scale Tip

Gleb D. Demin[1,*], Nikolay A. Djuzhev[1], Nikolay N. Patyukov[1], and Ilya D. Evsikov[1]

[1]National Research University of Electronic Technology (MIET), Shokin sq., bld. 1, Russia

*Contact: gddemin@gmail.com, phone +7-499-720-6907

Abstract— **Prediction of electron field-emission current of nano-scale devices is of significant importance in the field of vacuum nanoelectronics. Nowadays, modified Fowler-Nordheim model (F-N) being widely used to estimate electron field-emission current of various systems. Nevertheless, it was shown that experimental values of the emission current from nano-scale emitters can differ from such values obtained within F-N theory. Moreover, the F-N formalism is not entirely suitable for describing field emission from silicon nanostructures, where it is important to take into account the interaction of electrons with impurities, phonon scattering, temperature effects (Joule heating), and the penetration depth of the electric field in the semiconductor. For the accurate simulation of electron field emission in the case of nano-scale silicon tip, first-principle quantum-mechanical models based on Density Functional Theory (DFT) and Density Functional Perturbation Theory (DFPT) formalisms should be used. In this work, we calculated within first-principle approach the field-emission current in a system consisting of the silicon nanoscale tip (emitter) and collector of the same type, which are separated by a nanoscale vacuum gap. The calculations were performed using time-dependent perturbation theory for non-zero external electric field within the non-equilibrium Green functions (NEGF) formalism to describe the electron transport and inelastic scattering in the emitter, where pseudopotentials and Kohn-Sham equations are explored in a self-consistent manner. This theoretical approach can serve as an important step towards the consideration of field-electron emission process beyond the F-N theory to describe correctly field-emission experimental data at the nanoscale.**

I. INTRODUCTION

In recent years, there has been a rapid revival of vacuum nanoelectronics, which is associated with the emergence of the technological possibility of forming solid-state nanoscale structures, scalable to 7 nm and below. This opens the way to the fabrication of new electronic devices with a quasi-vacuum conduction channel of length below the mean free path of electrons in air (150-200 nm), which eliminates the need for additional vacuum packaging. The principle of operation of such devices is based on the effect of electron field emission into vacuum, where the ballistic transport of hot electrons through the vacuum gap ensures their high performance (with a frequency in the range from 0.35 to 4 THz), low power consumption (operating voltage is varied from 15 to 3 V and below), as well as stable operation under harsh conditions of high temperature and radiation.

Silicon field emission devices with a nanoscale vacuum channel, which can be fabricated using well-known CMOS (complementary metal-oxide-semiconductor) technology, seem to be the most preferable for progress in the field of vacuum nanoelectronics, since they have excellent reproducibility (in contrast to field emitters based on carbon nanotubes and some exotic materials), low cost and compatibility with most technological processes. For the above reasons, silicon-based nanoscale vacuum channel devices can be successfully applied to develop hybrid integrated circuits (ICs) that combine the advantages of both vacuum and silicon CMOS technology. These ICs can become especially in demand when performing long-term space missions to explore distant planets of the solar system, where reliable operation of the electronics is required under prolonged exposure to cosmic radiation. An important step towards the design and development of such hybrid ICs based on silicon field-emission nanodevices is to predict their operating parameters using correct theoretical models, which should be consistent with the experiment.

Currently, to interpret the results of experimental measurements of cold electron emission from nanostructures, most studies use the F-N theory with a number of corrections and modifications (Murphy-Good formalism [1], Forbes-Deane approximation [2]). Nevertheless, the F-N approach is well suited mainly for describing the field emission characteristics of nanostructures with a metal conductivity (carbon nanotubes, refractory metals (W, Mo, Au), heavily-doped semiconductors). At the same time, when measuring the metal-like type of field-emission structures, there is often a serious discrepancy between theory and experiment [3], where the theoretically predicted values of the field-emission current can be overestimated and strongly depend on the magnitude of the phenomenological constants chosen for a particular case. Calculations within F-N model will result in relatively accurate values only up to certain scale, where the impact of surface states at the atomic scale is not significant. Recent investigations clearly show that F-N calculations are incorrect for emitters with tip radius of less than 20 nm [4]. Also, even simple first-principles simulations demonstrate large deviation of field-emission characteristics from F-N behaviour at relatively high bias voltage.

The phenomenon of field emission from semiconductor nanoscale tips cannot be fully described within the framework of F-N theory, since it depends on the type and level of impurity

978-1-6654-2590-2/21 $31.00 © 2021 IEEE

concentration, temperature (as a result of Joule heating), and the penetration depth of the electric field into the semiconductor subsurface layer. These values determine both the potential profile at the semiconductor-vacuum interface and the field-emission current, which is not taken into account in the standard F-N formalism [5].

To describe the field emission from silicon (Si) nanoscale tip, we performed first-principle calculations of the ballistic electron transport in the system with five regions: 1) left Au electrode, 2) Si nanoscale tip (emitter), 3) nanoscale vacuum gap, 4) Si collector and 5) right Au electrode (see Fig. 1a). The calculations were performed in Quantum ESPRESSO package based on the DFT and DFPT using NEGF formalism, which is responsible for quantum electron transport in electrical contacts Au/Si. Such a theoretical analysis of electron field emission based on first-principle model, which allows one to take into account the influence of the electron-phonon interaction, temperature effects, and electron scattering by impurities in a field emitter, is necessary to form a correct theoretical concept of electron field emission from silicon nanotips at the nanoscale, which corresponds to the experimental results.

II. THEORY: FIRST-PRINCIPLE MODEL OF FIELD EMISSION

To calculate the field emission from a silicon nanoscale tip within DFT, a set of basis wave functions in the form of a linear combination of atomic orbitals (LCAO) was chosen, while transport through the vacuum gap was calculated based on the Kohn-Sham equation using NEGF to describe electron scattering in the electrode regions. With an increase in the thickness of the vacuum gap d_{vg} ($d_{vg} > 0.6$ nm) and a transition to the tunnelling regime, the transmission coefficient in first-principle calculations using the LCAO basis tends to zero.

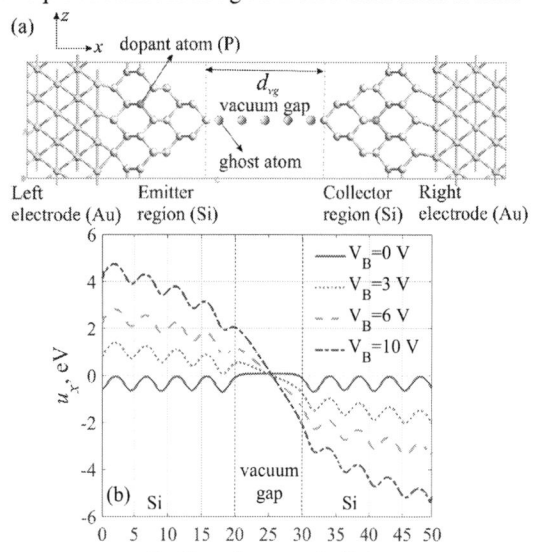

Fig. 1 (a) Schematic sketch of the Au/Si/vacuum/Si/Au field-emission (FE) structure. (b) Potential profile of the FE structure for different bias voltages.

The exponential decay of LCAO wave functions can be simulated using so-called ghost atoms, i.e. free orbitals that do not have a local distribution of electrons and pseudopotential. The description of a part of the vacuum region through ghost

atoms does not violate the electronic structure and improves the description of wave functions in the vacuum region. Placement of ghost atoms should occur within the boundaries of the orbitals of real surface atoms (Fig. 1a). For a qualitative estimation of the field-emission characteristics of a silicon nanotip, the field emission structure in Fig. 1a was considered with the vacuum gap thickness d_{vg} varied from 0.6 to 1.0 nm. The bias voltage V_B in the range from 0 to 10 V is applied. Based on the DFT using NEGF formalism and ghost atoms, the potential profile u_x of the structure (Fig. 1b) and the transmission coefficients $T_n(\varepsilon)$ through the scattering region were calculated, where the bias voltage $V_B = \mu_L - \mu_R$ was determined as a difference between electrochemical potentials μ_L and μ_R of the left and right electrodes, respectively.

III. RESULTS AND CONCLUSION

The current I_{fe} was calculated based on the Landauer formalism (Fig. 2a). The nonzero values of the current for 0.6 nm are explained by a partial transition to the contact mode within the framework of basic functions, namely, the contribution of the 2P orbitals of the nanoscale tip and collector atoms closest to the surface. In Fig. 2b, the I - V characteristics are plotted in the F-N coordinates for a visual representation of the cut-off and the linear region of field emission.

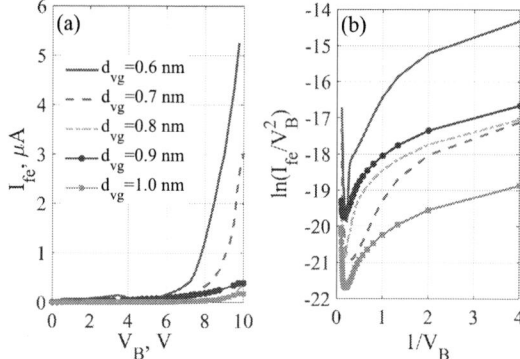

Fig. 2 (a) I-V characteristics and (b) F-N plot of the structure shown in Fig. 1a.

ACKNOWLEDGMENT

The work was performed in the R&D Center «MEMSEC» (MIET) and supported by the RF Ministry of Education and Science (grant No. 075-03-2020-216, 0719-2020-0017).

REFERENCES

[1] E. L. Murphy, and R. H. Good, "Thermionic emission, field emission, and the transition region," Phys. Rev., vol. 102, pp. 1464–1473, June 1956.

[2] J. H.B. Deane, and R.G. Forbes, "The formal derivation of an exact series expansion for the principal Schottky–Nordheim barrier function v, using the Gauss hypergeometric differential equation," J. Phys. A: Math. Theor., vol. 41, p. 395301, October 2008.

[3] R. G. Forbes, and J. H.B. Deane, "Fowler-Nordheim plot analysis: a progress report," Jordan J. Phys., vol. 8, p. 125, June 2015.

[4] G. D. Demin et al., "Comprehensive analysis of field-electron emission properties of nanosized silicon blade-type and needle-type field emitters," J. Vac. Sci. Technol. B, vol. 37, p. 022903, March 2019.

[5] J.A. Driscoll, S. Bubin, W. R. French, and K. Varga, "Time-dependent density functional study of field emission from nanotubes composed of C, BN, SiC, Si, and GaN," Nanotechnology, vol. 22, p. 285702, June 2011.

978-1-6654-2590-2/21 $31.00 © 2021 IEEE

The notional emission area for cylindrical posts and its variation with local electric field

Rajasree Ramachandran[1,2,*], Debabrata Biswas[1,2]

[1]Homi Bhabha National Institute, Mumbai 400 094, INDIA

[2]Bhabha Atomic Research Centre, Mumbai 400 085, INDIA

*Contact: rajasreer@barc.gov.in, phone 022-25590805

Abstract — **The dependence of the notional emission area on the apex electric field for a single field-emitter tip is investigated for cylindrical posts. It is found that a power-law dependence on the local field exists only when the end-cap on the cylinder is modelled as a hemi-ellipsoid with height $h_e \geq 3R_a$. Specifically, for a hemispherical end-cap, the notional area does not increase with local field as a power law. The power law behaviour is expected to hold for generic end-caps where the end-cap shape closely follows a parabola for $\rho \leq R_a$.**

I. INTRODUCTION

Field emitter based cathodes play a prominent role as electron sources in modern vacuum nano-electronic devices [1]. Such cathodes comprise of a planar substrate on which sharp emitter tips having apex radius of curvature R_a typically of the order of a few hundred nano-meters or even less are grown. When dealing with field emitter tips with $R_a \geq 100$ nm, conventional field emission theories for planar cathodes [2,3] generally hold while for smaller apex radius of curvature, the curvature corrected field emission formalism needs to be used [4,5].

It is a notable fact that most of the current I emitted from a single emitter tip can be attributed to a fraction of the entire surface area of the endcap. Conventionally [6], the net current is expressed as a product of the apex current density J_a and quantity referred to as the notional emission area A_n, defined as $A_n(E_a) = I(E_a)/J_a(E_a)$ where $E_a = \gamma_a E_0$ is the local electric field at the apex, γ_a is the apex field enhancement factor, E_0 is the macroscopic electric field, while the current density J_a is the zero-temperature Muphy-Good (MG) current density at the apex. At an arbitrary point $\mathbf{r}$ on the surface of the emitter,

$$J_{MG}(\mathbf{r}) = \frac{1}{t_F^2(r)} \frac{A_{FN}}{\phi} E(r)^2 e^{-B_{FN} \vartheta_F(r) \phi^{\frac{3}{2}}/E(r)} \quad (1)$$

Here, $A_{FN} \cong 1.541434 \, \mu A \, eV \, V^{-2}$ is the first Fowler-Nordheim constant while $B_{FN} \cong 6.830890 \, eV^{-\frac{3}{2}} V \, nm^{-1}$ is the second constant and ϕ is the work function. Here $\vartheta_F = 1 - f + \frac{1}{6} f \ln f$ and $t_F = 1 + \frac{1}{9} f + \frac{1}{18} f \ln f$ are potential barrier shape correction factor with $f = f(r) \approx \frac{c_s^2 E(r)}{\varphi^2}$ while $c_s^2 = 1.439965 \, eV^2 V^{-1} nm$ is the Schottky constant and $E(\mathbf{r})$ is the local electric field at that point. Thus, if we know the local field on the emitter surface, the net current $I(E_a)$ from the emitter tip can be obtained by integrating the current density (Eq. 1) over the surface.

Since the net current as well as the apex current density depends on E_a, it is natural to expect that the notional emission area will also be a function of the electric field. The net current is often expressed empirically as $I = A E_0^k \, e^{-\frac{B}{E_0}}$. Equivalently, it may also be expressed in terms of the applied gap voltage V_g, with $E_0 = V_g/D$ where D is the distance between the cathode and anode plates of a parallel plate diode. The exponent k incorporates the apex current density which contributes a term $2 - v$ where $v = \frac{\eta}{6} \approx 9.836 \, (eV)^{\frac{1}{2}} \phi^{-\frac{1}{2}}$. For $\phi = 4.5 \, eV$, $2 - v \approx 1.227$. Assuming that the notional area exhibits a power-law dependence $A_n \sim E_a^\alpha$, $k = 2 - v + \alpha$. Clearly, it is important to investigate the existence of the power-law behaviour as well as the value of α. For endcaps, well approximated by a parabola, it is known [7] that $\alpha \approx 1$.

In this study, we shall examine the existence of a power law for cylindrical posts with various endcaps numerically and also determine the value of α when a power law is observed.

II. METHODOLOGY

Consider the family of Hemi-Ellipsoid on Cylindrical Post (HECP) emitters, in which a hemiellipsoid end-cap is mounted on a cylindrical post. The total height h and apex radius of curvature R_a of the HECP emitter are kept constant while the height of the hemiellipsoid end cap h_e is varied from R_a to $5R_a$. In each case the surface field is evaluated using COMSOL in which the HECP is placed in a planar diode configuration in a 2D axis-symmetry domain (ρ,z) with right boundary having Neumann boundary condition (zero charge) and the anode kept at distance of about five times the total height of the emitter on which the surface charge specified.

Using the surface field thus obtained, the current density is evaluated using Eq.(1) and the total current emitted is calculated by integrating over the surface to obtain

$$I = \int J(\rho) 2\pi \rho \sqrt{1 + \left(\frac{dz}{d\rho}\right)^2} \, d\rho \quad (2)$$

for apex electric fields E_a in the range 1-10V/nm. The limit of the integration is chosen such that convergence is achieved for all values of the fields, which is typically from $\rho = 0$ (apex) to $\rho = 2R_a$. When $h_e = R_a$ (hemispherical endcap), the integration is extended over the cylindrical post till convergence is obtained. A plot of $ln\left(\frac{I}{J_a}\right) = ln(A_n)$ vs $ln(E_a)$ has been used to determine the dependence of notional emission area on electric field.

III. RESULTS

In the following, we shall investigate whether power law holds in the pre-exponential factor in the expression for the current emitted from a single emitter. For this, consider an HECP emitter of total height $h = 50\mu m$ and tip radius $R_a = 100$nm mounted in a parallel plate configuration. For E_a ranging from 1-10V/nm a plot of $ln\left(\frac{I}{J_a}\right)$ vs $ln(E_a)$ has been generated for different heights of the hemiellipsoid endcap h_e ranging from R_a to $5R_a$ keeping the total height of the emitter fixed. The following three cases are worth examining.

(i) $h_e = R_a$: This corresponds to a Hemisphere on Cylindrical post (HCP) emitter tip, which is a special category of emitter on which the parabolic approximation for the apex hardly holds. A typical fit of $ln\left(\frac{I}{J_a}\right)$ vs $ln(E_a)$ plot given by Fig. 1 shows that linear fit with slope 0.89 exists only in the low field region for small values of apex electric field $E_a < 1.5$ V/nm which is too low for a measureable current to detect. Hence for an HCP emitter, the dependence of the notional area A_n on E_a is not a power law and the relation $I = AE_0^k e^{-\frac{B}{E_0}}$ may not hold.

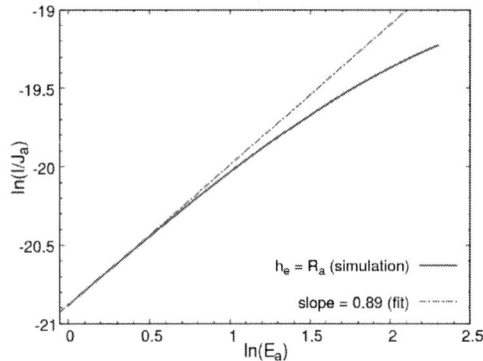

Fig. 1 Variation of notional emission area with apex electric field for HCP emitter. The linear fit holds for $E_a < 1.5$V/nm with slope 0.89.

(ii) $2R_a \geq h_e < 5R_a$: Here, as the height of the hemi-ellipsoidal endcap h_e increases from $2R_a$ to $4R_a$, the linear region in the plot of $ln\left(\frac{I}{J_a}\right)$ vs $ln(E_a)$ increases implying a power law dependence of the notional emission area on the apex electric field. This is expected because as h_e increases, the local parabolicity near the apex of the emitter increases and holds up to $\rho = 4R_a/5$. It is in agreement with the theoretical prediction that for a parabolic emitter tip, the notional area varies linearly with E_a in the low-moderate field range $E_a < 5$ V/nm. The results for $h_e = 3R_a$ are shown in Fig. 2 with the linear fit having a slope 1.03.

(iii) $h_e = 5R_a$: When the height of the hemiellipsoid end cap becomes five times the apex radius, substantial current is emitted from the end cap itself. In this scenario, the local parabolicity is valid upto $\rho \approx 2R_a$ and so the linear region in the $ln\left(\frac{I}{J_a}\right)$ vs $ln(E_a)$ plot is observed for $E_a < 7.4$V/nm. Thus, the validity of the linearity of the notional emission area with field is improved.

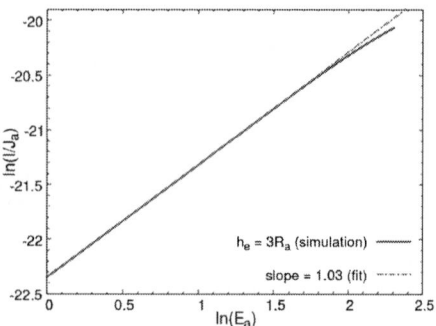

Fig.2 Variation of notional emission area with apex electric field for an HECP emitter with $h_e = 3R_a$. The linear region holds for $E_a < 4.5$V/nm with slope 1.03.

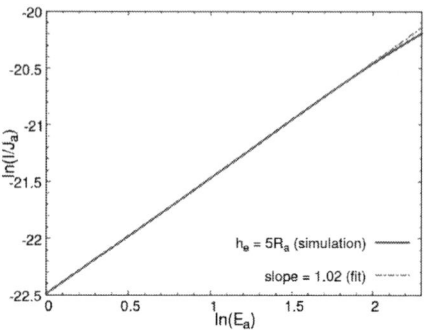

Fig. 3 Same as Fig.2 for $h_e = 5R_a$. The linear region holds for $E_a < 7.4$V/nm

IV. CONCLUSIONS

In this study we have investigated the field dependence on the notional emission area for HECP emitter tip. A plot of $ln(I/J_a)$ vs $ln(E_a)$ shows linear behaviour in a range which increases as h_e varies from R_a to $5R_a$. The slope of the linear region is found to be around 0.89 for $E_a < 1.3$ V/nm when $h_e = R_a$ while it takes a value close to unity for $h_e \geq 2R_a$. At $h_e = 5R_a$, the linear region is observed for $E_a < 7.4$ V/nm. These observations are in agreement with the theoretical calculations [7] of $A(E_a)$ which predict a linear dependence on E_a at low-moderate fields for tips that can be well approximated by the parabolic approximation and having apex radius of curvature $R_a \geq 100$nm.

The authors thank Dr. Raghwendra Kumar for discussions.

REFERENCES

[1] N. V. Egorov and E. P. Sheshin, "Carbon-based field emitters: Properties and applications," in *Modern Developments in Vacuum Electron Sources,* Topics in Applied Physics Vol.135, edited by G. Gaertner, W. Knapp, and R. G. Forbes (Springer, Cham, 2020).

[2] R. H. Fowler and L. Nordheim, Proc. R. Soc. A 119,173(1928).

[3] E. L. Murphy and R. H. Good, Phys. Rev. 102, 1464(1956)

[4] A. Kyritsakis and J. P. Xanthakis, Proc. R. Soc. London A 471, 20140811(2015).

[5] D.Biswas, R. Ramachandran, J. Appl. Phys. 129 194303(2021).

[6] R.G.Forbes, E.O.Popov, A.G.Kolosko and S.V.Filippov, R. Soc. Open Sci.8:201986(2021).

[7] D.Biswas, Phys. Plasmas 25, 043105(2018).

[8] D.Biswas and R.Ramachandran, J. Vac. Sci. Technol, B 37, 021801(2019)

Cascade Electron Source Based on Horizontal Tunneling Junction

Zhiwei Li
Key Laboratory for the Physics and Chemistry of Nanodevices

Department of Electronics, Peking University

Beijing, China
lzw111@pku.edu.cn

Xianlong Wei
Key Laboratory for the Physics and Chemistry of Nanodevices

Department of Electronics, Peking University

Beijing, China
weixl@pku.edu.cn

Abstract—On-chip electron source is an important trend for electron source and vacuum devices. As a mainstream alternative for on-chip electron source, tunneling electron source, which is usually in a vertical-stacked structure, is encountering bottleneck of low electron emission efficiency. Electron source based on horizontal tunneling junction has been proposed to improve the emission efficiency. Here, we proposed a cascade electron source by connecting several identical such electron sources in series to further improve the emission efficiency. Both theoretical and experimental results show that cascade electron source based on horizontal tunneling junction can greatly improve the emission efficiency to nearly 50%.

Keywords—electron source; cascade; on-chip; tunneling junction

I. INTRODUCTION

On-chip electron source is an important trend for electron source and vacuum devices. As a mainstream alternative for on-chip electron source, tunneling electron source, which is usually in a vertical-stacked metal-insulator-metal (MIM) or metal-oxide-semiconductor (MOS) structure, is encountering bottleneck of low electron emission efficiency. The emission efficiency, a benchmark for tunneling electron source, is defined as the current emitted into vacuum compared to that tunneled across the internal metal-insulator energy barrier. Apart from reducing the thickness of top electrode layer, electron source based on horizontal tunneling junction has been proposed to improve the emission efficiency.

Here, we proposed a cascade electron source by connecting several horizontal tunneling junctions in series to further improve the emission efficiency. [1] This cascade electron source removes the spatial restriction compared with vertical tunneling electron sources, so that we can extract emission current cumulatively step by step. Theoretical calculation predicts that the total emission efficiency increases approximately linear to the number of steps and an experimental result of 47.6% is obtained in a cascade electron source with three horizontal tunneling junctions.

II. RESULTS AND DISCUSSION

Figure 1a and 1b show the schematic structure and energy band diagram of a cascade electron source based on horizontal

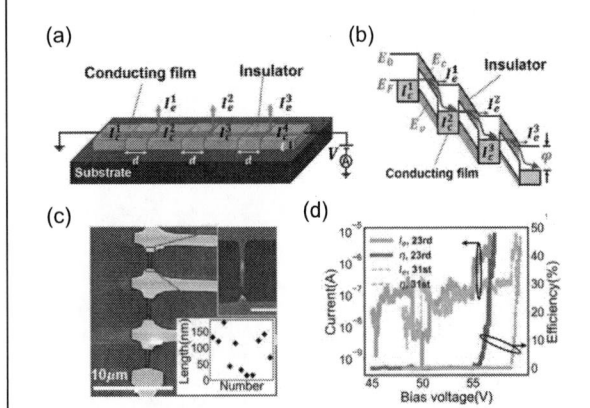

Figure 1 A cascade electron source based on horizontal tunneling junction (a) Schematic diagram **(b)** Energy band diagram **(c)** SEM image **(d)** Electrical characterization

tunneling junction. Several identical horizontal tunneling junctions are connected in series in a line. When a voltage bias is applied to two ends of the structure, electrons first tunnel from the side with lowest electric potential into the adjacent insulator, then travel along the rest metal and insulator layers, where electron emission into vacuum occurs in every junction, so that emission current can be collected additively.

Figure 1c shows a cascade electron source fabricated by using a metal nanowire with a width of 200 nm and height of 15 nm on a SiO_2 substrate. At each channel the nanowire is first broken by Joule heating, forming a silicon conducting filament between the electrical broken gap with the help of resistive switching phenomenon of SiO_2, then a bias voltage of over 7 V is applied across the silicon conducting filament to breakdown the filament into a $Si-SiO_x-Si$ nanogap, or a horizontal tunneling junction. An overall voltage bias is then applied across all the three tunneling junctions. Figure 1d shows the electrical performance of such a cascade electron source. When a bias voltage of 57 V is applied, the emission current reaches 5.0 uA. Since the total current from source electrode is 10.5 uA, the emission efficiency reaches 47.6%. This value is much higher than that of previously reported vertical tunneling electron sources.

978-1-6654-2590-2/21 $31.00 © 2021 IEEE

III. Summary

A cascade electron source based on horizontal tunneling junction is proposed to overcome the bottleneck of emission efficiency in tunneling electron sources. This novel vacuum device is designed and fabricated successfully, realizing a high emission efficiency of 47.6%.

References

[1] Z. Li and X. Wei, "A Cascade Electron Source Based on Series Horizontal Tunneling Junctions," *IEEE Transactions on Electron Devices,* vol. 68, no. 2, pp. 818-821, Feb 2021, doi: 10.1109/TED.2020.3044868.

Degradation of an emitter based on VACNT made by DC-PECVD during field emission

M.A. Chumak*, A.A. Rokacheva, L.A. Filatov,
I.S. Bizyaev

Peter the Great St.-Petersburg Polytechnical University,
Polytechnitscheskaya 29, St.-Petersburg,
195251, Russia
*Corresponding author: equilibrium2027@yandex.ru

E.O. Popov, S.V. Filippov, A.G. Kolosko

Ioffe Institute,
Polytechnitscheskaya 26, St.-Petersburg,
194021, Russia

Abstract—**This paper presents a technology for manufacturing LAFE based on VACNT by DC-PECVD on an Fe catalyst deposited by CVD on a SiO₂/Si substrate immediately in the form of nanoisland. The emission properties and the state of the cathode after emission tests were also investigated. Voltage-current characteristics tested for compliance with the cold field emission regime.**

Keywords — Field emission, LAFE, DC-PECVD, VACNT, catalyst, effective emission parameters, ferrocene.

I. INTRODUCTION

Carbon nanotubes (CNT) are promising candidates for production of various micro-sized devices. A special place is given to the development of sources of free electrons based on the field emission effect. The sources can be used to create electron nanolithography systems, photoelectric converters, amplifiers of electrical signals (traveling wave tubes), and devices for household and scientific research purposes: monitors, lamps, X-ray machines, gas sensors, microscopes, space telescopes, etc. [1-2].

The use of electron-beam lithography methods makes it possible to produce a regular and rather rarefied structure of catalyst islands [3]. Lithography methods are the most reliable in terms of reproducibility, but at the same time, the most expensive and ineffective for processing large-area field emitters (LAFE). Another, cheaper method is the coating of the substrate with a layer of catalyst, which is then subjected to heat treatment. As a result, the integrity of the layer is violated under the action of surface forces, and it turns into islands [4].

In this work we propose a new concept of the technology for creating LAFE based on nanotubes, using the CVD method for the controlled deposition of the catalyst metal immediately in the form of islands without subsequent thermal annealing.

II. FABRICATION PROCEDURE

The same setup was used for sequential formation of Fe layers by CVD and VACNT arrays using DC-PECVD on a Si/SiO₂ substrate. In the case of obtaining Fe layers, the reactor is organized according to the principle of a two-zone system according to the operating temperature. In the upper region, (the first in the gas flow), there is a diffusion evaporator with ferrocene (bis (η -cyclopentadienyl) iron) serving as a metal source. The evolved reagent vapors are transferred by an argon flow to the underlying deposition area, where a quartz pedestal with a built-in heater is located. There is a graphite washer 45 mm in diameter with substrates, which were polished KEF-5 (100) silicon wafers with a natural SiO₂ layer. For the sample presented in this work, deposition was carried out for 30 min at a pedestal temperature of 700 °C and a total pressure of 700 Pa. The consumption of argon and ferrocene was 50 and 0.27 ml/min, respectively. No morphological features were observed on the smooth surface of the obtained Fe layer (with a nominal thickness of $\approx$ 30 μg/m²). In the case of growing of VACNT arrays, the reactor was equipped with an electrode system. The above-mentioned graphite washer acted as the cathode. A stainless steel disc (ø 45 mm) served as an anode. The gap between the electrodes was 40 mm. The sample presented in this work was obtained by deposition for 10 min at a pedestal temperature of 740 °C and a total pressure of 300 Pa. The working medium was created from ammonia supplied with a flow rate of 200 ml/min, and acetylene - 100 ml/min. Samples obtained after iron deposition served as a substrate. The discharge was characterized by a current of 7.5 mA and an anode voltage of 480 V.

The SEM image in Fig. 1 (a) shows VACNTs discretely spaced from each other with the placement density is ~ 1.7 × 10¹⁰ cm². The vast majority of CNTs are oriented vertically and the CNT have a slightly conical shape and a small variation in height Fig 1 (b). Their average length is more than 200 nm. SEM images clearly show catalyst particles located at the free ends of CNTs.

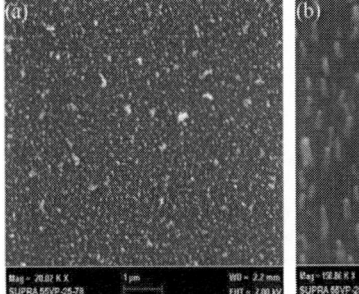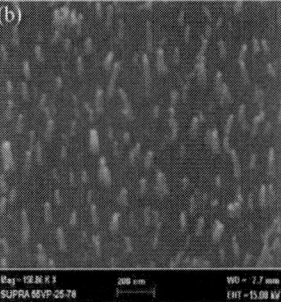

Fig. 1 SEM image of an as-grown VACNT made with DC-PECVD technology: a) top view in a wide area of the sample, b) detailed image of VACNTs.

III. RESULTS AND DISCUSSION

The field emission investigation was conducted using computerized method with multichannel data collection and online processing of the field emission data **Ошибка! Источник ссылки не найден.**. The method uses flat electrodes and fast high-voltage scanning regime (one current-voltage characteristic per 20 ms). The interelectrode distance was 300 μm and measuring chamber was under technical vacuum conditions (10^7 Torr).

At the first stage of measurements, the sample was subjected to high voltage training, while new emission sites were activated, and some of the most unstable, strongly protruding above the surface, were also destroyed (burned out in a vacuum discharge). Fig. 2 (a) shows time dependences of the amplitude of the voltage pulses and the corresponding amplitude of the emission current pulses (voltage and current levels). The vertical lines correspond to vacuum discharges that accompanied a stepwise increase in the applied voltage during the training of the emitter. After training, current levels below 250 μA were stable. Fig. 2 (b) shows the glow pattern at the maximum current, which shows the largest number of the sites in comparison with the pictures at lower current levels. However, there are an edge effects caused by the breakdown of the dielectric varnish that was used to cover the edges of the emitter.

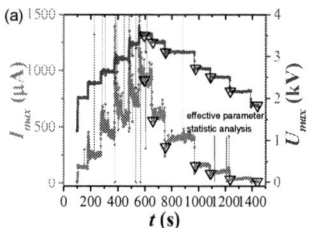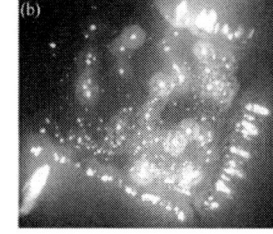

Fig. 2. a) Time dependence of current and voltage levels at stepwise change in voltage, b) The glow pattern of the VACNT sample.

The current-voltage characteristic (IVC) of the sample is shown in the Fig. 3 (a). To calculate parameters the standard IVC regression analysis in semi-logarithmic Fowler-Nordheim coordinates was applied. The average values: $\langle\gamma_{eff}\rangle = 1007$ and $\langle A_{eff}\rangle = 17620$ nm^2.

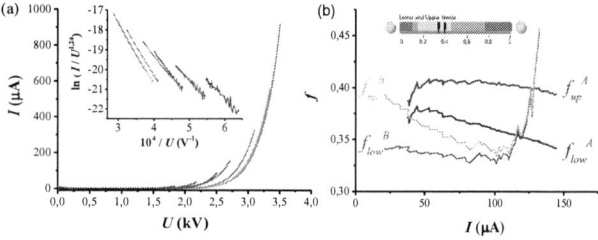

Fig. 3. a) IVC for different current levels. The inset shows the IVC characteristics in the Fowler -Nordheim coordinates, b) Dependences of the critical fields f_{low} and f_{up} on current level of the upper limit (method A) and the lower limit (method B).

Fig. 3 (b) shows the dependences of two critical dimensionless fields (f_{low} and f_{up}) on the choice of a fragment of the experimental IVC at a current level of $I = 150$ μA. The dependences are plotted on the corresponding level of the

emission current and nowhere go beyond the reliable range of electric fields. This method is described in more detail in [6].

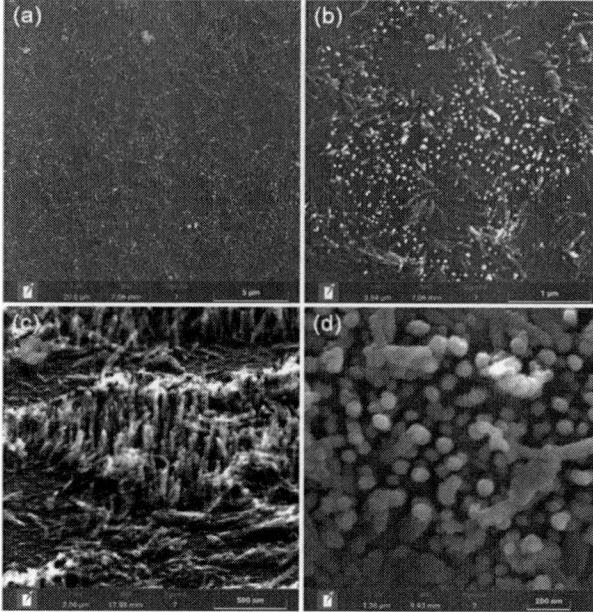

Fig. 4. SEM images of a VACNT sample on Si/SiO$_2$ after an emission experiment: a) general view of the emitter with the formed bald spots in the CNT array, b) more detailed image, the longest carbon structures are visible, c) a side view at an angle of 30 °, d) top view of a detailed image CNT.

Fig. 4 (a-c) shows the SEM images of the cathode after emission tests, from which it can be seen that, there are differences in the state of the cathode surface from the previous images, which demonstrated the initial array of CNTs (Fig. 1). More detailed images showed that CNTs lie on the substrate or are absent altogether (Fig. 4 (b-c). It is possible that such an inhomogeneity of the array is due to the detachment of nanotubes by electric field during the emission tests.

REFERENCES

[1] N. De Jonge, and J. M. Bonard, "Carbon nanotube electron sources and applications," Phil. Trans. Roy. Soc. London, vol. A362, pp. 2239–2266, November 2004.

[2] W. I. Milne, K. B. K. Teo, G. A. J. Amaratunga, P. Legagneux, L. Gangloff, J.-P. Schnell, V. Semet, V. Thien Binh, and O. Groening, "Carbon nanotubes as field emission sources," J. Mater. Chem., vol. 14 (6), pp. 933–943, February 2004.

[3] M. Jonsson, O. A. Nerushev, and E. E. B. Campbell, "Dc plasma-enhanced chemical vapour deposition growth of carbon nanotubes and nanofibres: in situ spectroscopy and plasma current dependence," Appl. Phys. A, vol. 88 (2), pp. 261–267, April 2007.

[4] K. Y. Jeong, H. K. Jung, and H. W. Lee, "Effective parameters on diameter of carbon nanotubes by plasma enhanced chemical vapor deposition," Trans. Nonferrous Met. Soc. China, vol. 22, pp. 712–716, December 2012.

[5] E. O. Popov, A. G. Kolosko, S. V. Filippov, E. I. Terukov, R. M. Ryazanov, and E. P. Kitsyuk, "Comparison of macroscopic and microscopic emission characteristics of large area field emitters based on carbon nanotubes and graphene," J. Vac. Sci. Technol. B, vol. 38 (4), pp. 043203-1–10, June 2020.

[6] E.O. Popov, A.G. Kolosko, and S.V. Filippov, "Test for Compliance with the Cold Field Emission Regime Using the Elinson–Schrednik and Forbes–Deane Approximations (Murphy–Good Plot)," Tech. Phys. Lett., vol. 46(9), pp. 838–842, October 2020.

Analysis of The Field Emission Current From an Array of Silicon Field Nanoemitters For Portable X-Ray Systems

Petr Yu. Glagolev [1,*], Gleb D. Demin [1], Nikolay A. Djuzhev [1], Ilya D. Evsikov [1], and Nikolay A. Filippov [1]

[1]*National Research University of Electronic Technology (MIET), Shokin sq., bld. 1, Russia*
*Contact: glagolev@ckp-miet.ru, phone +7-499-720-6907

Abstract—**Much attention in the development of portable X-ray tubes is paid to the fabrication of field emitter arrays as an electron source. First of all, this is due to the fact that such arrays of field emitters are capable of providing a high and uniform current density required to obtain a high intensity of X-ray radiation at the output, good stability of the X-ray tube operation over a long period of time, as well as a micrometer size of the focal spot for obtaining X-ray images with high resolution. One of the important aspects in the technology of the array nanofabrication is the analysis of the uniformity of the field enhancement factor and field emission current of each of the silicon emitter over the whole array, which is necessary for the stable operation of X-ray tubes (without reducing the number of active emission centres). In this work, we performed an analysis of the statistical dispersion of the field enhancement factor of the silicon field emitters in the fabricated array with a size of 300 x 300 and a distance of 3 μm between the array elements, where the maximum current from the array reaches values up to 393 μA. The results obtained can be used in the development of mobile microfocus X-ray sources.**

I. Introduction

Currently, in the context of a pandemic, medicine needs remote diagnosis of diseases without the need for patients to constantly visit polyclinics and hospitalize them in medical institutions where there may be a large number of people. For this purpose, it is important to develop portable X-ray systems based on transmission-type microfocus X-ray tubes capable of producing fast and high-quality radiographs of the skeleton, tissues, internal organs in the presence of hidden injuries or detection of diseases outside the clinic with a safe visit to the patient's home. An important role for this task is played by electron sources as the main functional part of X-ray tubes, which must demonstrate good stability, high currents and low power consumption [1]. Silicon field emitters, due to their high reproducibility and homogeneity of the array formed on their basis, nanometer radius of the tips (up to 5 nm and below), as well as the ability to achieve current densities up to 100 A/cm^2, stable in the temperature range from 25 to 400 ºC, can be perfectly suitable for solving the above task [2]. Over the past few years, a number of works have been devoted to the research of silicon field emitter arrays [2-4].

Nevertheless, in the case of large arrays of silicon field emitters, variation of the parameters of the technological process of their fabrication can lead to a spread of the emitter's aspect ratio h_e / r_e (the ratio of the emitter height h_e to the radius r_e of the emitter tip) over the array. This, in turn, causes an undesirable depolarization effect, that is, screening of the local electric field E_i at the tips of individual i-th emitters as a result of electrostatic interactions with neighboring j-th emitters, which have larger aspect ratio and, therefore, higher field enhancement factor (FEF) $\beta_j = E_j / E_m = f(h_j / r_j)$, where $i \neq j$, and E_m is the macroscopic electric field in the interelectrode space [5, 6]. As a result, not all field emitters within the array will take part in the formation of electric current in the X-ray tube. Moreover, the disappearance or redistribution of individual emission centers during X-ray tube operation will contribute to the current fluctuations, which negatively affects the size of the output X-ray intensity, as well as the size of the focal spot and the quality of the generated X-ray images [7]. For this reason, in order to optimize the technology of creating regular and homogeneous large-area arrays of silicon field emitters as an electron sources in portable X-ray tubes, a preliminary analysis of the statistical spread of the emitter's FEFs depending on the density of the array (lateral distance between neighboring emitters), its size and the distance to the anode with a transmission-type X-ray target should be carried out.

II. Statistical Analysis of The Electric Field Distribution In A Silicon Field-Emitter Array

To analyse the statistical dispersion of the FEF and current-voltage characteristics of the separate silicon field emitters, we fabricated the silicon field-emitter array with a size of 300 x 300 and a distance of 3 μm between the array elements (Fig. 1a), based on the developed technology from our previous work [8].

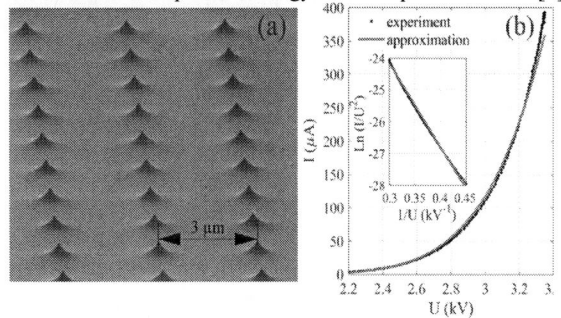

Fig. 1 (a) Fabricated array of silicon (Si) field emitters. (b) The measured current-voltage characteristic of a 300x300 array of Si field emitters (in the inset – the same plot in Fowler-Nordheim coordinates).

978-1-6654-2590-2/21 $31.00 © 2021 IEEE

The maximum current from the entire array, obtained in the experiment, reaches values up to 393 μA (Fig. 2b).

A fragment of a 300 × 300 field-emitter array containing 18 needle-type silicon field emitters was studied using scanning electron microscopy (SEM). Fig. 2 presents SEM image of the above-mentioned fragment with measured emitter heights. The inset to Fig. 2 shows a corresponding SEM image of a single silicon emitter with the measured height, lateral dimension of the bottom of the emitter, and the diameter of the emitter tip.

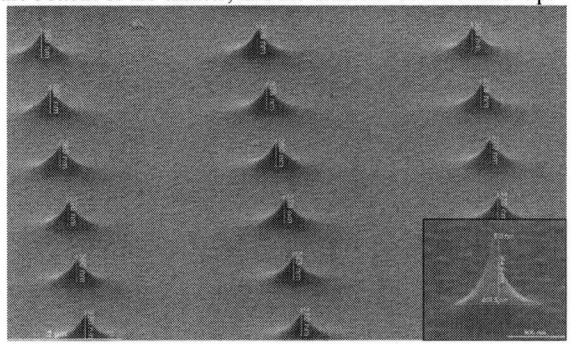

Fig. 2 SEM image of an array of needle-type silicon field emitters. Inset: SEM image of a single silicon emitter with measured geometric parameters.

Based on the experimental data obtained, a three-dimensional finite element model of an array of 18 silicon field emitters was developed in the COMSOL Multiphysics software package [9], where the geometric dimensions of individual emitters correspond to the dimensions of the selected emitters of the experimental sample. Fig. 3 presents the simulated distribution of the local electric field at the tips of silicon field emitters. The obtained electrophysical parameters made it possible to calculate the FEF of each individual emitter in the array, as well as to estimate the magnitude of the field-emission current from the selected emitter tips.

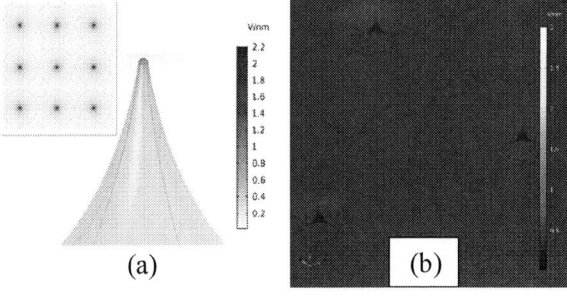

Fig. 3 (a) The electric field distribution of an individual tip from a (3x3) array of silicon (Si) field emitters. (b) The simulated electric field distribution in an array of Si field emitters.

III. CONCLUSIONS

Thus, the statistical dispersion of the emission characteristics of silicon needle-type field emitters within large-area field-emitter array (with a size of 300x300) has been investigated. Using the SEM method, the geometric parameters of several field emitters from the fabricated array were measured. A three-dimensional finite-element model of an array of silicon field emitters has been developed, which reproduces the distribution of geometric parameters within the array based on experimental data. The simulation of the electric field distribution over the array has been carried out, demonstrating the variation of the maximum values of the local electric field at the emitter tips (in the range from 2.2 to 3 V/nm).

ACKNOWLEDGMENT

The work was performed in the R&D Center «MEMSEC» (MIET) and supported by the RF Ministry of Education and Science (grant No. 075-03-2020-216, 0719-2020-0017).

REFERENCES

[1] R. Behling, Modern Diagnostic X-Ray Sources: Technology, Manufacturing, Reliability, 2nd ed., Boca Raton: CRC Press, 2021, pp.213–246.

[2] S.A. Guerrera, and A.I. Akinwande, "Nanofabrication of arrays of silicon field emitters with vertical silicon nanowire current limiters and self-aligned gatesNanofabrication of arrays of silicon field emitters with vertical silicon nanowire current limiters and self-aligned gates," Nanotechnology, vol. 27, p. 295302, June 2016.

[3] G.D. Demin, N.A. Djuzhev, N.A. Filippov, P.Yu. Glagolev, I.D. Evsikov, and N.N. Patyukov, "Comprehensive analysis of field-electron emission properties of nanosized silicon blade-type and needle-type field emitters," J. Vac. Sci. Technol. B, vol. 37, p. 022903, March 2019.

[4] R. Bhattacharya, N. Karaulac, W. Chern, A. I. Akinwande, and J. Browning, "Temperature effects on gated silicon field emission array performance," J. Vac. Sci. Technol. B, vol. 39, p. 023201, March 2021.

[5] R. Rudra, and D. Biswas, "Verification of shielding effect predictions for large area field emitters," AIP Adv., vol. 9, p. 125207, December 2019.

[6] J. Bieker, R.G. Forbes, S. Wilfert, and H.F. Schlaak, "Simulation-based model of randomly distributed large-area field electron emitters," IEEE J. Electron Devices Soc., vol. 7, pp. 997–1006, September 2019.

[7] C. Prommesberger et al., "Regulation of the transmitted electron flux in a field-emission electron source demonstrated on Si nanowhisker cathodes," IEEE Trans. Electron Devices, vol. 64, pp. 5128–5133, December 2017.

[8] N.A. Djuzhev et al., "Development of technological principles for creating a system of microfocus X-ray tubes based on silicon field emission nanocathodes," Tech. Phys., vol. 64, pp. 1742–1748, December 2019.

[9] COMSOL Multiphysics. v. 5.5. COMSOL AB (Stockholm, Sweden), https://www.comsol.com.

Experimental study of the multi-tip field emitter based on the array of silicon pyramidal microstructures

Ilya D. Evsikov[1,*], Gleb D. Demin[1], Tatiana A. Gryazneva[1], Maksim A. Makhiboroda[1], Nikolay A. Djuzhev[1], Oleg V. Pankratov[2], Eugeni O. Popov[3], Sergey V. Filippov[3], Anatoly G. Kolosko[3] and Maksim A. Chumak[4]

[1]National Research University of Electronic Technology (MIET), Shokina Sq., 1, Zelenograd, Moscow, 124498, Russia
[2]Ltd "Igla", Solnechnaya Al. 6, Off. 104, Zelenograd, Moscow, 124527, Russia
[3]Ioffe Institute, Polytechnitscheskaya St. 26, St.-Petersburg, 194021, Russia
[4]Peter the Great St.-Petersburg Polytechnical University, Polytechnitscheskaya St. 29, St.-Petersburg, 195251, Russia
*Contact: evsikov.ilija@yandex.ru

Abstract — **The paper presents a multiparameter experimental study of the multi-tip field emitter based on the array of silicon pyramidal microstructures. Emitter was created using standard silicon microfabrication process. Sample measurement was performed using computerized method with online processing of the field emission data. Obtained experimental current-voltage characteristics demonstrate good agreement with the expected theoretical estimates of the field-emission current.**

I. Introduction

Field emitters at this moment remain a promising alternative for the development of the electron sources. Electron sources based on field emitters are small in size, have a relatively low power consumption and small preparation time. These properties make them very attractive for the creation of electronic systems for compact X-ray sources in medical applications and security systems [1]. In addition, the sources of free electrons on field emission structures can find their application in the development of microfocus X-ray tubes for the maskless X-ray lithography. In terms of manufacturability, silicon is an attractive candidate for the field emitters. Silicon-based field emission structures created by means of the silicon microtechnology have a high degree of reproducibility and uniformity which is very important for multi-beam electron systems [2].

This paper describes the practical results of creating a multi-tip field emitter and obtaining experimental emission data of the developed structure.

II. Fabrication Process

Multi-tip field emitter based on the array of silicon pyramidal microstructures was fabricated using standard silicon technological operations. Pyramidal structures were formed in the bulk of the monocrystalline silicon substrate. The 150 mm n-type silicon wafers was oxidized to obtain a 100 nm SiO_2 layer. Deposition of Si_3N_4 layer (260 nm) was performed after wafer oxidation to form mask SiO_2-Si_3N_4 pattern. The photoresist was deposited by centrifugation on a rotating plate. The photolithography process included standard drying, exposure and strip operations. Anisotropic liquid etching at 80 °C with KOH 40% water solution was used to form the pyramidal microstructures on the surface of the silicon substrate. Fabricated silicon multi-tip field emitter is shown in Figure 1. Single silicon cathode was an octahedral pyramid of about 275 μm in height and the width of the cathode base was about 180 μm. Distance between cathodes in array was about 1 mm.

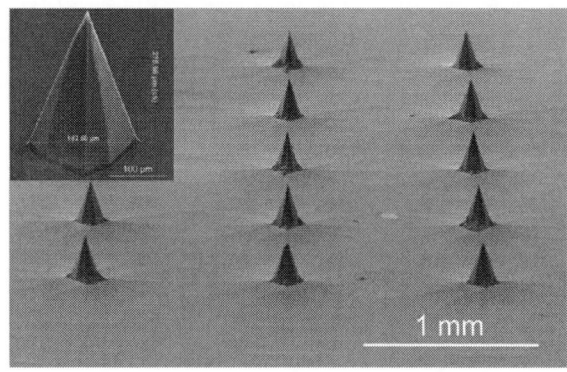

Fig. 1 SEM image of the silicon multi-tip field emitter based on the array of silicon pyramidal microstructures. In the inset: single silicon pyramidal field emitter.

III. Results and Discussion

The experimental data was obtained using computerized method with multichannel data collection and online processing of the field emission data [3]. This method uses flat electrodes and fast high-voltage scanning regime (one current-voltage characteristic per 20 ms). The interelectrode distance was 300 μm and measuring chamber was under technical vacuum conditions (10^{-7} Torr).

The current-voltage characteristic (IVC) of the sample is shown in the Figure 2. To estimate effective microscopic parameters (field enhancement factor γ and emission area A for a given work function φ = 4.5 eV) we used the modern theoretical concepts, which propose to use corrected Fowler-Nordheim coordinates (so-called Murphy-Good coordinates)

978-1-6654-2590-2/21 $31.00 © 2021 IEEE

for IVC analysis [4]. The values $\gamma = 1263$ and $A = 16.81$ nm^2 characterize the sample (see insert of Figure 2).

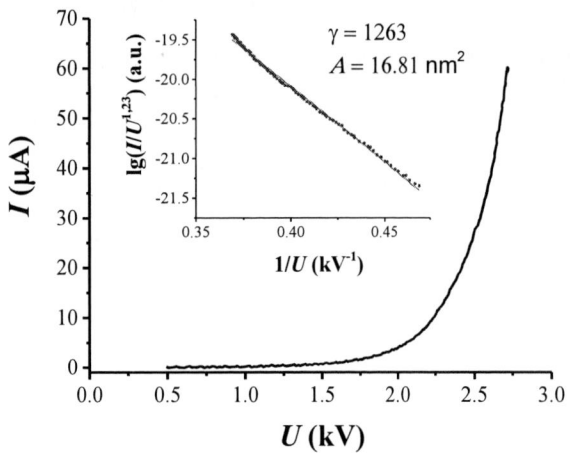

Fig. 2 Current-voltage characteristics of the fabricated silicon multi-tip field emitter in the standard and Murphy-Good (see insert) coordinates.

Maximum emission current level (amplitude of the pulses) that we were able to get from the cathode was 112 μA (Fig. 3). In general, emission in all voltage ranges was unstable, especially at currents above 50 μA. The Forbes orthodox cold emission test [4] showed that electric field on the emission sites apexes is too high for stable field emission.

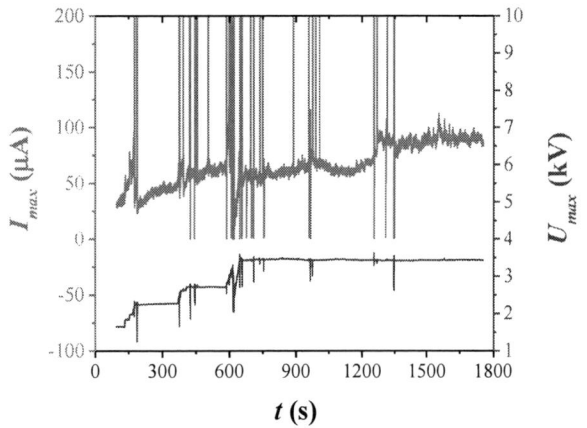

Fig. 3 Time dependences of the voltage and current levels.

In the second part of the investigation we used the computerized field emission projector which allows obtaining the distribution of the current load over the cathode emission sites and constructing the profile of the emission surface. Glow pattern analysis show that there are 18 emission sites (N) out of 28 possible, so emission area of one emission site can be estimated as $A_1 = A / N = 0.93$ nm^2. Solution of the Fowler-Nordheim equation with the found local current loads and corresponding voltage allowed to obtain local field enhancement factors. Obtained values were in the range from 1170 to 1300 (Fig. 4).

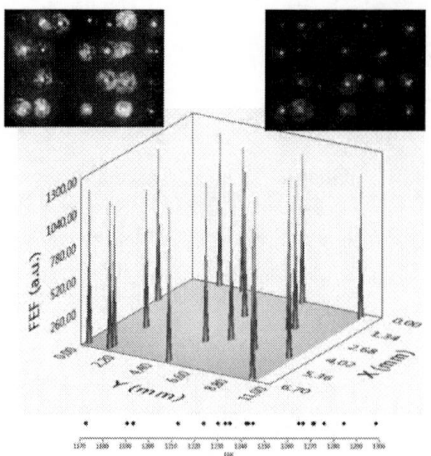

Fig. 4 Calculated field enhancement factors of the fabricated silicon multi-tip field emitter. Inset: glowing pattern of the emission surface.

IV. CONCLUSIONS

In this work we created the multi-tip field emitter based on the array of silicon pyramidal microstructures. Computerized system with multichannel data collection and field emission projector were used in the measurement of the prepared sample of the multi-tip field emitter. The investigated emitter occupies an intermediate position between single-tip macroscopic systems and large-area nanoscale field cathodes. It possesses both a distributed current load (this makes it possible to apply the system for determining local emission characteristics to its study) and the ability to clearly observe the shape of the emitting elements (this operation usually very difficult for emitters based on arrays of nanotubes). From a methodological point of view, the study of this kind of emitters should serve as a bridge for the transition from well-known macroscopic field emission calculations to understanding the nanoscale emission structures.

ACKNOWLEDGMENT

Authors I.D. Evsikov, G.D. Demin, T.A. Gryazneva, M.A. Makhiboroda and N.A. Djuzhev thank the RF Ministry of Education and Science for financial support (grant No. 075-03-2020-216, 0719-2020-0017) of the work performed based on the equipment of the R&D Center «MEMSEC» (MIET).

REFERENCES

[1] Basu A. et al., "A portable x-ray source with a nanostructured Pt-coated silicon field emission cathode for absorption imaging of low-Z materials," J. Phys. D Appl. Phys., vol. 48, no. 22, 2015, p. 225501.

[2] N.A. Djuzhev, et al., "Development of Technological Principles for Creating a System of Microfocus X-Ray Tubes Based on Silicon Field Emission Nanocathodes", Tech. Phys., vol. 64, no. 12, 2019, pp. 1742-1748.

[3] E.O. Popov, A.G. Kolosko, S.V. Filippov, E.I. Terukov, R.M. Ryazanov, E.P. Kitsyuk, "Comparison of macroscopic and microscopic emission characteristics of large area field emitters based on carbon nanotubes and graphene", J. Vac. Sci. Technol. B, vol. 38, 2020, p. 043203.

[4] E.O. Popov, A.G. Kolosko, S.V. Filippov, "Test for Compliance with the Cold Field Emission Regime Using the Elinson–Schrednik and Forbes–Deane Approximations (Murphy–Good Plot)", Tech. Phys. Lett. vol. 46, no. 9, 2020, p. 838.

Technology of the fabrication of Mo-based diode and triode structures with nanoscale vacuum gap

Tatiana A. Gryazneva [*], Nikolay A. Djuzhev, Gleb D. Demin, Nikolay A. Filippov, Ilya D. Evsikov and Maksim A. Makhiboroda

National Research University of Electronic Technology (MIET), Shokina Sq., 1, Zelenograd, Moscow, 124498, Russia

*Contact: gryazneva@ckp-miet.ru

Abstract — **In this work, experimental samples of Mo-based planar diode and triode field emission structures with a 150 nm vacuum channel were fabricated by electron beam lithography and plasma-chemical etching. The experimental current-voltage characteristics of the fabricated samples were obtained by means of the experimental set-up for the measurements in the high-vacuum conditions consisting of scanning electron microscope/focused ion beam system and semiconductor device analyzer. The experimental current-voltage characteristic confirmed Fowler-Nordheim mechanism of the electron transport in the fabricated molybdenum field emission structures.**

I. INTRODUCTION

Over the past ten years, there has been an active development of vacuum nanoelectronic devices associated with the emergence of the technological possibility of forming nanoscale (sub-100 nm) vacuum interelectrode gaps. It opens up prospects for the creation of THz field-emission devices with a quasi-vacuum (air) conduction channel of a length less than the electron mean free path in air (about 68 nm) [1] operating at voltage below the ionization potential of most atmospheric gases. Particularly attractive in this direction is the development of field-emission diodes and transistors with a nanoscale quasi-vacuum (air) channel based on various field emission materials – Si [2], SiC [3], VO$_2$ [4], W, Au, Pt [5], Al [6]. These field emission structures can operate in air at atmospheric pressure and do not need a special sealed package which facilitates their miniaturization and easy integration into conventional CMOS (complementary metal–oxide–semiconductor) integrated circuits. In turn, this makes it possible to create low-power hybrid electronics, combining both high-speed performance of field-emission structures (with a frequency in the range of hundreds GHz to several THz), due to the ballistic transport of electrons through a sub-100 nm quasi-vacuum channel, and the advantages of the technology of the state-of-the-art CMOS transistors [2], [7].

In this paper we present the practical results of creating a planar molybdenum diode and triode field emission structures as well as experimental investigation of emission characteristics of the fabricated structures.

II. FABRICATION PROCESS

The diode structure consists of two oppositely located flat electrodes. The distance between the electrodes is about 150 nm. The triode structure consists of four planar electrodes: triangle cathode and grid electrodes and flat anode. Distance between electrodes is about 150 nm as in the case of the diode structure (Figure 1).

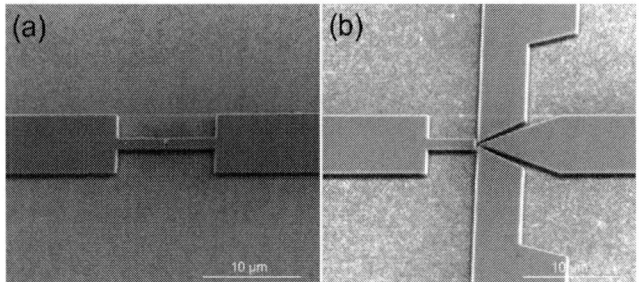

Fig. 1 SEM image of the Mo-based (a) diode and (b) triode structures with nanoscale vacuum channel.

Planar field emission structures were fabricated using electron beam lithography. Molybdenum layer (100 nm) was deposited by PVD process on silicon wafers which were previously oxidized in wet oxygen ambient to obtain SiO$_2$ layer (350 nm). The main advantages of molybdenum are its high melting temperature (2896 K), chemical inertness and high manufacturability. Al 50 nm layer was used as an additional mask for molybdenum dry etching. Al layer was coated with electron beam resist and the device structure was patterned by electron beam lithography. Plasma-chemical etching was performed on the Al and Mo layer and insulating SiO$_2$ layer was wet etched. Fabrication process and final field emission structure presented in the Figure 2.

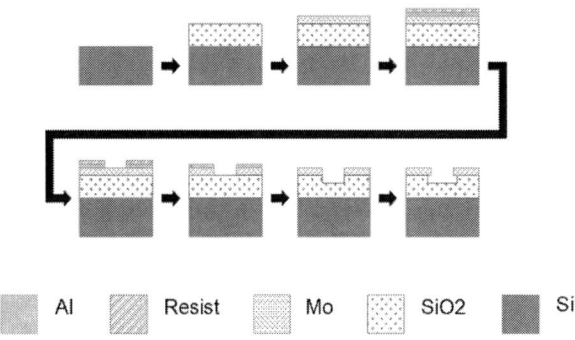

Fig. 2 Process flow for the formation of a planar Mo-based diode and triode structures.

978-1-6654-2590-2/21 $31.00 © 2021 IEEE

III. RESULTS AND DISCUSSION

The experimental current-voltage characteristics of Mo-based diode and triode field emission structures were obtained using developed experimental set-up consisting of dual-beam scanning electron microscope/focused ion beam (SEM/FIB) system FEI Qaunta 3D FEG and semiconductor device analyzer Agilent B1500A (Figure 3a). SEM/FIB system was modified with vacuum electrical inputs thus fabricated field emission structures can be investigated *in situ* under high-vacuum conditions. Special sample stage with movable measuring probes was created to install experimental samples into vacuum chamber of the SEM/FIB system (Figure 3b).

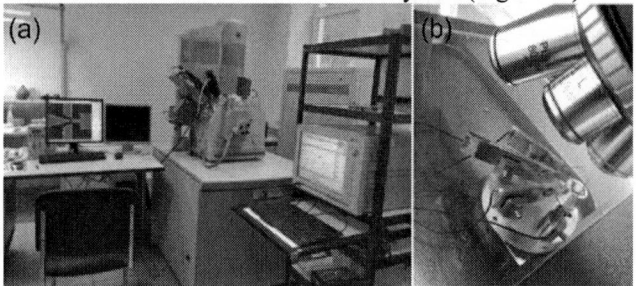

Fig. 3 (a) Experimental set-up with scanning electron microscope (on the left) and semiconductor device analyser (on the right) (b) Sample stage for the measurement of field emission structures.

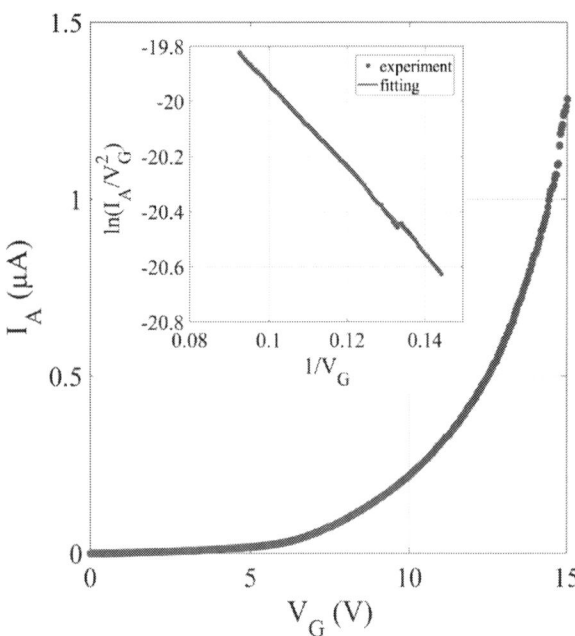

Fig. 4 Current-voltage characteristic for the planar Mo-based triode field emission structure. Anode voltage in the measurement was fixed at 15 volts. In the inset: current-voltage characteristic for the field emission region in the Fowler-Nordheim coordinates.

All the measurements were carried out in a vacuum chamber of the SEM/FIB system under pressure of about $5 \cdot 10^{-4}$ Pa or $3.5 \cdot 10^{-6}$ Torr. The experimental current-voltage characteristic of the Mo-based triode field emission structure is shown in the Figure 4.

In the experiment voltage of the anode electrode was set to the level of 15 V and voltage of the grid electrode was varied from 0 to the 15 V. Current-voltage characteristics were obtained in the square pulse mode of the voltage source (semiconductor device analyzer).

Maximum emission current that was obtained from the experimental sample of triode structure was about 1.28 μA. Raising the grid voltage above the 15 V caused significant instability in emission characteristics of the experimental sample. The derived value for the field enhancement factor γ was 637.

IV. CONCLUSIONS

In this work we developed a technology for fabrication of Mo-based diode and triode structures with 150 nm vacuum gap. Experimental set-up consisting of SEM/FIB dual-beam system and semiconductor device analyzer was used to examine fabricated molybdenum field emission structures and collect experimental data in the high-vacuum conditions. Maximum value of the emission current that we managed to obtain was about 1.28 μA with voltages on the anode and grid electrodes 15 V. The results of the study can be used in the development of the technology for fabrication of the planar field emission devices with nanoscale vacuum and quasi-vacuum (air) interelectrode gaps.

ACKNOWLEDGMENT

This work was performed based on the equipment of the R&D Center "MEMSEC" (MIET) and supported by the RF Ministry of Education and Science (grant No. 075-03-2020-216, 0719-2020-0017).

REFERENCES

[1] S.G. Jennings, "The mean free path in air", J. Aerosol Sci. vol. 19, no. 2, 1988, pp. 159-166.

[2] J.-W. Han, J.S. Oh, and M. Meyyappan, "Co-fabrication of Vacuum Field Emission Transistor (VFET) and MOSFET", IEEE Trans. Nanotech., vol. 13, no. 464, 2014, pp. 464-468.

[3] J.-W. Han, M.-L. Seol, D.-I. Moon, G. Hunter, and M. Meyyappan, "Nanosacle vacuum channel transistor fabricated on silicon carbide wafers", Nature Electronics vol. 2, no. 9, 2019, pp. 405-411.

[4] M. Liu, W. Fu, Y. Yang, T. Li, and Y. Wang, "Excellent field emission properties of $VO_2(A)$ nanogap emitters in air", Appl. Phys. Lett. vol. 112, 2018, p. 093104.

[5] S. Nirantar, et al., "Metal-Air Transistors: Semiconductor-Free Field-Emission Air-Channel Nanoelectronics", Nano Lett. vol. 18, 2018, pp. 7478-7484.

[6] S. Srisonphan, Y.S. Jung, and H.K. Kim, "Metal-oxide-semiconductor field-effect transisotr with a vacuum channel", Nature Nanotech vol. 7, no. 8, 2012, pp. 504-508.

[7] J.-W. Han, J.S. Oh, and M. Meyyappan, "Vacuum nanoelectronics: Back to the future?–Gate insulated nanoscale vacuum channel transistor", Appl. Phys. Lett., vol. 100, 2012, p. 213505.